Innovation Behavior of Older Consumers

Marktorientierte Unternehmensführung

Herausgegeben von Hermann Freter

Band 36

PL ACADEMIC RESEARCH

Armin Tank

Innovation Behavior of Older Consumers

An Empirical Analysis of Older and Younger Consumers' Determinants of Consumer Electronic Products Ownership

PL ACADEMIC RESEARCH

Bibliographic Information published by the Deutsche Nationalbibliothek

The Deutsche Nationalbibliothek lists this publication in the Deutsche Nationalbibliografie; detailed bibliographic data is available in the internet at http://dnb.d-nb.de.

Zugl.: Siegen, Univ., Diss., 2013

D 467
ISSN 0178-8183
ISBN 978-3-631-64149-1 (Print)
E-ISBN 978-3-653-03180-5 (E-Book)
DOI 10.3726/978-3-653-03180-5

PL Academic Research is an Imprint of Peter Lang GmbH.

Peter Lang – Frankfurt am Main · Bern · Bruxelles · New York · Oxford · Warszawa · Wien

This book is part of the Peter Lang Edition list and was peer reviewed prior to publication.

www.peterlang.com

Für meine Eltern und meine Geschwister

I would like to dedicate this doctoral dissertation to
my parents, my sister and brother

Vorwort des Herausgebers

Die Schriftenreihe „Marktorientierte Unternehmensführung“ enthält Arbeiten, die sich mit der Führung von Unternehmen und Organisationen vom Markt her widmen. Dabei kann es sich um Fragen der strategischen und taktischen Planung von Marketing-Instrumenten, um das Marketing einzelner Branchen sowie um Fragen der Informations- und der Organisationsseite des Marketing handeln. Die Themen beziehen sich sowohl auf theoretische Fragen als auch empirische Analysen.

Die vorliegende Arbeit greift zwei Mega-Trends auf. Einerseits verdeutlichen die aktuellen und prognostizierten Bevölkerungszahlen eindrucksvoll, in welchem strukturellen Umbruch sich die Gesellschaft in Deutschland befindet. Dabei hat der demographische Wandel vielfältige Auswirkungen – auch auf Unternehmen und ihre Kunden. Andererseits ist die Mehrzahl der Märkte von stetig kürzer werdenden Produktlebenszyklen, stagnierender Nachfrage und globalem Wettbewerb geprägt. In dieser Situation stellt die Vermarktung innovativer Produkte oder Dienstleistungen eine Möglichkeit dar, um in diesem Umfeld bestehen zu können. Trotz der Relevanz dieser beiden Trends sowohl in der Forschung als auch in der Praxis, bestehen Defizite in einer gemeinsamen, ganzheitlichen Betrachtung.

Die Arbeit von Armin Tank verknüpft wichtige Aspekte dieser beiden Themengebiete und leistet als erste Untersuchung dieser Art in Deutschland einen wesentlichen Beitrag, Unterschiede im Kaufverhalten älterer und jüngerer Konsumenten von innovativen Unterhaltungselektronikgeräten – einer Produktkategorie mit außergewöhnlicher Dynamik und dementsprechend hoher Innovationsgeschwindigkeit – zu erklären.

Der Verfasser konzeptualisiert und operationalisiert das innovative Verhalten von Verbrauchern und ihre verhaltenswissenschaftlichen Determinanten zunächst auf Basis einer geeigneten theoretischen Grundlage und überprüft diese in einer eigenen empirischen Untersuchung. Er verfolgt dabei einen interdisziplinären Ansatz, bei dem sowohl Erkenntnisse aus der Gerontologie, Demographie, Biologie, Psychologie, Soziologie, Ökonomie als auch aus der Innovationstheorie einfließen. Der Autor ermittelt somit umfassend, welche Unterschiede im Innovationsverhalten jüngerer und älterer Konsumenten bestehen. Die Untersuchung lässt Aussagen über die Relevanz einzelner Determinanten zu und identifiziert solche Einflussfaktoren, die eine altersspezifische Komponente aufweisen.

Die vorliegende Arbeit bietet mit den gewonnenen Erkenntnissen zahlreiche Ansatzpunkte für weiterführende Untersuchungen. Angesichts der weitreichenden Veränderungen, die sich durch den demographischen Wandel ergeben, wünsche ich der Arbeit eine lebhafte Aufnahme und hohe Aufmerksamkeit in Wissenschaft und Praxis. Die Relevanz der Thematik ist umso bedeutender, als es sich beim demographischen Wandel nicht um ein auf Deutschland beschränktes Phänomen handelt. Die Mehrzahl der westlichen Industrienationen ist hiervon in gleicher Weise betroffen.

Prof. Dr. Hermann Freter

Editor's Preface

The series "market oriented business management" ("Marktorientierte Unternehmensführung") consists of publications that focus on corporate business management and organizational management from a market perspective. The topics of the publications reach from the strategic and tactical planning of the marketing instruments to marketing issues of specific sectors or research questions dealing with the informational and organizational side of marketing. The covered topics relate to theoretical questions as well as empirical analyses.

The present book picks up two mega trends. On the one hand actual and predicted population figures demonstrate the current structural transition of the German society. This demographic change has manifold consequences – also on companies and their clients. On the other hand most of the markets are characterized by increasingly shorter product life cycles, a continuing stagnant demand and global competition. In this situation the commercialization of innovative products and services represents one possibility to remain competitive in this environment. Despite the fact that both of these trends are highly relevant in academic research as well as in practice, deficits exist in terms of a joint interdisciplinary, holistic consideration.

As the first study of its kind in Germany, the dissertation by Armin Tank connects important aspects of these two subject matters and substantially contributes to explaining differences in the consumer behavior of older and younger individuals regarding innovative consumer electronic products. These type of products are part of a product category that is characterized by an exceptional dynamic and therefore an exceptionally high pace of innovation.

The author conceptualizes and operationalizes innovative behavior of consumers and its determinants – initially on a theoretical basis which is subsequently validated by his own empirical analysis. He follows an interdisciplinary approach by considering findings from diverse research areas such as gerontology, demography, biology, psychology, sociology, economics as well as innovation theory. Thus, the author comprehensively identifies existing differences in the innovative behavior between older and younger consumers. The study makes statements on the relevance of specific determinants and identifies those influencing variables that exhibit an age-specific component.

The findings of the present study present numerous starting points for further research. In light of the far-reaching changes that result from the demographic change, I hope for a vigorous reception of and a high degree of interest in this work in the academic and practical realm. The relevance of this issue is even more significant as demographic change is not a phenomenon exclusively limited to Germany. The majority of western industrial nations is affected in a similar way.

Prof. Dr. Hermann Freter

Table of Content

Figures

Tables

Abbreviations

ANCOVA	Analysis of covariance
ANOVA	Analysis of variance
CEP	Consumer electronic products
CEP_i	Innovative CEP
CEPO	CEP ownership
CEPO _i	Innovative CEPO
CEPO_ni	Non-innovative CEPO
CEPO_w	Weighted CEPO
CFA	Confirmatory factor analysis
Ch.	Chapter
CLV	Centered leverage values
Diss.	Dissertation
DV	Dependent variable
DVD	Digital versatile disc
DVR	Digital video recorder
ed.	Edition
Ed.	Editor
EFA	Exploratory factor analysis
EVS	German income and expense study (Einkommens- und Verbrauchsstichprobe)
FA	Factor analysis
FMCG	Fast moving consumer goods
G	Group
GLM	General linear model
H	Hypothesis
hp	Help potential / contact quality
ibid.	Ibidem

imp	Imputation
IV	Independent Variable
KMO	Kaiser-Meyer-Olkin
Misc.	Miscellaneous
MSA	Measure of sampling adequacy
n.a.	Not applicable
OECD	Organisation for Economic Co-operation and Development
p.	Page
PAF	Principal axis factoring
PC	Personal computer
PCA	Principal component analysis
pp	Percentage points
pp.	Pages
pt.	Point
RQ	Research question
SD	Standard deviation
sdr_abs	Absolute studentized deleted residuals
SE	Standard error
SEM	Structural equation model
si	Social integration / contact frequency
SME	Small and medium enterprises
TAM	Technology acceptance models
VCR	Video cassette recorder
VIF	Variance inflation factor
vs.	Versus
WLS	Weighted least-square
WOM	Word-of-mouth
wtr	Wrong tendency of relationship

ZPRED	Standardized predicted values
ZRESID	Standardized residuals

Statistical Abbreviations and Symbols

α	Intercept of the regression line
β	Slope coefficient of the regression line
ε	Error term
σ_X	Standard deviation of the independent variable X
σ_Y	Standard deviation of the dependent variable Y
a	Estimator for α
b_j	Coefficient b (estimator for β / Partial regression coefficient of the independent variable j)
$\hat{b}_j$	Coefficient beta / Beta (standardized coefficient of b_j)
e_i	Unexplained error/residual of observation i
i	Number of the particular observation
j	Specific independent variable
k	Number of independent variables
n	Number of cases / observations
p	Specific CEP
t_{pop}	Partial regression coefficients in the total population
w	Index weights
y_i	Observed value of the dependent variable Y for x_i
$\hat{y}_i$	Estimated value of the dependent variable Y for x_i
$\bar{y}$	Mean of the dependent variable
D	Cook's Distance
J	Number of independent variables
R	Multiple correlation coefficient
R^2	Coefficient of multiple determination

R^2_{pop}	Coefficient of multiple determination in the total population
SS_e	Error sum of squares
SS_r	Regression sum of squares
SS_t	Total sum of squares
X	Independent variable
Y	Dependent variable
$\hat{Y}$	Estimation of the dependent variable Y
%	Percent / percentage

1 Introduction

1.1 Aging Societies and the Importance of Older Consumers

In the nearer and farther future many countries will be affected by profound demographic changes. To be more precise, most of the concerned countries are already in the middle of this process. The average age of most industrial nations has already increased and will grow even further. For example the median age[1] of the German population increased from 32 years (men) and 37 years (women) in 1950 to 44 years (men) and 46 years (women) in 2010. It is forecasted to further increase by nearly seven years (men) and eight years (women) in the next 50 years, so that by 2060 the median age of the male German population will be around 51 years and that of the female population around 54 years (s. Figure 1). Another remarkable growth-rate can be observed for people aged 60 or older. This age group will grow by an impressive 22% from 21.5 million in 2010 to 26.2 million in 2060 while the group of persons who are younger than 60 years will experience a sharp drop by 36%.[2] As will be shown later, demographic trends are relatively stable and hardly changeable. Therefore, it can be expected that these numbers portray a realistic scenario presenting major challenges for the future.[3]

So far, the general discussions about the demographic change centered mainly around negative aspects of the aging society. In politics, financing the social security systems has for a long time been the dominating topic (without any breakthrough long-term solutions) – spreading the negative image that old people "cost too much" and are some sort of a "burden" to society.

---

1 The median age divides the population into two groups of the same size. 50% of the population are younger and 50% of the population are older than the median age. See for example Bundesinstitut für Bevölkerungsforschung (2011a).

2 Own calculations on the basis of the data provided by the Statistisches Bundesamt (2009).

3 A detailed review of the German and international demographic developments, the reasons for the demographic shift, and the resulting consequences are discussed in Chapter 2.2, pp. 20.

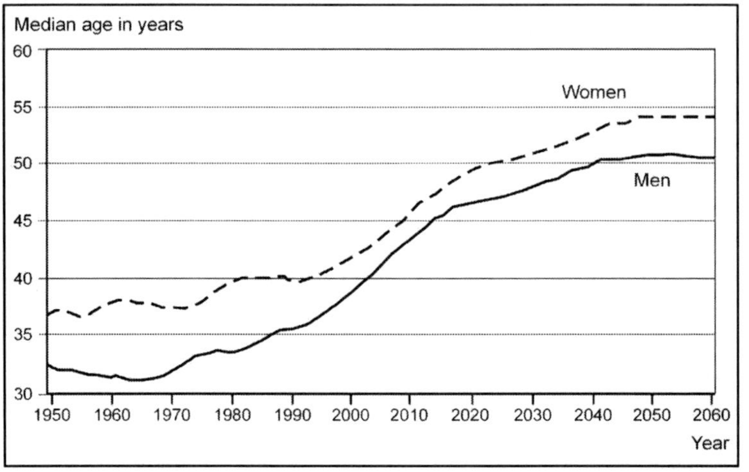

Figure 1: Development of the male and female median age from 1950 to 2060 (Source: **Bundesinstitut für Bevölkerungsforschung** (2011b), online; according to Statistisches Bundesamt (2009).)

In the course of time other aspects also started to receive the attention of the general public and the media. Books such as "Die deformierte Gesellschaft" (2002), "Das Methusalem-Komplott" (2004), or "Der Generationenpakt" (2004) recently introduced the topic to the general public.[4] In the following, the public discussion broadened as increasingly social and psychological aspects of aging were added. The positive aspects of aging and the potentials of aged individuals became more apparent to the public. As examples, the huge potentials existing due to the tremendous know-how of elderly retirees, their willingness to be actively engaged in the community (e.g. through voluntary work), or their support of own children or grand-children can be named. The demographic changes impact all areas of life which is why many other research areas picked up the demographic influences as well, discuss eventual future problems, but also offer solutions to these problems and point to the potentials of old age.[5] Lately, another positive trend emerged as more and more literature on the life of elderly persons was published. Self-help literature on "successful" aging supports older people in adapting to the new situations which are arising with their retirement (e.g. supporting to find activities for the newly gained time). Other topics that

4 Cf. Opaschowski, H. (2004); Schirrmacher, F. (2004); Miegel, M. (2002).

5 See the overview of gerontological research findings in Chapter 2, pp. 17.

are usually covered by this type of books are health status changes or changes of the social network with higher age.[6]

The demographic changes will also have far reaching consequences for companies. Although it is known that the age structure of most companies' customers is going to change in the future, the older consumer group is still widely neglected or not actively considered by the overwhelming part of businessmen. This is dangerous because the needs of consumers are likely to change when people age. Furthermore, it can be expected that older persons have different expectations towards products or services compared to younger consumers. If companies are not able to effectively adapt their business to the changing needs and expectations of their customers, they jeopardize the existence of the company in the worst case. Even though the mature market starts receiving more attention, many companies continue to direct all their marketing efforts towards the common 14-49 year old target group which is, however, a target group that will continue to shrink in the future. A study conducted by Capgemini (2005) demonstrates that managers view the demographic change more like a threat than like an opportunity: Only 11% believe that it has a positive impact on their business whereas 34% expect negative consequences and 55% think that it has no influence at all.[7]

Schmidt (1997) names three reasons why a large number of companies have disregarded aged consumers as a target group so far:[8]

- Product managers and creative workers/employees are young and are therefore more apt to identify with younger than elderly people. The needs of older citizens are overlooked because of the widespread orientation towards youth and young customers
- In many societies *old-age* is negatively stereotyped and elderlies are stigmatized
- Many clichés about older consumers are still predominant, including the imagination that this target group is futureless, as older people are at the end of their lives and are soon "going to die anyway;" that elderlies are physically and mentally restrained, what makes them reluctant to consummate; that they are resistant to promotional campaigns and adver-

6 See for example Kuntze, S. (2011); Bethke-Brenken, I./Brenken, G. (2010); Stumpf, H. (2008); Nuland, S. B. (2007).

7 Cf. Capgemini (2005).

8 Cf. Schmidt, I. (1997), pp. 36.

tisement in general; and that targeting old consumers has a negative influence on the image of the brand.[9]

The awareness of the "older consumers" and the "senior market" increased since the Capgemini study in 2005.[10] The sound financial situation of older persons and the supposed preference of older consumers to purchase high quality products and to spend considerable amounts of their money led to a real boom of consultants and authors of non-scientific books and articles talking about the "mature market" in a very euphoric way.[11] In this context a countless amount of names have been given to the targeted consumer segment (e.g. golden ager, best ager, generation gold, master consumers). All this to express that this consumer segment is highly attractive and that the market of affluent, "young" seniors is ready to be exploited. One has to give these consultants and publishers some credit that they created an awareness for the neglected consumer group, however, oftentimes lacking a *deeper understanding* of older individuals.

Many studies merely contain generally formulated strategies and recommendations that neglect the heterogeneity of older consumers and therewith overlook for example a great part of elderlies – namely those who are not as affluent as described. Often enough analyses are only of superficial nature and draw conclusions which are solely based on plausibility judgments. As a consequence, many companies still avoid dealing with the impact of the demographic trend on their business. Moschis (2003) strikingly explained the hesitation of companies with a combination of incomplete and anecdotal knowledge on elderly consumers. He also identified the high complexity of the mature market as a key reason why so much incertitude exists at the managerial level about choosing the right strategy to successfully market products to the aged consumers.[12] In consequence, this aspect needs to be added to the already high requirements on the

---

9 For this last statement see also Schönert, W. (1997), pp. 114.

10 While earlier books were for example published by Hens, M. (2004); Meyer-Hentschel, H./Meyer-Hentschel, G. (2004); Hock, E.-M./Bader, B. (2001); Krieb, C./Reidl, A. (2001); Härtl-Kasulke, C. (1998); Kaupp, P. (1997), more recent publications come for example from Fehlau, E. G. (2011); Hunke, G. (2011); Kohlbacher, F./Herstatt, C. (2011); Meyer-Hentschel, G./Meyer-Hentschel, H. (2010); Osl, P. (2010); Schorb, B. (2009); Otten, D. (2008); Rößing, A. (2008); Ness, O. (2007); Reidl, A. (2007); Gassmann, O./Reepmeyer, G. (2006); Haimann, R. (2005).

11 Cf. Heinze, R. G./Naegele, G./Schneiders, K. (2011), p. 62.

12 Cf. Moschis, G. P. (2003). Creating a specific strategy towards the mature market would, however, not be necessary, if older consumers were a homogeneous group, and if there were no significant differences between younger and older shoppers at the same time. Targeting innovations especially to the elderly would then be needless.

marketing to reach the best possible coordination of the marketing mix variables.[13]

Different to the non-scientific publications, academic research has been striving for long time to perform theory-grounded and empirical research in order to achieve research advances. While general research related to "age" and "aging" has more than 2,000 years of history in academic research (see the Gerontology Chapter 2.1, pp. 17), it was not until the late 1970s and early 1980s that first articles in marketing research addressed the special needs of older consumers. The article by Bartos (1980) is deemed to be the first publication in this area turning away from regarding elderlies as a "disadvantaged" group and, instead, advanced a positive view of the aged *consumer*.[14] Since then, numerous studies focused on this issue. Contrary to the diversified U.S. research, relatively little attention has been devoted to marketing aspects and older consumers by *German* researchers. Only a few authors have conducted scientific research in this area so far. Table 1 gives an overview on German dissertations dealing with older persons as consumers that are related to marketing. As can be seen many dissertations are theoretical works or merely refer to secondary data. Conclusions are derived from plausibility. Only more recent publications offer empirically founded results. But even the empirical studies reveal significant shortcomings. They lack comparisons between older and younger consumers for example (in most cases only older consumers are considered) or data analyses are only of descriptive nature. More comprehensive research designs are only provided by Landschulze (2010) and Hupp (1999).

Autor (year)	**Research focus (branch / industry)**	**Empirical research type**	**Considered age groups**	**N**	**Inferential statistics***
Gefromm (2010)	Luxury goods (in Japan)	qualitative (exploratory interviews in pre-study) + quantitative	50-59; 60-64; 65+	309	no, descriptive
Landschulze (2010)	FMCG** (margarine / shampoo)	quantitative	50-; 50-64; 65+	1.189 (586 / 603)	yes
Eitner (2008)	Reactions of retailers	qualitative + quantitative	-	(168 retailers)	no, descriptive
Möhlen	Automobiles	qualitative	-	-	-

13 Cf. Freter, H. (1987).

14 Cf. Bartos, R. (1980).

(2007)					
Peschmann (2005)	Automobiles	quantitative	30-54; (55-75) 55-64; 65-74; 75+	469	yes
Schulz (2005)	Banking	qualitative (expert interviews)	-	-	-
Enslin (2003)	Advertising analysis	qualitative (content analysis)	50+	134	no, descriptive
Peskes (2001)	Internet	qualitative / quantitative (content analysis)	-	-	-
Reitzler (2001)	Insurance	qualitative (pre-study) + quantitative	45-59; 60-75	4.421	no, descriptive
Kaufmann (2000)	Banking	quantitative	20-; 21-30; 31-40; 41-50; 51-60; 60+	300 [N (60+) = 15]	no, descriptive
Hupp (1999)	FMCG**	quantitative (two studies)	40-49; 50-59; 60-74; 75+ / In second study also 20-39.	798 / 207	yes
Brünner (1997)	Older consumers in general	secondary statistics	-	-	-
Kölzer (1995)	Customer orientation of retailers	quantitative	40-50; 60+	660	no, descriptive
Federsel-Lieb (1992)	Information behavior of older consumers	secondary data	-	-	-
Meyer-Hentschel (1990)	Product and retail store design	secondary data	-	-	-

* = Inferential statistics include methods that allow concluding from characteristics of a particular survey sample to characteristics of the population from which the sample was taken[15]
** = Fast moving consumer goods

Table 1: Overview of German dissertations dealing with older consumers

This short overview of the demographic situation and the current practical and academic knowledge about elderly consumers has shown that numerous gaps in this area still exist. To gain a deeper understanding about eventual differences between older and younger consumers, a lot of research still needs to be done. Before being able to identify a promising reseearch field that bears enough potential to be analyzed within the scope of this dissertation, it is indispensable to define who "older" and "younger" consumers are.

1.2 "Older Consumer" Definition

It is essential to concretize the terms "old" and "young," "older" and "younger," or "elderly" that have been used without further explanation so far. The discussion about defining the exact age of an "older" consumer is as diverse as the various names that have been given to the older consumer group. In some books the targeted "mature market" already starts at the age of 45 or 50 years of age.[16] Other authors account people older than 55, 60 or 65 years to this segment.[17] A definition of older consumers is, hence, indispensable.

This dissertation follows the orientation of the gerontopsychological research literature that focuses on the 60 years and older age group.[18] Similar to the dissertation of Kölzer (1995) the retirement status will also be considered.[19] However, while the author describes persons aged 60 years or older *and* retired, this study defines older consumers as persons who are retired *or* 60 years or older. That means that older non-retired individuals are also explicitly integrated. This is reasonable as the transition into retirement is a fluent process. It is likely that individuals in the pre-retirement state already look into issues regarding the time after their paying job.[20] On the other hand some retired persons stay active and

15 Cf. Allen, M. P. (1997), p. 50.

16 See for example Otten, D. (2008); Dobbelstein, T./Pflaum, M. (2007); Laumann, C. (2007); Pompe, H.-G. (2007); Gassmann, O./Reepmeyer, G. (2006); Härtl-Kasulke, C. (1998).

17 See for example Hunke, G. (2011); Neßlböck, R. (2008); Hock, E.-M./Bader, B. (2001); Brünner, B. O. (1997); Meyer-Hentschel, H. (1990).

18 Cf. Martin, M./Kliegel, M. (2005), p. 28.

19 Cf. Kölzer, B. (1995), p. 41.

20 See also the explanations given by Brünner, B. O. (1997), pp. 20.

become involved in voluntary work. A lower age limit seems to make no sense in this context. It cannot be expected that people aged 50 significantly differ in time spending or consumption related behavior from persons who are 47 years of age for example. It is also very likely that at this age people do not yet start dealing with the details of their retirement. The terms old/young persons, older/younger persons, or elderly persons are used interchangeably in the following.

1.3 Research Objective

Some general and also specific aspects (e.g. focus on a branch) of elderly have previously been researched in the marketing context as the overview of dissertations showed. As early studies have already done, this dissertation will not research the *general* implications of the demographic shift on marketing. By looking at the current market situations of companies, a promising research focus is attempted to be identified which will add to the state of knowledge about elderly consumers.

The intensifying global competition, the shortening product life-cycles, and stagnating demands are characteristics of many of today's markets.[21] In this difficult environment, companies have to actively manage their business strategies if they do not want to jeopardize their profits, market shares or, in last consequence, their existence.[22] *Innovations*[23] represent one way to counteract these market tendencies. They allow companies to replace old products that are at the end of their product life-cycle, adapt to changed customer needs, and break out of saturated markets by expanding to new markets or by occupying promising niches.[24] This is especially important for small and medium-sized enterprises

21 Cf. Breitschopf, B./Haller, I./Grupp, H. (2005), p. 43; Leker, J. (2005), p. 569; Sood, A./Tellis, G. J. (2005); Cooper, R. G. (2000), p. 38; Golder, P. N./Tellis, G. J. (1997); Böcker, F. (1996), p. 215.

22 Cf. Cooper, R. G. (2000), p. 38; Wöhe, G. (1996), p. 637; Zahn, E. (1991), pp. 116. For example, the number of insolvent companies in Germany rose by almost 50% from 1996 until 2005. While from 2006 until 2008 market tensions eased off a little, the fierce economic situation led to a recurring increase in 2009 and 2010. Cf. Institut der deutschen Wirtschaft (2011), p. 8; Institut der deutschen Wirtschaft (2006), p. 2; Institut der deutschen Wirtschaft (2004), p. 8.

23 The "innovation" term is defined in Chapter 3.1, pp. 67.

24 Cf. Trommsdorff, V./Schneider, P. (1990), p. 2.

(SMEs)[25] that are confronted with the market power of big companies.[26] Other advantages of introducing innovative products include a temporal monopoly position that allows the company to demand higher prices than for products where only marginal differences to competitor products exist. Furthermore, innovations can create a positive image at retailers and in the mind of the consumers.[27] Therewith companies are able to maintain their competitive advantage and expand or, at least, protect their profits.[28] This explains why innovating constitutes a central business function and plays such an important role for many managers.[29]

However, there are significant production and market-specific risks associated with product innovations that need to be taken into consideration.[30] Companies undergo substantial financial risks due to the fact that the process of innovating often requires large investments into tangible assets such as new production lines and materials as well as recruiting and training of employees.[31] Today, more money must be invested because of higher research and development efforts which result from the necessary specialization and increased complexity.[32] In addition, marketing expenses increase in order to position and advertise the product in this competitive environment.

Whether or not these investments pay off depends on the performance of the newly developed product or service – whether or not the market introduction was achieved at all. Due to limitations in predicting a new product's success there are high levels of uncertainty on the company side which are accompanying the innovation development.[33] The relevance of market risks becomes apparent when looking at flop rates of recently introduced products. For example Weis (2004), referring to a Madakom study, reports of flop rates ranging from

---

25 The flexibility and market proximity are characteristics of SMEs that are helpful to pursue this innovation and/or niche-strategy. Cf. Freter, H. (1982), p. 114 and pp. 120; Freter, H./Keßler, R./Wangen, E. (1980), p. 50.

26 Cf. Freter, H. (1983), pp. 33.

27 Cf. Witt, J. (1996), pp. 1; Wahle, P. (1991), p. 157.

28 Cf. Geroski, P./Machin, S./Reenen, J. V. (1993); Cravens, D. W./Shipp, S. H. (1991), p. 57.

29 Cf. Heinze, R. G./Naegele, G./Schneiders, K. (2011), pp. 33; Brockhoff, K. (1999), p. 1; Schumpeter, J. A. (1912), p. 177.

30 Cf. Koppelmann, U. (2001), pp. 102.

31 Cf. Hauschildt, J. (2004), pp. 168 and p. 498.

32 Cf. Fischer, M. (2001), p. 9; Kogler, A. (1991), pp. 13.

33 Cf. Gerpott, T. J. (2005), pp. 438; Hippel, E. v. (1992), pp. 3.

48.7% to 74.5% depending on the chosen product class.[34] Unsuccessful new product launches are subsequently withdrawn from the market, or, if the innovation project fails due to internal disagreements or negative assessments because of poor sales expectations, it is terminated even prior to commercialization.[35] Many other studies come to similar results, affirming the supposed high levels of risks related to innovations.[36] As a consequence, it is crucial to carefully select markets, innovations, and product launch strategies.

Despite the high levels of competition and despite the fact that markets are portrayed as being highly saturated, it is surprising to observe a very low interest in targeting innovations to the constantly growing group of older consumers. Mainly companies *exclusively* targeting this segment are among the small fraction of companies that pay attention to older consumers. It can be assumed that companies offering products like denture cleaners, stair lifts, or services such as meals on wheels or nursing already possess a deeper knowledge about the needs of their customers. Even if to some extent deficits in the marketing and conception of these products and services may exist, innovations in this area and their commercialization are more advanced. Therefore, innovations exclusively targeting older persons do not constitute the research focus of this dissertation. Instead it is more interesting to take a closer look at "regular" innovations that do not have an exclusive "elderly focus." The conception of these everyday products or services usually disregards the needs of the older clients and purely focuses on younger consumers.

In academic research the issue of older consumers *and* innovations has also found little consideration so far. This is, according to Strutton/Lumpkin (1993), due to the fact that "they are usually among the last to adopt innovations."[37] Other studies similarly argue that – although in good financial situation – older persons only spend little money on shopping, are undemanding customers, are reluctant to try out new products and are the last persons in the diffusion process to adopt innovations.[38] Instead, the typical innovator is described as younger

---

34 Cf. Madakom cited in Weis, H. C. (2004), p. 302. See also the overview given by Robertson, T. S. (1971), p. 17.

35 For an overview of existing product test approaches see Freter, H. (2004), p. 72.

36 For an overview of studies analyzing success and failures of innovations see for example Panne, G. v. d./Beers, C. v./Kleinknecht, A. (2003), p. 310; Montoya-Weiss, M. M./Calantone, R. (1994); Crawford, C. M. (1979).

37 Cf. Strutton, H. D./Lumpkin, J. R. (1993), p. 124.

38 Cf. Heinze, R. G./Naegele, G./Schneiders, K. (2011), p. 62; Kölzer, B. (1995), pp. 184; Federsel-Lieb, C. (1992), p. 65.

than the later adopters.[39] These prevailing opinions about elderlies resulted in only few theory-based studies dealing with the specific issue of innovations and older consumers. In fact, no dissertation was identified in the research process for this dissertation having addressed this subject in greater detail. Therefore it seems promising to research the innovation behavior of older consumers.

Besides the fact that the research of the older persons' innovation behavior has nearly been neglected, the review of relevant dissertations (generally dealing with elderlies and marketing, independent from any innovation focus) showed that a great part of these scientific publications lack solid empirical foundations. While some researchers do not empirically verify their hypotheses and make assumptions of plausibility, others only refer to secondary statistics. Accepting or rejecting the concrete research hypotheses of this study will, thus, occur on the basis of empirical evidence.

Although it is repeatedly mentioned that longitudinal studies in age and aging research are superior to cross-sectional studies[40] they also bear disadvantages. Besides the systematic respondent drop-out that occurs over time (the test and retest groups differ systematically)[41] the main handicap is that the time frame of dissertations is too short to gather enough data in order to answer longitudinal research hypotheses. Longitudinal research projects usually extend across decades – e.g. Berlin Aging Study (BASE), Bonn Longitudinal Study of Aging (BOLSA) in Germany, or the U.S. Seattle Longitudinal Study (SLS).[42] In consequence, with cross-sectional study designs not age changes but age *differences* are analyzed.[43]

To unveil eventual age differences it is essential to compare young with old consumers. Often age related studies exclusively research the behavior of older *or* younger consumers, so that no conclusions regarding older *versus* younger consumers can be made. In 1977 Phillips/Sternthal already criticized the large number of studies which solely examine the consumption behavior of elderly persons and do not consider other age groups.[44] The lacking comparison between young and old is a major drawback of these studies as they only allow state-

---

39 Cf. Robertson, T. S. (1968), pp. 362; Rogers, E. M. (1962), pp. 172.

40 See for example Carr, D./Bodnar-Deren, S. (2009), p. 715; Schaie, W. K./Hofer, S. M. (2001), pp. 61.

41 Cf. Phillips, L. W./Sternthal, B. (1977), pp. 452.

42 A overview over current longitudinal studies is given by U.S. National Institutes of Health (2011), online.

43 Cf. Phillips, L. W./Sternthal, B. (1977), p. 453.

44 Cf. Phillips, L. W./Sternthal, B. (1977), p. 444.

ments about elderly consumer specifics, but provide no insights whether their behavior even (significantly) differs from that of younger consumers. This however is essential in order to conceptualize whether marketing must adapt innovations to particular needs or whether an "ageless" market is prevalent.

As the discussion has shown, various research areas related to older consumers have not yet been regarded by studies. The literature review further revealed a broad neglect of innovations in this context. No German dissertation was found to address any details of the innovation behavior of elderly consumers. For this reason it is the principal objective of this dissertation to expand the marketing knowledge about the innovation behavior of elderly consumers in Germany.

This study will, in a cross-sectional analysis, research the innovation behavior of younger and older individuals. Its results will lead to statements on eventual age *differences*. More precisely the following central questions are addressed here:

- What are the relevant research findings concerning aging and old age?
- Are there any differences between the innovation behavior of younger and older consumers?
- What are the explanatory variables/determinants of the innovation behavior and how strong is their influence?
- Does age significantly influence the identified determinants?

This study does not aim to deliver concrete implications for the design of the companies' marketing mixes. This study much rather addresses the preceding step: before being able to research the elderly consumers' sensitivity to marketing activities, the differences between innovative consumer behavior of older and younger consumers must be identified.

Despite having limited the scope of the dissertation to "young and old consumers" and "innovation behavior" the research topic is still very broad in nature. It is likely that such a general research would not produce any meaningful results which would add new insights to the state of knowledge on consumer behavior. Therefore it is necessary to farther limit the perspective regarding the relatively broad terms of "innovations" and "innovation behavior." The exact focus on specific product or service innovations is identified in the course of the dissertation and will be discussed in more detail later (s. Ch. 4.1, pp. 117). The same holds true for formulating the concrete hypotheses. The research deficit in this area requires *developing* hypotheses, which is possible only after a detailed literature review of the associated disciplines.

1.4 Dissertation Structure

The structure of the dissertation, which is depicted in Figure 2, follows the logic of the general research questions that have been posed in the previous chapter. Therefore the study starts with a chapter devoted to "gerontology" – a science field which deals with all aspects of age and aging (s. Ch. 2, pp. 17). Findings from this chapter dealing with the general knowledge about elderlies are important to a great extent. They help to understand the situation of elderly consumers and evaluate various aspects throughout the dissertation. In addition to a brief description of the history of gerontology and the dimensions of aging, this chapter includes research results of different disciplines such as demography, biology, or psychology (Ch. 2.1 – Ch. 2.7).

Before being able to answer the three other research questions, it is necessary to take a look at theoretical aspects of innovation research (s. Ch. 3, pp. 67). Therefore, in a first step, the term innovation is defined (Ch. 3.1). The diffusion and adoption[45] of innovations are closely related to each other. Thus, before talking about innovation adoption research (Ch. 3.3) and explaining the single steps of the innovation decision process in more detail (Ch. 3.4), aspects of innovation diffusion research (Ch. 3.2) are briefly touched upon. The last aspect relating to innovation theory that is discussed in Chapter 3 deals with adoption determinant conceptualization (Ch. 3.5). The extensive literature review performed in this context helps to clarify one central aspect of this study: identifying potential factors influencing innovation adoption.

As mentioned in the last section, "innovations" represent a research object that is too broad in this context. For this reason Chapter 4 (pp. 117) commences with a discussion on which innovation(s) are reasonable to be focused on (Ch. 4.1). After the decision in Chapter 4.1 to regard the product category of consumer electronic products (CEP), the determinants identified earlier in Chapter 3.5 are assessed in the light of this category focus. At the same time age related knowledge supports formulating the concrete research hypotheses from Chapter 4.2.1 to 4.2.4.

45 At this point, the term innovation adoption is synonymously used as "purchase" of an innovation.

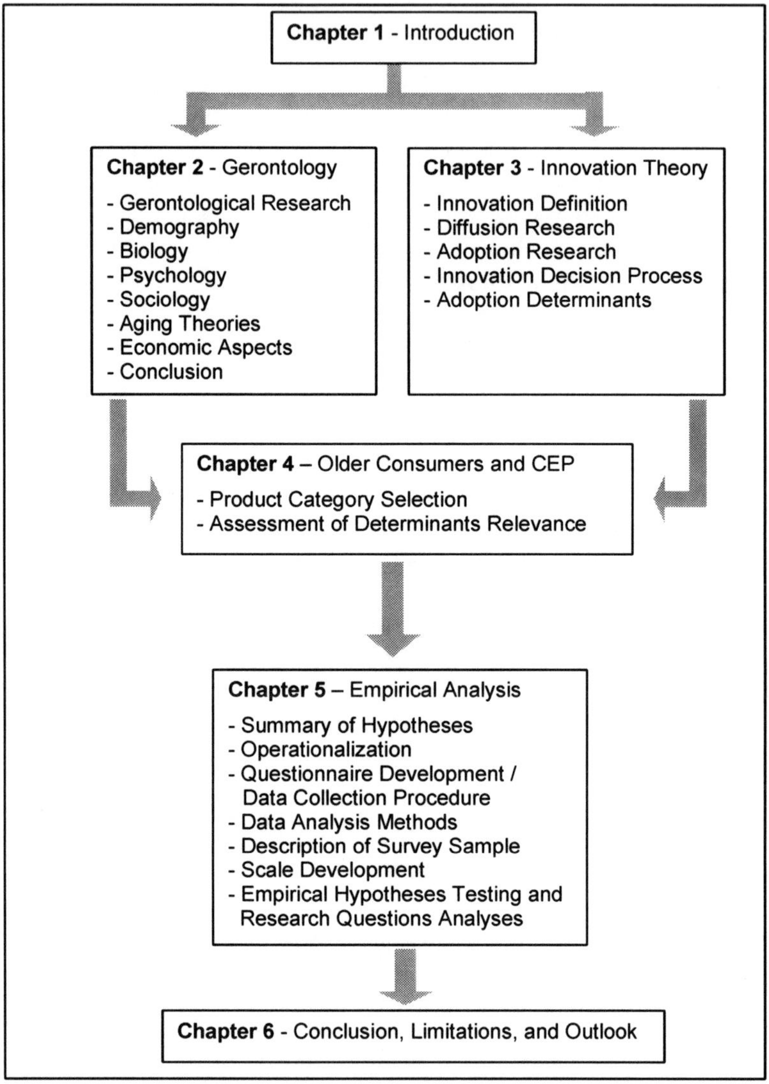

Figure 2: Schematic representation of the dissertation structure

As the literature review will later show, no adequate data is available which allow answering the research questions that have been posed. Therefore an empirical study needs to be set up and the statistical methods which are required to either accept or reject the concrete research hypotheses need to be explained. This is done in Chapter 5 “Empirical Analysis” (pp. 149). Its introductory chapter summarizes the concrete research hypotheses and the "diverse" research questions, and explains the research strategy (Ch. 5.1). With the help of the conceptualized adoption determinants (Ch. 3.5) and by considering aspects of consumer electronic products and "old age" (Ch. 4), the theoretical constructs are operationalized (Ch. 5.2). Therewith they become measurable and permit a formulation of survey questions. Chapter 5.3 describes the final questionnaire development and the data collection procedure. Then, in Chapter 5.4 the concrete statistical data analysis methods are discussed. The theoretical background as well as the assumptions of the employed methods of Exploratory Factor Analysis (Ch. 5.4.1), Analysis of Variance (Ch. 5.4.2), and Regression Analysis (Ch. 5.4.3) are presented. The analyses of the collected data begin with Chapter 5.5 where the survey sample is described. Then, in Chapter 5.6 the final scales are developed that measure the adoption determinants. The central research hypotheses are tested in Ch. 5.7 of the empirical part by use of the collected data (Ch. 5.7.1 to 5.7.4). The "diverse research questions," that are covering aspects which are not central to this study, are also considered here (Ch. 5.7.5).

The dissertation ends with Chapter 6 (pp. 273) and a summary of and critical discussion on the obtained results of this study. A synopsis and interpretation of the study results are given in Chapter 6.1. The critical discussion picking up the study's limitations follows (Ch. 6.2). Finally areas of further research are identified and a research outlook is given (Ch. 6.3).

2 Gerontology – Aging from the Perspective of Different Sciences

Unlike the rather unspecific definition of the National Institute of Aging that defines gerontology as "the study of aging from the broadest perspective," Baltes/Baltes (1994) concretize the scope of gerontology the following way:[46] "Gerontology deals with the description, explanation and modification of physical, psychological, social, historical and cultural aspects of aging and age, including the analysis of age-relevant and age-constituting environments and social institutions. This already indicates that gerontology is not an independent science – it can more precisely be called a science *field*."[47]

As gerontology refers to methods and results from various disciplines, it is only possible to discuss the most important findings next. After a short look at the historical background and development of gerontology as a science field (Ch. 2.1), the actual demographic situation and future developments will be discussed (Ch. 2.2). After that, some aspects of the age related biological changes are picked up (Ch. 2.3). Findings in the area of psychology (Ch. 2.4) and sociology (Ch. 2.5) broaden the perspective of aging and help to understand the situations that elderly individuals are confronted with. As many approaches exist trying to explain why and how aging occurs, one chapter is devoted to the most pertinent aging theories (Ch. 2.6). Finally, the economic situation of older persons will be analyzed (Ch. 2.7).

2.1 History and Objective of Gerontology

The precursors of gerontology can be found in the ancient Greek-Roman history. Far more than 2,000 years ago, philosophers and writers like Plato, Aristotle and Cicero already occupied themselves with the myths of age. They observed and described the elderly of their time and the process of aging with their own body.[48]

The phenomenon of aging was first evaluated in a scientific and systematic way in the period between the First and Second World War. But, it was not before the 1960s that researchers turned away from the dominating biologic-medical focus and gerontology became institutionalized. Birren (1961) called this the

46 Cf. Baltes, P. B./Baltes, M. M. (1994), p. 8.

47 Wahl H.-W./Heyl, V. (2004), pp. 35 (translated literally from the original German source).

48 Cf. Lehr, U. (2007), pp. 21.

"stage of expansion," which came along with a growing number of more sophisticated research studies.[49] During the 1980s, gerontology established itself in research and teaching. Since then, differentiation and specialization has led to an almost unmanageable number of different theories and approaches. Considering only the number of aging theories that will be discussed later, the wide variety of research results becomes evident. An exponential rise of gerontological publications can be observed.[50] Important research results have already been achieved, definitions have been established, and theories have been set up which will be explained in more detail in the following.

Wahl/Heyl (2004) name the following three questions as the basic ones in the research of aging:[51]

1. How does aging progress?
2. Why does aging occur – in general and specifically – the way we can observe it?
3. Is it possible to systematically influence this process?

Apart from describing and explaining the phenomenon of age and aging, it is also interesting to forecast intra- and interindividual differences.[52] Furthermore, gerontology aims at finding ways to improve and optimize age-specific changes – that means to positively influence the aging process.[53]

By means of twelve so-called "essentials" Table 2 illustrates the dimensions of gerontology. In the same way it shows the interdisciplinarity of this field. In this context, it should moreover be pointed out that – when answering research questions about aging – a holistic and theory-grounded analysis is necessary. Without such an approach, only partial knowledge would be the outcome of research efforts.[54]

	Aging can be seen as a...
1.	Dynamic process
2.	Biologically and medically determined process
3.	Lifelong and biographically anchored process
4.	Socially determined process
5.	Product of personal and spatial environment

49 Cf. Wahl, H.-W. (2004), pp. 29.
50 Cf. Birren, J. E./Schroots, J. J. (2001), p. 22.
51 Cf. Wahl H.-W. /Heyl, V. (2004), p. 31.
52 Cf. Kruse, A./Martin, M. (2004), p. 9.
53 Cf. Hoyer, W. J./Rybash, J. M./Roodin, P. A. (1999), p. 3.
54 Cf. Wahl H.-W. /Heyl, V. (2004), pp. 41.

6.	Economically determined process
7.	Gender-specific process
8.	Differential process
9.	Multidimensional process
10.	Multidirectional process
11.	Process that needs to differentiate between objectivity and subjectivity
12.	Plastic process with limitations

Table 2: The 12 dimensions of gerontology
(Source: **Wahl, H.-W./Heyl, V.** (2004), p. 42.)

Essential (1) expresses, that aging is not a condition, but a dynamic process that comprises both losses and gains. Essentials (2), (4), (5) and (6) underline the above mentioned interdisciplinary, when aging is discussed from the perspective of biology/medicine, social science and physical environment, as well as from the economic perspective. Essential (3) refers to the fact that situations or processes that occur at higher age may have their origin in earlier years. Therefore, the biographic development plays a significant role which needs to be considered carefully. Keeping in mind that women have a higher life-expectancy, the female part of the population strongly increases with age compared to the total population and consequently makes it necessary to respect gender differences (essential 7). Furthermore, gerontological research understands aging as an interindividually differing (essential 8), and multidimensional – as on different levels occurring – process (essential 9) that is multidirectional because it includes both losses and gains (essential 10). Due to the discrepancy between objective facts and their subjective interpretation, essential (11) plays a fundamental role in gerontological research. The last essential (12) considers the plasticity of aging and therewith the ways of modifying and shaping the organism in the course of a lifespan.[55] This characteristic allows the individual person to adapt to changes in his or her environment and to compensate losses for example.[56]

The goal of this brief introduction into gerontology was to point out the research field's multidimensionality and interdisciplinarity. The following study attempts to incorporate as many of the named aspects as possible which have been described in the essentials. Research findings from demography, biology, psychology, sociology, and economy dealing with the issue of aging and especially findings that allow an understanding of the situation of elderly consumers are at the center of interest.

55 Regarding the term "plasticity" see also Singer, T./Lindenberger, U. (2000), pp. 41; Hoyer, W. J./Rybash, J. M./Roodin, P. A. (1999), pp. 5.

56 Cf. Wahl H.-W./Heyl, V. (2004), pp. 41.

2.2 Demographic Aspects of Aging

Demography can be described as "the scientific study of human population, including its size, distribution, composition, and the factors that determine changes in its size, distribution, and composition."[57] The population statistics – the development in the past as well as the estimations of future developments – impressively illustrate the kind of structural change affecting the German society at the moment and also indicate what future challenges will arise. Scientists in the field of demography use different ways of projection to estimate future populations.[58] As the underlying assumptions for their calculations differ, there is a discrepancy between the results of projections. But, the estimations of the upcoming population development only vary slightly. They still point in the same direction and deliver a clear indication for a rapidly aging German society.

In Germany, the number of people who are 60 years or older more than doubled from 10.1 million in 1950 to 21.2 million in 2009.[59] The proportion compared to the overall population also increased. While in 1950 only about one of seven inhabitants was 60 years or older, in 2009 already one out four Germans passed that age (s. Table 3). Even more strikingly is the growth of that part of the population which is sometimes referred to as the "old" olds. The group of individuals aged 80 years and older increased by the factor 6 in the same period (from 0.7 million in 1950 to 4.2 million in 2009). And while in 1965 there were barely any individuals who entered the group of centenarians (276), today more and more persons reach this respectable age. In 1994 as many as 4,122 persons celebrating their 110th birthday already lived in Western Germany – the total number of persons who are 100 years or older was of course much larger.[60] The drastic changes are also visible when looking at the development of life expectancy which continuously increased in the last 130 years. While in 1871/1881 the average life expectancy was as low as 35.6 years (boys) and 38.4 years (girls),[61] in 2002/2004 newborns had an average life expectancy of 75.9 years (boys) and 81.5 years (girls).[62] Even in recent years an increase could be observed. In a period of 15 years (1991/1993 - 2006/2008) the life expectancy still augmented by

---

57 Swanson, D./Siegel, J. S. (2004). p. 1.

58 Cf. Statistisches Bundesamt (2011e); Statistisches Bundesamt (2009), pp. 37.

59 Both, the 1950 and the 2009 numbers include the people from whole Germany (new and old federal states).

60 Cf. Lehr, U. (2007), p. 37.

61 Due to the high infant mortality rate at that time, the further expected life expectancy of boys and girls at the age of five years was already 49.4 years (boys) and 51 years (girls).

62 Cf. Statistisches Bundesamt (2006), pp. 36.

6.6 years to 79 years (men) and by 5.2 years to 82.4 years (women).[63] Due to this sharp increase in life span, it is assumed that the life expectancy will cease to grow at such rates in the future. Thus, forecast scenarios also respect this slow down. Still, it is assumed that an augmentation of the life expectancy in the more conservative scenario of 7.8 years (men) and 6.8 years (women) until 2060 is possible – under this condition men will have a life expectancy of 85 years and women even 89.2 years.[64]

Year	People aged 60 years or older (in thousands)	% of the total population
1950	10,122	14.6
1960	12,732	17.4
1970	15,567	19.9
1980	15,174	19.4
1990	16,263	20.4
1995	17,209	21.0
1998	18,362	22.4
1999	18,881	23.0
2000	19,412	23.6
2001	19,871	24.1
2002	20,102	24.4
2003	20,336	24.6
2004	20,565	24.9
2005	20,540	24.9
2006	20,603	25.0
2007	20,770	25.3
2008	20,958	25.6
2009	21,210	25.9

Table 3: Population development of Germans who are 60 years or older (1950-2009) (Source: **Statistisches Bundesamt** (2011b), p. 9.)

On the other hand, the birth rate in Germany has steadily declined in the last decades. At the beginning of the 1960s the birth rate was at a high level of around 2.5 children per woman. From then on, a drastic drop followed until the mid 1970s to as few as 1.4 children per woman.[65] In the last ten to 15 years it leveled off at around 1.4 (2000: 1.38; 2005: 1.34; 2008: 1.38).[66] Not only were fewer kids born per woman, the decrease of the total number of women in the childbearing age (defined as women aged 14 to 49) led to a lower total number of births: The number of women in that age bracket was 404,000 lower in 2006

63 Cf. Statistisches Bundesamt (2011f).

64 Cf. Statistisches Bundesamt (2009), pp. 30.

65 Cf. Statistisches Bundesamt (2006), p. 27.

66 Cf. Statistisches Bundesamt (2009), p. 24.

compared to 1997.[67] The birthrates in Germany are actually so low that the total population will shrink. There will be more deaths than births,[68] so that the total population is expected to shrink from 81.5 million in 2010 to 64.7 million in 2040 (s. Figure 3). This corresponds to a relative reduction by more than 20% compared to 2010.

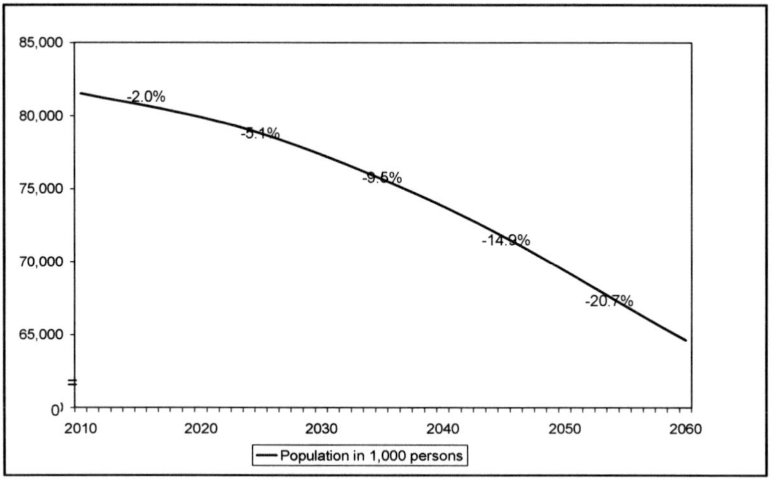

Figure 3: Population forecast for Germany until 2060
(Source:**Statistisches Bundesamt** (2009), p. 46.)

The decrease of childbirths in combination with the longevity results in an increase of the population's average age. In the period from 1990 to 2009 alone, the median age in Germany climbed from 39.3 to 43.4 years.[69] Another increase to an average age of more or less 50 years is expected in 2050.[70]

This also implies that researches do not see any sign that this trend will stop let alone reverse. The projections of the demographic situation in 2060 give a good impression of the future age structure. Compared with the age structure in 1910 the actual population structure cannot be described as a pyramid any more. While first deviations from the pyramid structure could already be observed in 1950, today's visualized age structure has no similarity with a pyramid at all. The form shift is even more drastic in the 2050 forecast. As the trend of only few newborn children is expected to continue, as well as the increase in life ex-

67 Cf. Pötzsch, O. (2007), p. 6.
68 Cf. Statistisches Bundesamt (2010a), p. 55; Statistisches Bundesamt (2009), p. 13.
69 Cf. Krack-Roberg, E./Grobecker, C. (2011), p. 429.
70 Cf. Statistisches Bundesamt (2006), p. 17.

pectancy, the form of the German age class composition for the year 2050 can best be described as a cigar and starts to even show tendencies of a mushroom form (s. Figure 4).

Applied to the single age classes this indicates for example that persons younger than 20 years will decrease from 15.6 million (2008) to 13.6 million (2020) and further to only 10.1 million (2060) (s. Table 4).

Age in years	**2008**	**2020***	**Difference (vs. 2008) in millions**	**2060***	**Difference (vs. 2008)**
0-19	15.6	13.6	-2.0	10.1	-5.5
20-29	9.9	8.5	-1.3	6.1	-3.8
30-49	24.3	19.8	-4.4	14.6	-9.7
50-64	15.5	19.2	3.7	11.9	-3.6
65-79	12.7	12.6	0.0	12.9	0.3
80+	4.1	6.0	1.9	9.0	5.0
total	82.0	79.9	-2.1	64.7	-17.4
in %					
0-19	19	17	-13	16	-35
20-29	12	11	-14	9	-38
30-49	30	25	-18	23	-40
50-64	19	24	24	18	-23
65-79	15	16	0.0	20	2
80+	5	8	48	14	123
total	100	100	-3	100	-21
*= forecast version "mittlere Bevölkerung, Untergrenze" with the following assumptions: constant birth rate of 1.4 children per woman; increase of life expectancy of newborn boys (eight years) and girls (seven years) in 2060, and a net immigration of 100,000 persons per year starting from 2014.[71]					

Table 4: Population development in Germany of selected age groups (Source:**Statistisches Bundesamt** (2009), p. 17.)

The forecasted group total in 2060 represents a decrease of 35% compared to 2008. Opposed to that a strong growth of the group of individuals who are 80 years and older has been calculated. This age class will more than double from 4.1 million (2008) to 9 million (2060). Its relative importance (compared to the total population) will boost from 5% in 2008 to 8% in 2020 and to 14% in 2060 (s. also Figure 5).

71 Cf. Statistisches Bundesamt (2009), p. 11.

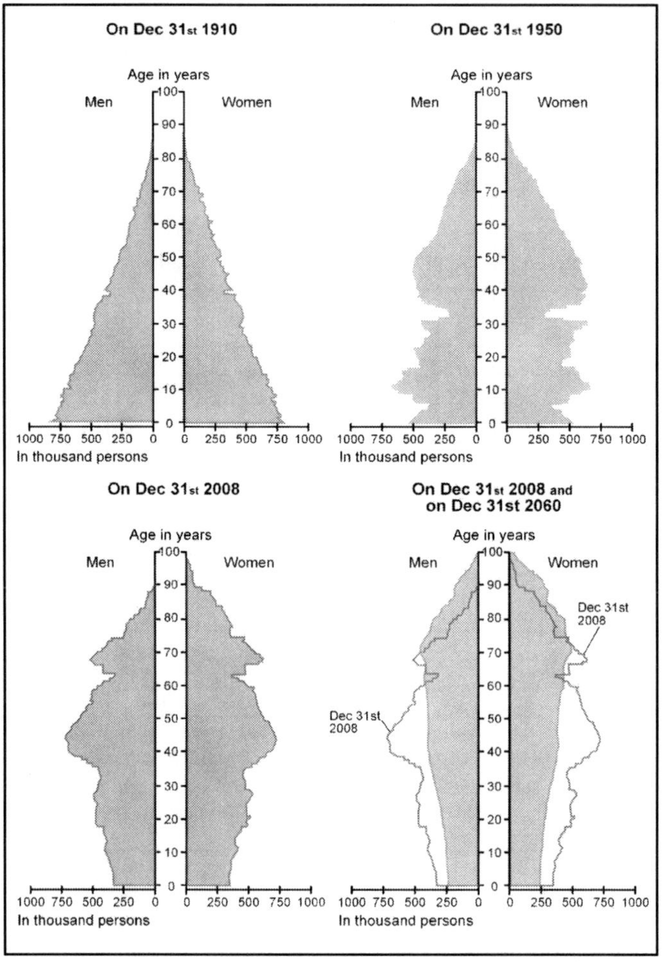

Figure 4: Age structure of the German population in 1910, 1950, 2008, and forecast of the 2060 population

(Source: According to **Statistisches Bundesamt** (2009), p. 15.)

In *explaining* the demographic situation of a nation as well as future changes in the population structure, singular reasons need to be separated from factors that have a continuous influence. The two major dents which can be observed for example in the German population pyramids of the years 1950 and 2008 can be explained by exceptional incidences (World War I and II).

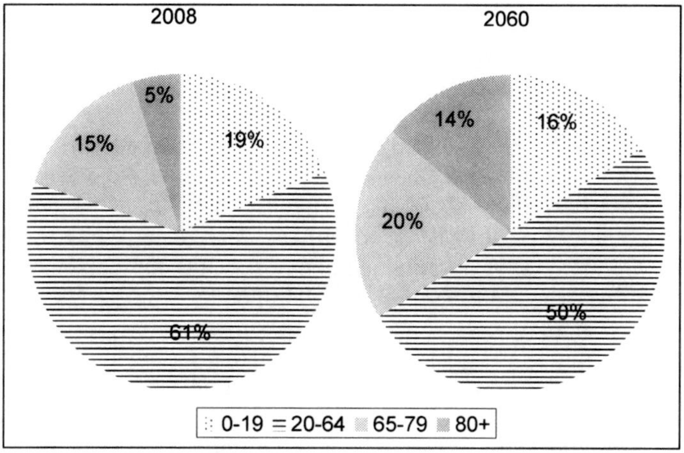

Figure 5: Age structure comparison 2008 vs. 2060 in Germany (Source:**Statistisches Bundesamt** (2009), p. 16.)

Other occurrences can be epidemics or natural catastrophes. Different to these singular events, three main factors with long-term effects that are responsible for the demographic composition of a country exit:

- Mortality rates
- Fertility rates
- Migration rates

Thanks to the advances in the health care system, improvements in hygiene, nutrition, housing conditions, and working conditions, the mortality rate could be drastically reduced. At the end of the 19th century, especially the prevalence of infant mortality and infectious diseases has been reduced which caused the average life expectancy to increase. In the second half of the 20th century it was predominantly the reduced mortality rate of elderly individuals which was responsible for this trend.[72]

As mentioned above, the birth rate has experienced a serious decline since the 1960s. The baby bust can be explained with the invention of the birth control pill in the early 1960s. Birth control was convenient for the lifestyles of societies. Longer education, more personal freedom in career planning, in private lifestyle, and in leisure time activities as well as higher levels of required and desired mobility and flexibility have led to the fact that couples have their first

72 Cf. Statistisches Bundesamt (2006), p. 36.

child at later age, if at all.[73] Even political decision to control the birth rate, such as the one-child policy that was introduced in 1979 in China, can have a very sustainable effect on the population development.[74]

The third aspect which influences the population structure of a society is migration. In the beginning of the 1990s, each year more than one million (in 1992 even 1.4 million) people moved to Germany.[75] On the other hand between 600,000 and 800,000 inhabitants left Germany each year. While the net immigration into Germany always remained above the 300,000 person line in these years, the number dropped in the following. In 2008 it was the first time a negative balance (-56,000) appeared, meaning more people migrated from Germany than immigrated into the country. As will be shown briefly later, migration can help to absorb the aging of a society to some extent.[76]

Aging is not just a phenomenon which can be observed in the German society – aging is a global phenomenon. Among the top ten of the world's oldest populations, measured in terms of percent population aged 60 years and older, and in terms of median age, nine countries are from Europe (s. Table 5). In both categories, however, a non-European country (Japan) is ranked first.

Country	Population aged 60+ (in %)	Country	Median age (in years)
Japan	27.9	Japan	42.9
Italy	26.4	Italy	42.3
Germany	25.3	Germany	42.1
Sweden	24.1	Finland	40.9
Greece	23.4	Switzerland	40.8
Austria	23.3	Belgium	40.6
Bulgaria	22.9	Croatia	40.6
Belgium	22.9	Austria	40.6
Latvia	22.8	Bulgaria	40.6
Portugal	22.8	Slovenia	40.2
Switzerland	22.7	Sweden	40.1

Table 5: Countries with the worldwide oldest populations
(Source: **Victor, C.** (2010), p. 62; according to numbers of the United Nations (2007).)

Taking a look at the age structure of selected nations demonstrates the different aging effects on countries. The data of the U.S. Census Bureau allows compar-

73 Cf. Lehr, U. (2007), p. 39.

74 Cf. Wu, X. (2008), p. 1.

75 See in the following Krack-Roberg, E./Grobecker, C. (2011), pp. 422; Statistisches Bundesamt (2010a), p. 67.

76 See for example the actual situation of the age structure and forecasts for the United States of America.

ing the structural changes of Germany (s. Appendix 1), Japan (s. Appendix 2), and USA (s. Figure 6) as examples for some leading industrial nations with those of two countries which exemplarily represent some less developed countries (Afghanistan s. Appendix 3 and Ethiopia s. Figure 7). For the latter it can be observed that even in 2050 a pyramid shape of the age structure is forecasted. As expected, the Japanese as well as the German age structure do not resemble a pyramid. In fact, the age structure of the USA also deviates from the pyramid form. Nevertheless, the society does not age to a similar extent. The U.S. population will, different to Japan and Germany, increase (by 127 million or +42.9% by the year 2050 compared to 2005). Likewise, the fertility rate (2.0) and net immigration (+1.8 million) will be significantly higher.[77]

At first sight the presented data gives the impression that aging is most pertinent in developed countries. However, with medical advances and social changes also less developed countries societies will age. Their aging process will even be more distinctive compared to industrial nations.[78] But, also here, great regional differences will be prevalent. For example in countries that are particularly affected by HIV/Aids, the average life expectancy may be reduced five years or more compared to similar countries which are less affected.[79]

---

77 Cf. U.S. Census Bureau (2011a), online.

78 Cf. Hoffmann, E./Menning, S./Schelhase, T. (2009), p. 22.

79 Cf. Pohlmann, S. (2002), p. 5.

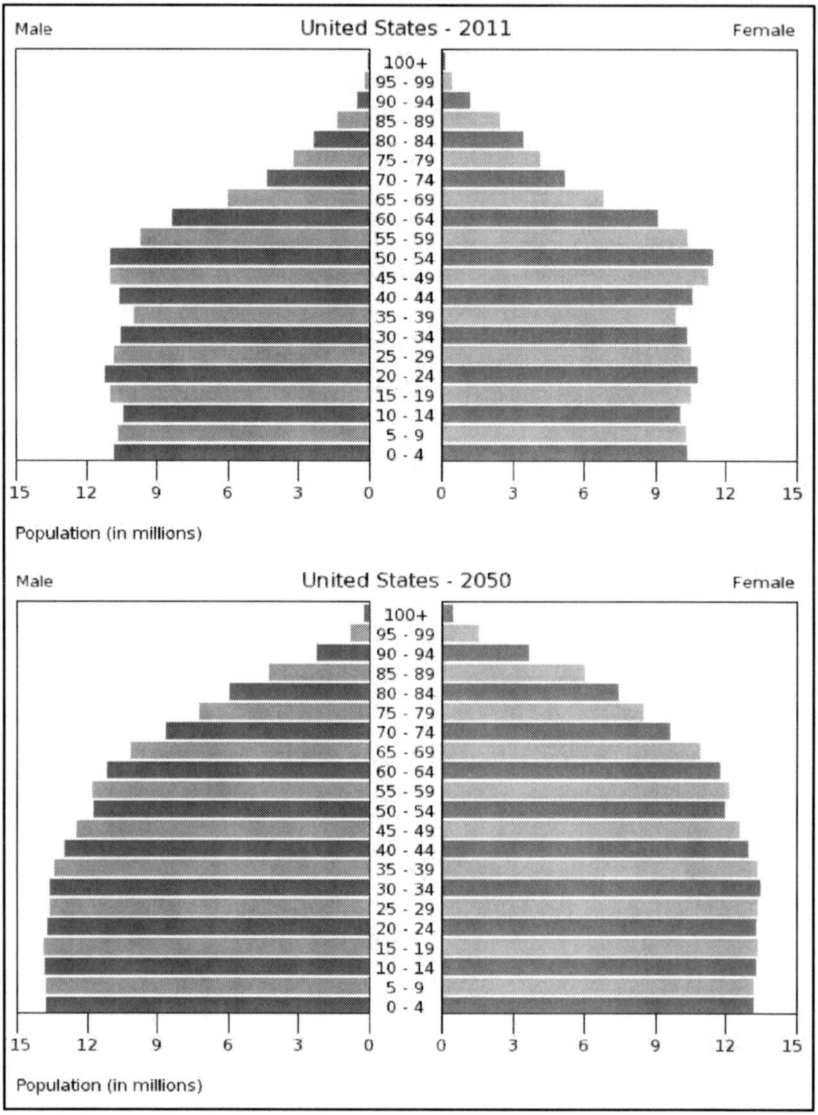

Figure 6: Age structure of USA
(Source: **U.S. Census Bureau** (2011b), online.)

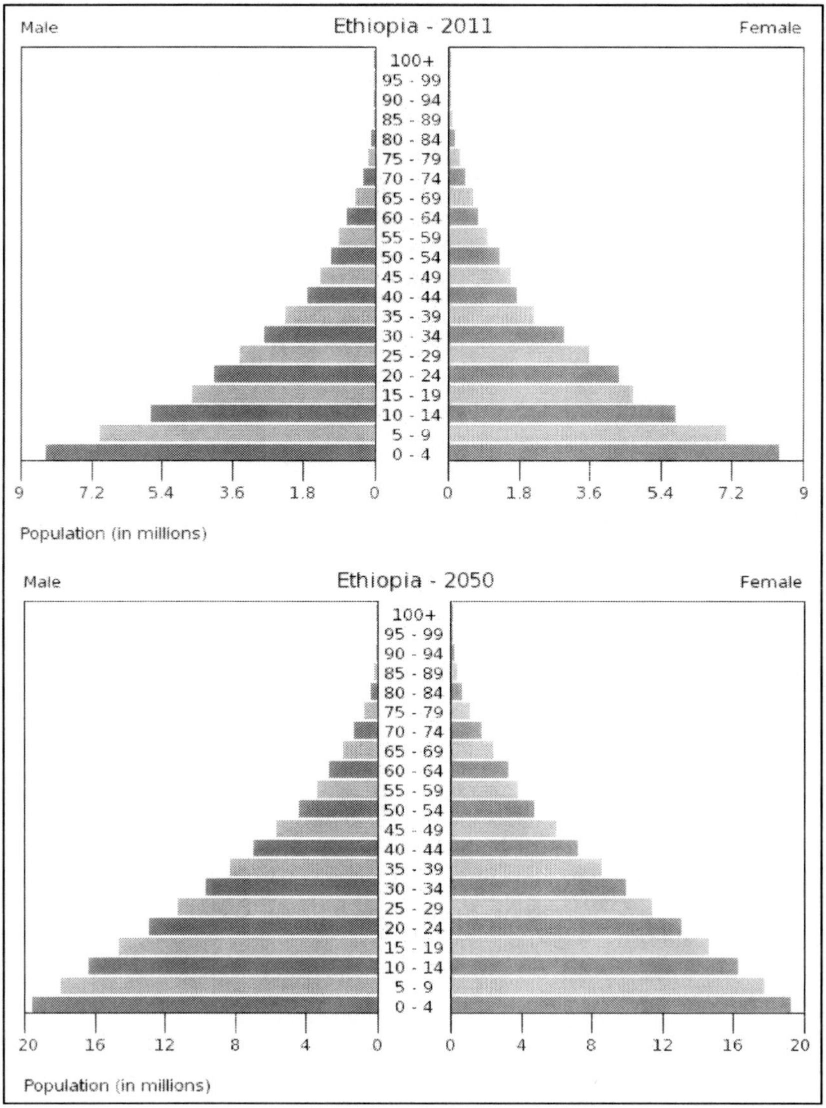

Figure 7: Age structure of Ethiopia
(Source: **U.S. Census Bureau** (2011b), online.)

Finally, one further demographic aspect should be mentioned in this context: the development of the household structure in Germany. Despite the decreasing total population in Germany, no similar trend is observable for the number of households (s. Figure 8). The development of the latter has decoupled from the population development and shows its own dynamic. The increasing number of divorces, breakups, and blended families, but also the growing life expectancy and fewer multi-generation households are the causes.[80] Smaller household sizes cannot only be found in younger age classes. At old age more couple and single households can be observed, and an increasing number of retirees deliberately decide to live alone.[81] Due to their longer live expectancy, the large majority of single households consists of older women (s. Figure 9).

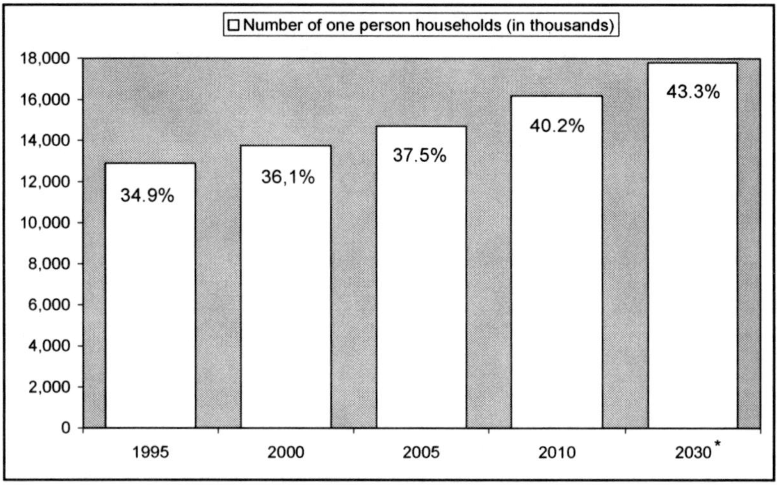

Figure 8: One person households in Germany
* = estimation of 2030 household situation
(Source: **Statistisches Bundesamt** (2011c), p. 15; **Statistisches Bundesamt** (2011d), p. 17; **Statistisches Bundesamt** (2010a), p. 46.)

80 Cf. Heinze, R. G./Naegele, G./Schneiders, K. (2011), pp. 43.
81 Ibid., p. 44.

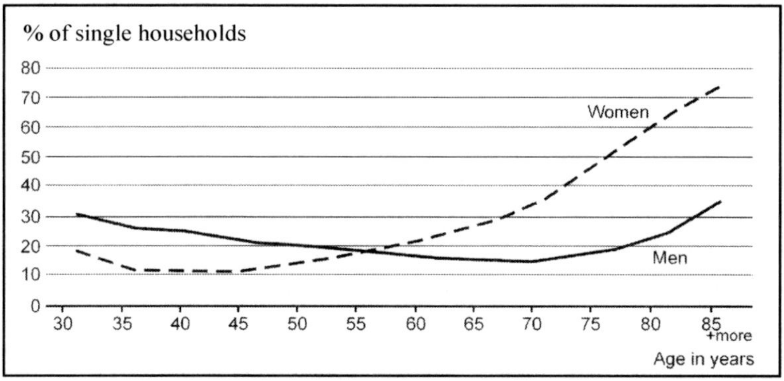

Figure 9: Male and female single households in Germany
(Source: **Statistisches Bundesamt** (2011a), p. 20.)

This sub-chapter provided an overview on actual demographic research findings. The current demographic situation, its development and future outlook clearly demonstrates the relevance of researching the consequences of an aging society.

2.3 Biological Aspects of Aging

The aging process is associated with biological changes. Of the many biological changes that occur in later life, the changes of external appearance (Ch. 2.3.1), sensory organs (Ch. 2.3.2), locomotor system (Ch. 2.3.3), and changes in the general health status of elderly persons (Ch. 2.3.4) will be discussed in further detail in the following chapters.

2.3.1 Changes of External Appearance

The advanced age of a person can first be noticed based on the external appearance. Changes of external appearance that are accompanied by the process of aging, include for example turning gray due to pigment loss in the hair follicles, emergence of face wrinkles, or the decline in muscles mass with simultaneous fat proportion increase. The changes appear gradually and at different rates, however they are universal and unavoidable.[82] When taking a look at the emergence of face wrinkles for example it can be stated that this is a very complex process that consists of several stages. However, researchers have also found that the aging process of the skin can be influenced to a great extend by each person itself, as it is predominantly the intensity by which the skin is exposed to

82 Cf. Hooyman, N. R./Kiyak, H. A. (2008), pp. 77; Cavanaugh, J. C./Blanchard-Fields, F. (2002), pp. 34.

ultraviolet light that has a major influence on the aging process of the skin and therewith the emergence of face wrinkles.[83]

2.3.2 Changes of Sensory Organs

The sensory organs are responsible for the connection of an organism with its surrounding physical world. Every sensory organ consists of receptors and neural pathways which transmit the sensory information to the brain where it is further processed.[84] The human sensory organs include eyes (vision), ears (hearing and equilibrium), nose (olfaction/smelling), tongue (gustation/taste),[85] and skin (somatosensory system/touch).[86]

Regarding the visual capacity it was discovered that no significant increase of blind people prevails at higher ages. Nonetheless the visual acuity decreases and defects of vision are more prevalent than in the younger parts of the population.[87] The parts of the eye experiencing the greatest change when people age are the cornea, pupil, rods and cones in the retina, and the lens.[88] The most common visual impairments resulting from physical changes of the eye are the presbyopia, cataract, macular degeneration, glaucoma, and retinopathy.[89] The presbyopia is a degenerative condition which is caused by an increasing inflexibility of the lens and results in a blurring of near vision. Similar to the presbyopia, the senile cataracts also relates to structural changes of the lens. These lead to a clouding of the lens and reduce the visual acuity. The most common visual impairment among older individuals is macular degeneration. While the peripheral visual field is not affected by damages of the macular, the central vision can eventually be lost. Opposed to that, the glaucoma causes a loss of the side vision while the central vision remains unaffected. The cause for glaucoma can be mainly attributed to an increased intraocular pressure which has an effect on the visual filed but which also may damage the optic nerves. The retinopathy usually emerges as a consequence of diabetes. The disease causes changes in the blood vessels of the retina which, in turn, can blur or distort the visual informa-

---

83 See the literature cited in Cavanaugh, J. C./Blanchard-Fields, F. (2002), pp. 35.

84 Cf. Tittel, K. (2003), p. 353.

85 More precisely, other organs also contribute to the sensation of taste (e.g. the smell of food detected by the olfactory system).

86 Cf. Csillag, A. L. (2005).

87 Cf. Hooyman, N. R./Kiyak, H. A. (1999), p. 73.

88 Cf. Hooyman, N. R./Kiyak, H. A. (2008), p. 94.

89 See in the following Jamison, J. R. (2007), pp. 65; Crews, J. E./Whittington, F. J. (2000), pp. 90.

tion that is sent from the retina to the brain.[90] The mentioned defects of vision are likely to have a significant negative effect on the elderlies' daily activities. Problems are expected for example when reading small prints, with recognizing objects in the periphery of the vision field, with bright light, or in situations where the eye needs to adapt from light to darkness or vice versa.[91]

The auditory system also undergoes several changes in the aging process. The ear consists of three components: the outer ear, the middle ear, and the inner ear.[92] The function of the outer ear is to capture sound waves. It is separated from the middle ear by a sensitive membrane, the ear drum. Three little bones in the middle ear transmit the vibrations to the inner ear. Hair cells in the inner ear convert the pressure waves to nerve impulses which are further transmitted to the brain where they are translated into meaningful sounds. In the large majority of cases (approximately 80%) where people suffer from hearing loss, the impairments are due to damaged hair cells of the inner ear or due to the loss of nerve cells of the auditory nerve.[93] Consequences are, among others, an elevation of the audible threshold or a lowered audible frequency range. Problems furthermore arise for many older individuals, when several audio sources exist as it is the case in group conversations, or when background noise interferes.[94]

Age related decreases also exist in the functioning of taste-, smelling-, and touch-senses. Regarding the decline of the taste sense it has been found in medical studies that this process occurs for each person individually – no general taste loss takes place at older age. Still, chances increase that the taste thresholds of one or more taste sensations (sweet, sour, bitter, salty, umami) change with consequences for the appetite or food choices. This may lead to suboptimal nutritional states or malnourishment.[95] It is not just taste buds on the surface of the tongue that are responsible for tasting. The olfactory system also plays a crucial role in food consumption e.g. as appetite stimulant. The odor sensations are further important to detect dangers like fire smoke or leaking gas.[96] Decreases of the efficiency of the tactile sense can be traced back to changes of the skin. With age, the number of nerve endings reduces. Especially fingertips, palms, and

---

90 For further analysis see Cavallotti, C. A./Cerulli, L. (2008).

91 Cf. Schiffman, S. (2009), p. 90.

92 See in the following ibid., pp. 91; Hooyman, N. R./Kiyak, H. A. (2008), pp. 99; Masoro, E. J. (1999), pp. 108.

93 Cf. Schiffman, S. (2009), p. 91.

94 Cf. Craik, F. I./Salthouse, T. A. (2000), pp. 157.

95 Cf. Hooyman, N. R./Kiyak, H. A. (2008), pp. 104. For further information see Schiffman, S. (2009), pp. 78 and the cited literature.

96 Cf. Robnett, R. H./Chop, W. C. (2010), p. 77 and pp. 160.

lower extremities are affected by reduced touch sensitivity. Compared to younger persons, older adults are less able to discern among levels of painful stimuli which can have serious injuries as a consequence.[97]

Even though it can be assumed, that impairments in the areas of taste, smelling, or touch are not as fatal as impairments in the auditory- and visual system, they nonetheless represent limitations in the day-to-day living of individuals at old age. The impact of age related decreases of the organ functionality is intensified especially in those situations when they occur in combination – that means when difficulties are experienced with impairments of several sensory organs.

2.3.3 Changes of Locomotor System

When analyzing age related changes of the locomotor system, especially issues such as movement variability, coordination, posture and body balance, strength control, reaction time and movement speed are of interest.[98] Declines in motor control at old age have their cause in changes of the brain-structure, the muscle composition, and the sensorimotor integration. For instance the reaction time, defined as the time needed to initiate a motor response to a stimulus, significantly slows with age. In this context for example Masoro (1999) discovered that it is mainly the central processing which is responsible for slowing down the reaction – even more so when elderly persons find themselves in a situation where they are confronted with several choice alternatives.[99] As mentioned, also the posture, body balance and coordination are of interest in gerontological research. Changes in bones (loss of bone tissue), muscle weakness (loss of up to 40% of muscle tissue by the age of 80), but also diminished vision or damages in the inner ear account for walking difficulties of the elderly and the fact that they need more spatial cues to orient themselves.[100] These physical changes can in the worst case lead to "coordination catastrophes" – tripping and falling, which often have serious consequences, be it a direct impact on the health status of aged individuals or an indirect impact on the quality of life due to fear of falling.[101]

97 Cf. Hooyman, N. R./Kiyak, H. A. (2008), pp. 82.

98 Cf. Ketcham, C. J./Stelmach, G. E. (2001), pp. 313; for an overview of the musculoskeletal changes see Hoyer, W. J./Rybash, J. M./Roodin, P. A. (1999), p. 72.

99 Cf. Masoro, E. J. (1999), p. 114.

100 Cf. Hooyman, N. R./Kiyak, H. A. (2008), p. 82; Cavanaugh, J. C./Blanchard-Fields, F. (2002), pp. 37.

101 Cf. Jamison, J. R. (2007), pp. 66.

As has been briefly shown above, restrictions on the movement apparatus can have far-reaching consequences for the affected persons. Depending on the specific illness and the severeness of the symptoms, elderlies may experience the whole range from only a slight limitation in their usual activities in the best case to a total dependency on help of others in the worst case. External help can prevent those stressful and high-pressure situations. Still, possibilities to arrange daily activities, leisure time, or the cultivation of social contacts are limited. In the end, these indirectly affecting factors also have a negative influence on the perceived quality of life.

2.3.4 Changes in Health and Epidemiology

Scientists in the field of epidemiology collect data of acute and chronic illnesses in the population and study the factors affecting public health. Gathered knowledge is used to adapt capacities in hospitals or in the area of care, to analyze the effectiveness of recently introduced therapies, as well as to identify risk factors that are modifiable in order to develop new treatment approaches.[102]

The aging process is accompanied by an increasing vulnerability of the organism. These higher chances of fragility of the organism enhance the occurrence of illnesses (see the overview of the most frequent chronic diseases in Table 6). The likelihood of reaching a pathological condition increases which in turn can shorten the individual lifespan.[103]

In addition, illnesses usually do not occur separately. Older people more frequently suffer from different diseases simultaneously – this state is called multimorbidity.[104] As an example, the prevalence of having five or more diseases rises from 3% at the age group of 40-54 year old persons to 8% for the 55-69 year old individuals. A sharp increase can be detected for the oldest age group of the survey respondents (70-85 years of age) where 21% suffer from five or more illnesses simultaneously (s. Figure 10). In some cases the simultaneously occurring illnesses develop independently, while in other cases illnesses influence each other and become an even more serious matter as correct diagnoses and therapies are by far more difficult.[105]

102 Cf. Weyerer, S. (2000), pp. 27.

103 Cf. Ding-Greiner C./Lang, E. (2004), p. 195.

104 Cf. Ibid., p. 197.

105 Cf. Steinhagen-Thiessen, E./Wrobel, N./Borchelt, M. (1999), p. 283.

Rank	Total	40-54 years	55-69 years	70-85 years
1	Hypertension 34.3%	Arthrosis 22.5%	Hypertension 41.2%	Hypertension 52.7%
2	Arthrosis 33.0%	Hypertension 18.9%	Arthrosis 37.0%	Arthrosis 46.5%
3	Hightened cholesterol 24.1%	High cholesterol 16.1%	High cholesterol 28.9%	High cholesterol 32.2%
4	Arthritis / rheumatism 15.0%	Arthritis / rheumatism 10.0%	Arthritis / rheuma-tism 15.6%	Arthritis / rheu-matism 23.1%
5	Circulatory disorder 9.8%	Mental illness 6.5%	Diabetes 10.9%	Cardiac insuffi-ciency 20.6%
6	Diabetes 9.5%	Circulatory dis-turbance / dis-turbed blood flow 4.7%	Circulatory distur-bance / disturbed blood flow 9.2%	Circulatory dis-turbance / dis-turbed blood flow 19.6%
7	Cardiac insuf-ficiency 8.7%	Chronic lung disease 3.9%	Cardiac insuffi-ciency 8.1%	Diabetes 19.1%
8	Osteoporosis 6.0%	Diabetes 3.2%	Cancer 6.3%	Osteoporosis 14.5%
9	Mental illness 5.7%	Gastric / intesti-nal ulcer 2.9%	Osteoporosis 5.4%	Cancer 10.3%
10	Cancer 5.6%	Cardiac insuffi-ciency 2.7%	Mental illness 5.2%	Glaucoma / mac-ula degeneration 9.4%

Table 6: Prevalence of self-reported diseases

(Source: **Wurm, S./Schöllgen, I./Tesch-Römer, C.** (2010), p. 97; based on the data of the DEAS 2008.)

It is important to understand that old age is not necessarily related to illness.[106] Drawing on the numbers of older persons that are in need of care support that statement. The average age of persons who are in inpatient care is 81.8 years: while for example only 5% of the people who need inpatient care in Germany are 65-69 year old, 20% are between 85 and 90 years of age and 25% are 90 years and older.[107] A similar statement can be made for nursing home residents.

106 Cf. Baltes, P. B./Baltes, M. M. (1994), p. 16.

107 BMFSFJ (2006), p. 12 (based on numbers of the Infratest-Heimerhebung 2005).

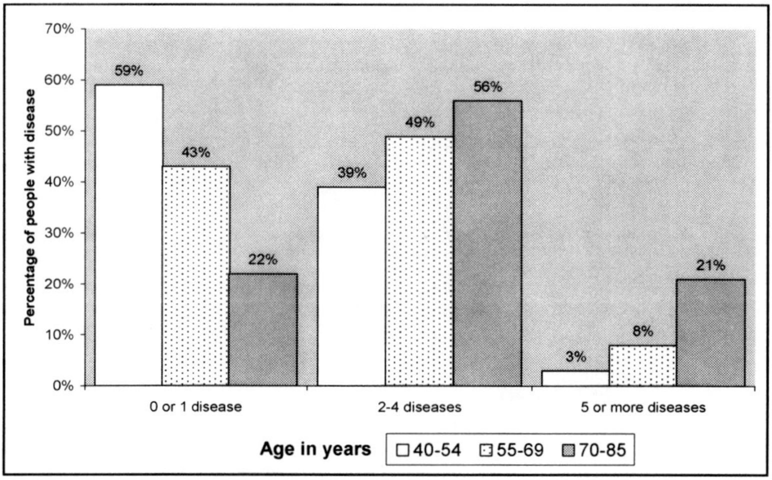

Figure 10: Multimorbidity for selected age groups
(Source: **Wurm, S./Schöllgen, I./Tesch-Römer, C.** (2010), p. 99; based on the data of the DEAS 2008.)

A significant part of the respective age group becomes nursing home residents only at very old age (s. Figure 11). Even if the rate of institutionalization and the likelihood of multimorbidity increases when people age, it is important to consider that the objective judgment of physicians and researchers usually differs from the subjective perception of illnesses and its severeness by the person concerned.[108] The subjective well-being, for instance, of the age groups between 40 and 85 years is at a comparable high level.[109]

It is important to point out that the biological changes that have been discussed in this chapter represent average results. Deviations from the described phenomena exist due to the heterogeneity of the analyzed group of elderly persons. This means that on an individual basis, diseases or biological changes may occur to a stronger or weaker extent, or even not at all. Still, they can be expected at certain tendency.

108 Cf. Zank, S. (2000), pp. 44.

109 Cf. Tesch-Römer, C./Wiest, M./Wurm, S. (2010), pp. 263.

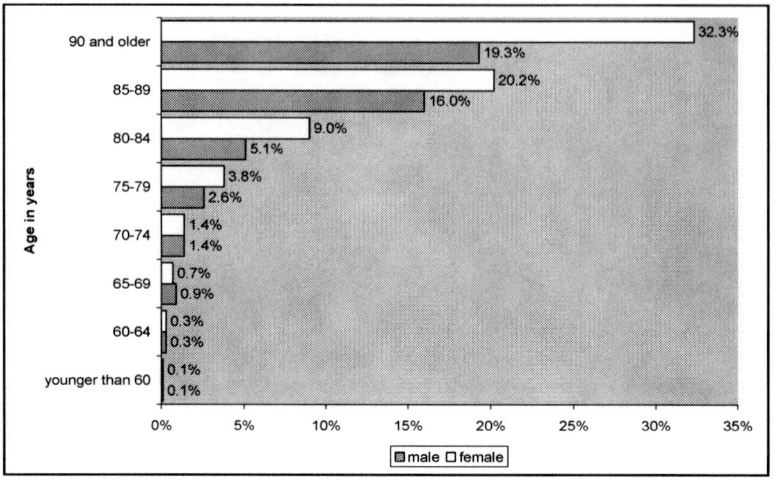

Figure 11: Percentage of people per age group living in nursing homes
(Source: **BMFSFJ** (2006), p. 13; according to the 2005 Infratest-Heimerhebung.)

2.4 Psychological Aspects of Aging

Findings about psychological changes in the aging process add important knowledge to gerontological research. Researchers especially deal with the issues of age related changes of learning and memory (Ch. 2.4.1), intelligence (Ch. 2.4.2), and personality (Ch. 2.4.3) which is why these issues are picked up here as well and will be discussed in the next chapters.

2.4.1 Learning and Memory

The general information processing model represented in Figure 12 shows the flow of information from perception of the stimulus by sensory organs via sensory memory to the primary memory form where the information passes to central processing. After having processed the information, the individual gives a response and/or the information is stored through learning in secondary memory from where the information can be recalled at a later point in time. The model shows that the changes in the sensory organs which have been discussed earlier can already have an effect on information processing.

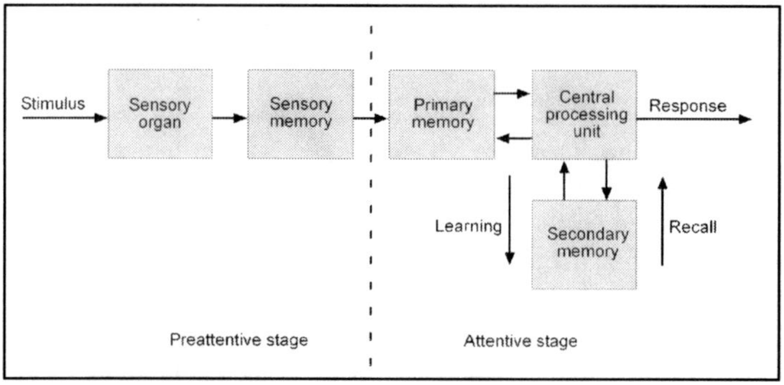

Figure 12: Schematic representation of the information processing model
(Source: **Hooyman, N. R./Kiyak, H. A.** (2008), p. 184.)

Various studies found evidence that the ability to learn and remember (recall information) decrease at higher age.[110] There seems to be no area of the brain that remains unaffected by age related decrements. It is notably the episodic memory that shows the strongest impairment.[111] The episodic memory relates to the past and stores information about specific events or episodes of the more recent past (e.g. remembering whether one has just turned on the stove) or more distant past (e.g. remembering the first day of school).[112] Different to that, all non-episodic forms of memory are oriented toward the present or future when the information is retrieved.[113] The procedural memory (e.g. remembering how to ride a bike) and the semantic memory (containing knowledge about the general world as for example language, facts, symbols) can be accounted to the group of non-episodic memories.[114] The individual is usually not aware of the contents of these memory types. The existence of knowledge in these areas is most likely to be identified indirectly when an individual *performs* a task – then, it can be derived that the necessary knowledge is existing.[115] While on average the episodic memory is strongly affected by age, the performance of *non-episodic* memory tasks indicates no or just little decrease of learning and remembering at older age when these parts of the memory are addressed.[116] More specifically, recognition tasks are well performed by most elderly persons so

110 See the studies cited in Zacks, R. T./Hasher, L./Li, K. Z. (2000), p. 293.
111 Cf. Wahl, H.-W./Heyl, V. (2004), pp. 170; Bäckman, L./Small, B. J./Wahlin, A. (2001), p. 349.
112 Cf. Robnett, R. H./Chop, W. C. (2010), p. 123.
113 Cf. Bäckman, L./Small, B. J./Wahlin, A. (2001), p. 350.
114 Cf. Robnett, R. H./Chop, W. C. (2010), pp. 123.
115 Cf. Wahl, H.-W./Heyl, V. (2004), p. 170.
116 Cf. Robnett, R. H./Chop, W. C. (2010), pp. 124.

that only small age differences can be observed, whereas the largest differences are uncovered for free-recall tasks when for example unrelated words are presented and no retrieval cues exist.[117]

The lower performance in memory tasks have been attributed to slowed neural transmissions, lower working memory capacities, and negative changes of attention (elderly are more prone to distraction by irrelevant stimuli).[118] Although the discussion shows that people at old age do not achieve results as good as younger persons when doing a memory test, it needs to be taken into consideration that oftentimes some basic test conditions may negatively impact their test performance. In other terms, the results of training related studies can be improved, among others, when materials are used with which the elderly are familiar in everyday life, the learning pace can be adapted by the test person, instructions are well-defined and unambiguous, or the learning materials are adapted to sensory deficits of elderly (e.g. font size, contrasts, adequate light). The learning performance of older persons can best be improved, however, when the concerned test participants are highly motivated and have an own interest in improving their performance as for example when learning improvements will have a positive effect on their daily life.[119]

2.4.2 Intelligence

When it is intended to measure intelligence, a differentiation is typically made between "fluid" and "crystallized" intelligence.[120] Fluid intelligence for example comprises the ability to quickly and accurately process information, to perform multiple activities, and to perform tasks of discrimination, comparison, or categorization. Different to that, stored information that has been acquired in the course of a lifespan, well-learned skills, as well as language comprehension rely on crystallized intelligence.[121]

Studies on adult intelligence in the 1950s report of a decline of intelligence when people age.[122] However, this holds true only for one type of intelligence.

---

117 See the overview of studies in Wilmoth, J. M./Ferraro, K. F. (2007), pp. 94; Wahl, H.-W./Heyl, V. (2004), p. 171; and Zacks, R. T./Hasher, L./Li, K. Z. (2000), p. 295.

118 Cf. Sigelman, C. K./Rider, E. A. (2012), pp. 267; Rogers, W. A./Fisk, A. D. (2001), pp. 271.

119 Cf. Martin, M./Kliegel, M. (2005), pp. 164.

120 Cf. Cattell, R. B. (1987), pp. 87.

121 Cf. Stuart-Hamilton, I. (2011), p. 175; Robnett, R. H./Chop, W. C. (2010), pp. 126; Wilmoth, J. M./Ferraro, K. F. (2007), p. 96.

122 Cf. Erber, J. T. (2010), pp. 176.

While geronto-psychological studies aiming to test fluid intelligence showed that old participants score lower than their younger counterparts, old persons performed as good as or even better than young participants in tests which are focusing on crystallized intelligence.[123] Figure 13 illustrates these different developments of fluid and crystallized intelligence. The mentioned longitudinal studies can be credited to have delivered important insights. Cross-sectional measurement methods have, in contrast, often favored young participants.[124] In order to bypass biases, utilizing abstract testing materials which equally disadvantage all participants should be preferred.[125]

As shown in the graphic, decreases in fluid intelligence occur incrementally. But, declines are offset partly by increases in crystallized intelligence. As a consequence, the effects of changes in intelligence become noticeable only at a later point in time and complicate the interpretation of research results.[126] Evolutionary and genetically reasons have been postulated to be responsible for intelligence changes in later life, especially for fluid intelligence.[127] Great interindividual differences of intelligence development in the aging process, however, indicate the complexity of this research issue. The complexity is caused by the sophisticated interplay of biological changes of the sensory organs, sociobiographic variables (e.g. education, social class, income), and the cognitive changes. This interplay makes it difficult to find clear evidence of age related differences and to isolate the relevant determinants.[128]

---

123 See for example Robnett, R. H./Chop, W. C. (2010), pp. 126; Baltes, P. B. (1990), pp. 5 and the cited literature.

124 Cf. Sorce, P. (1995), pp. 468.

125 Cf. Stuart-Hamilton, I. (2011), pp. 175.

126 Cf. Wilmoth, J. M./Ferraro, K. F. (2007), p. 96.

127 Cf. Robnett, R. H./Chop, W. C. (2010), p. 127.

128 Cf. Wahl, H.-W./Heyl, V. (2004), pp. 168.

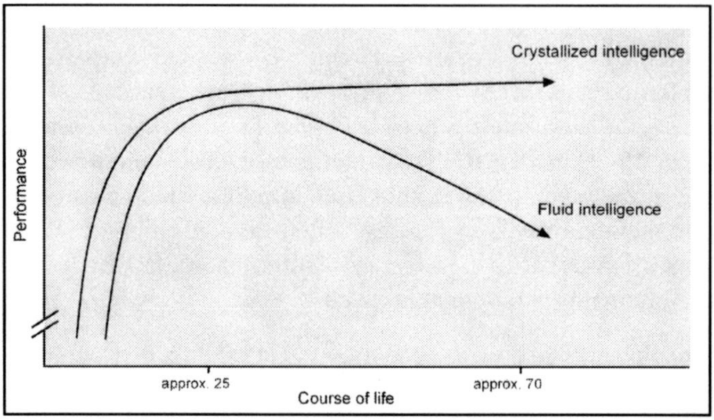

Figure 13: Schematic representation of the development curves of different intelligence types over life-span
(Source: **Baltes, P. B.** (1990), p. 5.)

2.4.3 Personality

Personality refers to "more or less stable, internal factors that make one person's behavior consistent from one time to another, and different from the behavior other people would manifest in comparable situations."[129] In other words, it is assumed that personality is stable, internal, consistent, and different.[130] Although one of the assumptions of personality is stability, this does not mean that a personality does not undergo *any* changes. But, these changes occur rather gradually and in the long term than in the short term. Therefore, one of the central questions in gerontology relating to personality is whether a personality changes in the aging process and, if yes, whether the whole personality is affected or whether it is just specific areas of the personality that alter over time.[131]

Hoyer/Rybash/Roodin (1999) differentiate between three broad approaches that attempt to describe and explain the personality development of older adults: the stages approach, the trait approach, and the life events approach.[132] While the life events approach will be discussed in more detail later in the context of sociological aspects of aging (s. Ch. 2.5.4, pp. 49), the stages approach, which is dividing the lifespan into eight major periods, is omitted. It has been, among others, criticized for its abstract character and dichotomous style of the stages

129 Child, I. (1968), p. 83.

130 Cf. Hampson, S. E. (1992), pp. 1.

131 Cf. Wahl, H.-W./Heyl, V. (2004), p. 173.

132 Cf. Hoyer, W. J./Rybash, J. M./Roodin, P. A. (1999), pp. 377.

(e.g. intimacy vs. isolation in the early adulthood stage, or integrity vs. despair in the late adulthood stage) which is neglecting the individual differences of the persons.[133] Therefore, the focus will be on a brief description of the trait approach in this chapter.

The personality definition refers to internal factors that are more or less stable. The ensuing discussions in the research community dealt with the question of how to conceptualize these factors.[134] One approach that has received much attention was the trait approach, which suggested that several broad dispositions exist which account for the personality.[135] Cattell (1979) puts it as follows: "A trait may be defined as that which defines what a person will do when faced with a defined situation."[136] Thus, the individuals' behaviors serve as a way to determine their personality traits.

Many of the studies that identify personality differences between young, middle-aged, and old adults come to this result because they detect existing differences between various generations (cohorts), not related to age per se.[137] Quite the contrary, Costa and McCrae employed cross-sectional as well as longitudinal research in the trait approach.[138] The five traits (also called "the big five") that have been found to remain stable throughout adulthood are:[139] neuroticism, extroversion, openness to experience, agreeableness, and conscientiousness.[140] Instead, other personality traits such as self-confidence, cognitive commitment, outgoingness, and dependability show some change when they are compared over a 30 to 40 year period.[141] Other longitudinal studies of dispositional traits, such as The Berkeley Studies, unveiled that personality is more likely to show variability when the considered time period contains transitions from for example young adulthood to parenthood or from employment to retirement.[142] Overall, it can be doubted that the personality stops changing with a certain age and

133 Cf. Franz, C. E./White, K. M. (1985), pp. 22.
134 Cf. Hampson, S. E. (1992), p. 3.
135 Cf. Hirschberg, N. (1978), p. 45.
136 Cattell, R. B. (1979), p. 14.
137 Cf. Ruth, J.-E./Coleman, P. (1996), p. 314.
138 Cf. McCrae, R. R./Costa, P. T. (2003); Costa, P. T./McCrae, R. R. (1994), pp. 28.
139 See also the description summary of the five factors given by Cavanaugh, J. C./Blanchard-Fields, F. (2002), pp. 332.
140 Cf. Costa, P. T./McCrae, R. R. (1988), pp. 855.
141 Cf. Jones, C. J./Meredith, W. (1996).
142 Cf. Haan, N./Millsap, R./Hartka, E. (1986), pp. 226.

stays stable from then on throughout life. It is more likely that "some traits change, opening the door to personality development in adulthood."[143]

2.5 Sociological Aspects of Aging

After having discussed biological and psychological findings that play a major role in gerontological research, this section is dedicated to important sociological research results. With the assessment of the development of German household structures one sociological aspect has been dealt with already in the demography chapter. In the following, the issues of daily activities of elderly individuals (Ch. 2.5.1), the development of social networks (Ch. 2.5.2), the family life cycle (Ch. 2.5.3), and critical life events and daily hassles (Ch. 2.5.4) will be looked at more closely.

2.5.1 Daily Activities

To better understand older people, it is helpful to take a closer look at their social life. While the development of social networks in old age will be discussed in further detail later (s. Ch. 2.5.2, pp. 46), it is looked at the elderlies' daily activities here: How is their day structured? How much time do they spent for specific tasks or activities? What are preferred leisure time activities?

The combination of increasing life expectancy, the ongoing trend of early retirement in the last decades, and the structural changes at higher age has brought forward the term "deprofessionalization"in old age.[144] Daily work (e.g. household work), regeneration and leisure activities – in general all activities that can be defined as distractions from commitments and stereotyped behavior patterns[145] – dominate and form an area of life which is free from paid work.[146] Figure 14 shows the average time spent by four age classes (20-39; 40-59; 60-69 and 70 years and older) in six activity categories (sleeping; eating/personal hygiene; paid work/further education or qualification; unpaid work; social contacts/entertainment/events; and sport/hobby/use of mass media).[147] It elucidates the sharp increase of both, unpaid activities and leisure time activities at higher age.

143 Cavanaugh, J. C./Blanchard-Fields, F. (2002), p. 336.

144 Cf. Tews, H. P. (1990), pp. 484.

145 Cf. Kolland, F. (2000), p. 179.

146 Cf. Engstler, H. et al. (2004), pp. 218.

147 Similarly, see Weick (2004) who illustrates the categories of weekdays/working days activities with a finer gradation of 13 age classes. Cf. Weick, S. (2004), p. 414.

With regard to leisure time activities, it is especially time watching television which increases in old age – from 126 minutes per day (25 to 45-year) to 163 minutes per day[148] (65-year and older).[149] Looking at activities where older persons help or assist in other households, taking care of children ranks number one.[150]

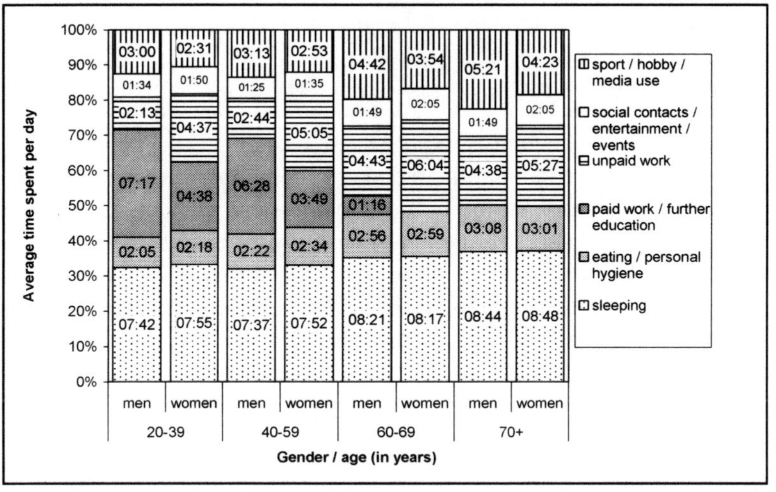

Figure 14: Time spending patterns of men and women of different age classes by activity category (in hours:minutes per day)
(Source: **Engstler, H. et al.** (2004), p. 221; based on numbers of the Zeitbudgeterhebung 2001/02.)

Compared to former cohorts, today's generation of elderlies possesses largely more possibilities of personal development. They can more easily pursuit their interests and organize their spare-time. These possibilities emerge thanks to their higher average income and assets, higher degree of mobility, better education, and better state of health.[151] Since the state of health has been developing positively, the mobility remains high up to old age – especially when compared to earlier generations. This increased mobility allows elderly people, among others,

148 The indicated minutes refer to the average TV-watching time for individuals who used this medium for at least five or ten minutes.

149 Cf. Jäckel, M./Wollscheid, S. (2004), p. 381.

150 Cf. Engstler, H. et al. (2004), pp. 230.

151 Cf. Clemens, W./Naegele, G. (2004), p. 394; Tews, H. P. (1993), pp. 34.

to benefit from a wider range of leisure time activities that would have been impossible to perform in the case of poorer health status and reduced mobility.[152]

These time spending analyses indicate averages for specific age classes. Although they do not reveal the fluctuation margins which are expressed in certain leisure activities,[153] they provide a first insight into this issue and point to the differences existing between older and younger individuals.

2.5.2 Social Network

The way the process of aging is experienced substantially depends on the support elderlies receive from family members, neighbors or friends, ergo from their social network. The possibility to maintain social contacts is influenced by various factors. One important moment that changes the structure of the social network is reaching the retirement age. Health limitations, a decreasing social net (through widowhood, death of close relatives or friends, moving/relocation of children, etc.), decreasing income, or living-conditions that enhance isolation are all variables that negatively influence the chance of maintaining social contacts.[154]

While gender or regional differences play a minor role in determining the network size, partnership and family is more important. Persons having a life partner and children possess a larger social network than persons without life partner and without any children.[155] But, evidence supporting the hypothesis that network sizes decrease when people age has also been found. A review of related research studies discovered that people at age 65 and older only have five to 15 important or close social contacts whereas mid-aged adults have an average of 15 to 35 social relationships.[156] The reduction of social contacts, however, is not necessarily a result of external factors and constraints, but can be due to a self-intended disengagement. Friendships involving greater distances between individuals are particularly given up voluntarily. Furthermore, personal contact is often reduced or completely abandoned in case of relationships that are perceived as one-sided or unbalanced. Opposed to this, contacts to family members are oftentimes intensified. If available, it can be the spouse, sisters, brothers, own children, or grandchildren with whom interactions are intensified. Besides this group of persons, long friendships are also carried on, and it is not rare that

---

152 Cf. Ibid., p. 397; Mollenkopf, H. et al. (2004), p. 347.

153 Cf. Kolland, F. (2000), p. 181.

154 Cf. Clemens, W./Naegele, G. (2004), p. 393.

155 Cf. Huxhold, O./Mahne, K./Naumann, D. (2010), pp. 220.

156 See the studies referred in Lang, F. R./Carstensen, L. L. (1998), pp. 56.

those friendships similarly experience intensification with higher age. Crucial to maintaining relations is – besides the already mentioned balance of giving and receiving – a meaningful and positive emotional experience.[157]

The social network represents an important resource for life quality and takes over many different functions which have only briefly been mentioned here. As the interpersonal relationships at old age are expected to play an important role in the context of this study, they will also be discussed in further detail at different sections in this dissertation (s. Ch. 3.5.2.3, pp. 111; and s. Ch. 4.2.3.3, pp. 144).

2.5.3 Family Life Cycle

The family life cycle is based on the idea that people who are living in their own household run through different stages in their life. Attributes that are employed to define those stages entail for example the martial status, number and age of children, age of spouse, and employment or retirement status of the spouse.[158] Different to the more general term "life cycle" which is used in various disciplines for the origin, maturation, and death of animals, plants, or products,[159] the term "*family* life cycle" was shaped in American sociology of the 1930s.[160] The intention was to segment the family life into stages, determine their demographic structure, unveil sociological changes that occur at the different stages, and prove the existence of a pre-determined progression through the defined family life cycle stages.

In the following, the life cycle concept was applied in many other research areas for example to describe the financial situation of families.[161] In marketing research the family life cycle concept received a lot of attention. There, the concept was used to explain consumer expenditures and to predict consumption patterns due to specific needs of the people at a certain stage.[162] In an early study Wells/Gubar (1966) differed between nine life cycle stages:[163]

157 Cf. Lang, F. R. (2000), pp. 142.

158 Cf. Freter, H. (2008), pp. 101.

159 Cf. Kamenz, U. (1987), p. 10.

160 See for example Loomis, C. P. (1936); Loomis, C. P./Hamilton, C. H. (1936); Sorokin, P. A./Zimmerman, C. C./Galpin, C. J. (1931).

161 Cf. Dorbritz, J./Gärtner, K. (1995), pp. 262.

162 See for example Kroeber-Riel, W./Weinberg, P. (2003), pp. 449; Wilkes, R. E. (1995); Kamenz, U. (1987), pp. 87; Müller-Hagedorn, L. (1986), pp. 175; Lansing, J. B./Kish, L. (1957).

163 Cf. Wells, W. D./Gubar, G. (1966), p. 362.

1. Bachelor stage (young single people not living at home)
2. Newly married couples (young, no children)
3. Full nest I (youngest child under six)
4. Full nest II (youngest child six or over six)
5. Full nest III (older married couples with dependent children)
6. Empty nest I (older married couples, no children living with them, head in labor force)
7. Empty nest II (older married couples, no children living at home, head retired)
8. Solitary survivor (employed / self-employed)
9. Solitary survivor (retired)

Each stage has its own characteristics that influence consumption. Starting a family, children leaving their parent's house, reaching the retirement age etc. change the structure of needs and therewith the consumption patterns which are leading to a shift of expenses into different areas. When children move out of the family's home and become independent, the parents' disposable income increases and can be used for own consumption purposes. When looking at consumer behavior of furniture for instance, married couples show the greatest interest for new purchases and then again when the children have aged or move out of the household. And while young families focus more on practicability, in later stages the attractiveness and representativeness becomes more important in the purchase decision.[164]

In gerontology, the dynamic life stages approach helped to get away from a too narrow view of merely analyzing people at "old age" (old age as one stage of the life cycle). The differentiation of "old age" into several phases came along with the hypothesis that people of different stages differ from each other, while similar social characteristics were attributed to those people at the same life stage. More precisely, it was assumed that people at a certain life stage take on similar roles, are in a situation where they have to adapt their self-concept, or have a similar view of their identity.[165]

164 Cf. Freter, H. (2008), pp. 102.
165 Cf. Backes, G. M./Clemens, W. (1998), pp. 148.

Problems and limitations of the family life cycle concept refer to the definition and operationalization of the stages as well as to the interpretation of the results.[166] Since household structures are becoming increasingly diverse, the early approaches have been criticized for their too simplistic analysis of family situations. While traditional models started with as few as three stages, more recent models consist of ten, twelve, or even up to 24 stages.[167] These approaches try to integrate the non-traditional households and the changing situation of family structures caused by the decline in average family size, delay in time of first marriage, delay in parenting, increasing divorce, rising number of blended families, etc. Further, modern approaches are more flexible and do not hold on to the orderly progression of stages as they allow skipping stages or reverting to a prior stage.[168]

With the reflection of current trends the models have become more precise in considering the actual living situations of people, but they have also become increasingly complex. Another consequence of the differentiation of the family life cycle models is that it becomes increasingly difficult to make universally valid statements: Research results are exceedingly fragmented and each newly identified segment that is added to the concept represents a smaller group of people.

2.5.4 Critical Life Events and Daily Hassles

The exploration of critical life events has an important place in the socio-psychological gerontology. Critical life events can be defined as situations in which a burdening discrepancy between a desired and expected state is perceived or anticipated by the concerned person. On the one hand these situations include foreseeable events (as for example the children's moving out of the parental home, retirement etc.) and unexpectedly, unforeseeable, or rarely occurring events (as for example death of the spouse, severe illness etc.). On the other hand the critical life events can be punctually (e.g. moving out, acute illness) or permanently burdening (e.g. chronic illness, need for care).[169] The definition of

166 Cf. Müller-Hagedorn, L. (1984), pp. 562; Derrick, F. W./Lehfeld, A. K. (1980), p. 214.

167 See the overviews given by Kamenz, U. (1987), pp. 15 and Müller-Hagedorn, L. (1984), pp. 564.

168 Cf. Schaninger, C. M./Danko, W. D. (1993), pp. 581; Danko, W. D./Schaninger, C. M. (1990), p. 887; Gilly, M. C./Enis, B. M. (1982); Murphy, P. E./Staples, W. A. (1979), pp. 15.

169 Cf. Martin, M./Kliegel, M. (2005), pp. 91; Wentura, D./Greve, W. (2000), pp. 49; Filipp, S.-H. (1999), pp. 116; Backes, G. M./Clemens, W. (1998), p. 162; Filipp, S.-H. (1990).

critical life events already shows why especially gerontologists have dealt with this issue. Many of the inevitable crisis situations are more likely to occur at higher ages. Therefore, gerontological research seeks to understand the effects of such events as well as to find intervening variables and coping strategies.[170] Coping can be defined as the "constantly changing cognitive and behavioral efforts to manage specific external and/or internal demands that are appraised as taxing or exceeding the resources of the person."[171]

In order to identify the severeness of critical life events, they have been classified according to their "stress value." For this reason Holmes/Rahe (1967) carried out a study following the "Social Readjustment Rating Scale." The study results are shown in Table 7 and serve as one approach for the classification of life events. Whether or not an event has a "critical" effect largely depends on the concerned person and the possibility or impossibility to influence the event.[172] The extent to which the critical event

Rank	Critical life event	Stress value
1.	Death of a spouse	100
2.	Divorce	73
3.	Marital separation	65
4.	Imprisonment	63
5.	Death of a close family member	63
6.	Personal injury or illness	53
7.	Marriage	50
8.	Dismissal from work	47
9.	Marital reconciliation	45
10.	Retirement	45
11.	Change in health of family member	44
.	.	.
.	.	.
.	.	.
40.	Change in eating habits	15
41.	Vacation	13
42.	Christmas	12
43.	Minor violation of law	11

Table 7: Rank order of critical life events with the highest stress value (Source: **Holmes, T. H./Rahe, R. H.** (1967), p. 216 CBib70.)

170 Cf. Lehr, U. (2007), pp. 179; Hoyer, W. J./Rybash, J. M./Roodin, P. A. (1999), pp. 401.
171 Lazarus, R. S./Folkman, S. (1984), p. 141.
172 The life event is the more critical the less the person has an influence on its occurrence/happening.

can be anticipated, whether it questions the values of the concerned individual, as well as at what life-stage this critical event happens is also of major significance.[173] Assessing these situations merely in an objective way does not lead to a correct analysis because subjective interpretation can intensify or relativize effects.[174] On the other side, critical events should not only be of subjective but also of objective relevance.[175] In this context, Olbrich (1995) explicitly points to the fact that critical life events – besides the negative aspects most studies have their focus on – may work as a stimulus initiating or supporting a positive development.[176] In the negative case, however, these life events are able to threaten the continuity and integrity of the concerned adult's self-concept and self-esteem.[177]

Due to the extensive research efforts in the field of severe life events which have a negative effect on the concerned person, the relevance of many obstacles and little problems of the daily life have long been neglected. These obstacles are subsumed under the keyword "daily hassles." Among others, problems related to work, family or financial stability but also losing things, planning meals, having to wait, or filling out forms can be accounted to the daily hassles which are potential sources that create stress.[178] Though not so drastic when observed separately, all these small problems can add up and finally become larger obstacles reducing the quality of life.[179] The consequences of daily hassles can have serious consequences on the (subjectively) perceived well-being as well as on the (objective) health – sometimes these consequences are even more serious than the effect of critical life events.[180] However, research shows that it is extremely difficult to measure the extent and consequences of daily hassles. Furthermore, a clear separation of daily hassles and critical life events is not always self-evident, due to the interdependences between these two concepts.[181]

2.6 Aging Theories

To get a better understanding of the research about aging, different gerontological theories are presented. The objective of these theories is to provide *explana-*

173 Cf. Filipp, S.-H. (1999), pp. 116.
174 Cf. Ibid.
175 Cf. Backes, G. M./Clemens, W. (1998), p. 162.
176 Cf. Olbrich, E. (1995), pp. 133.
177 Cf. Diehl, M. (1999), p. 175.
178 Cf. Hooyman, N. R./Kiyak, H. A. (2008), p. 220; Wheaton, B. (1999), p. 284.
179 See Filipp, S.-H. (1990), p. 295 and the cited studies.
180 Cf. Lazarus, R. S./Folkman, S. (1984), p. 13 and pp. 311.
181 Cf. Filipp, S.-H. (1990), p. 296.

tions for the aging process. Therefore, the most important theories will be picked up in the following which are structured according to the categorization suggested by Wahl/Heyl (2004).[182] The authors differentiate between theories of change (theories that emphasize the changes that occur in later life) and theories of continuity (theories that emphasize factors that function continuously and do not change over time). Both categories are subdivided depending on whether they rather stress differential or universal elements. Theories which predominantly focus on individual *differences* (between persons) in later life are associated with the differential theories. The other category consists of theories focusing on *universal* aspects of aging which mainly apply for all aging individuals. The authors even introduce a third form of classification by assigning the theories into rather thematically oriented approaches and approaches taking the meta-perspective. The meta-perspective theories are not limited to a concrete issue of aging (e.g. the development of intelligence or social networks) but investigate global issues about general aspects of aging (e.g. the relationship between the development of intelligence and social integration).[183] To reduce the complexity of the theory categorization, no distinction is made between thematically oriented and meta-perspective theories, as this is beyond the scope of this thesis.

2.6.1 Change Theories

Change theories with focus on universal elements: Biologically oriented theories can be attributed to the change theories which partly focus on universal elements and partly on differential ones. In the case of biological aging theories, the aging process is examined on a cellular and molecular level. One approach that attempts to explain the decrease in the functionality of organs and organ systems at higher age draws on stochastic processes on the one hand and deterministic processes on the other hand.[184] Stochastic processes argue that aging occurs because of incidental damages e.g. through little injuries, pollutants or free radicals. The decreasing effectiveness of repair mechanisms leads to weaker remedies for those little damages.[185] As a consequence the organs "age" whereas all individuals are affected more or less in the same way. Different to that, the theories that focus on deterministic processes hold genetic influences responsible for

182 Cf. Wahl H.-W. /Heyl, V. (2004), pp. 136.

183 Cf. Wahl H.-W. /Heyl, V. (2004), pp. 137.

184 Cf. Danner, D. B./Schröder, H. C. (1994); Dandekar, T. (1999).

185 Cf. Danner, D. B./Schröder, H. C. (1994), p. 111; Vömel, T. (1991), pp. 47.

the aging process and therewith focus on the individual differences of the process.[186]

The "deficit model" was based on the published results of Yerkes (1921) who developed an intelligence test with other researchers in order to assess U.S. army recruits.[187] The model belongs to the early psychological aging theories that can be accounted to the change theories with a universal focus. The author assumed that aging inevitably includes loss: a general decline of competence as well as a deterioration of social, emotional and affective capabilities. However, following research studies could not confirm this unavoidable process of losses. Quite in contrast, the limited design of the proposed deficit model has unequivocally been proven wrong.[188]

With the so called "disengagement theory" Cumming/Henry (1961) introduced another aging theory.[189] Their theory picks up individual as well as social aspects of aging. At higher age, the number of contact persons is reduced, e.g. through death of relatives or friends. Apart from this undesired reduction, it has been additionally hypothesized that aged people desire to withdraw from social relationships leading to social isolation.[190] This statement has been criticized in the following equally as the postulate that societies should not engage in activity programs for elderlies as this would "disturb" the disengagement which is a necessary adaptation process to higher age.[191]

One theory which can be categorized as focusing on differential elements has already been mentioned in the beginning of this chapter in the context of biological theories (the approach of deterministic processes). One further theory belonging to that category is the "cognitive theory of aging" by Thomae (1970).[192] His approach differs from the deficit and disengagement view as he puts an emphasis on the subjective perception and interpretation of the aging process instead of persisting on objective conditions in the life of elderlies. De-

---

186 Cf. Dandekar, T. (1999), pp. 243; Backes, G. M./Clemens, W. (1998), pp. 90. For an overview of further biological theories see for example Schachtschabel, D. O. (2004), pp. 167; Danner, D. B./Schröder, H. C. (1994), pp. 109.

187 Cf. Yerkes, R. M. (1921) cited in Lehr, U. (2007), pp. 47.

188 Cf. Brünner, B. O. (1997), p. 25.

189 Cf. Cumming, E./Henry, W. E. (1961).

190 Cf. Ibid., pp. 224.

191 Cf. Wahl, H.-W./Heyl, V. (2004), p. 133.

192 Cf. Thomae, H. (1970), pp. 1.

pending on the biography and the particular motivational situation, similar life conditions can be experienced in extremely different ways by the individual.[193]

2.6.2 Continuity Theories

The continuity theories emphasize on factors that do not substantially change when people age. One of them is the "activity theory" that can be traced back to Havighurst et al. (1970) and Tartler (1961). This theory basically corresponds to the opposite of the disengagement approach as it postulates that it is not the withdrawal from activities, but *higher* levels of activity that lead to a superior quality of life. For this reason, the authors deem the preserving of roles and activity patterns which had been practiced during middle age/mid to late adulthood as meaningful. It has not only been proposed to keep these activities up as long as possible, but to compensate lost roles and activities through new ones.[194] Yet, it has been criticized that no significant correlation has been found between more activity and higher levels of life-satisfaction. It has been argued that the desired degree of activity depends on particular personality traits which in turn have been established in earlier years.[195]

The continuity theory of aging by Atchley (1989) combines the basic ideas of disengagement and activity theory presented above. Central to this theory is the assumption that no single, universally applicable level of more or less activity exits which represents the optimum for *all* elderly persons. According to authors supporting the continuity theory, successful aging means to keep life-patterns intact. People who have been rather passive in younger years must be offered possibilities to withdraw from certain activities. Opposed to that it is important to create varying activities for people who have always been more active when they were younger.[196] The continuity theory stresses on universal elements in the same way as the activity theory. The approach of biographical aging determinants comprises both elements of change-oriented and also of continuity-oriented theories. It can, therefore, not be unequivocally linked to a single of the two categories. This theory basically assumes that there is a strong influence of biographic factors on the aging process. Here, the different courses in life are responsible for interindividual variations inside the group of older people,

---

193 Cf. Lang, F. R. (2000), pp. 147; Backes, G. M./Clemens, W. (1998), pp. 93.
194 Cf. Havighurst, R. J et al. (1970); Tartler, R. (1961).
195 Cf. Rosenmayr, L. (1976), pp. 360.
196 Cf. Atchley, R. C. (1989), p. 184.

whereas intraindividual changes are relevant to the same extent as the stability of some factors.[197]

One aging theory that is not considered in the categorization of Wahl/Heyl (2004) is the context-oriented model. Among others, researchers such as Papalia et al. (2007) understand the aging process as a continuous exchange between individual and environment.[198] Thus, aging can occur in extremely different ways. Cavanaugh/Blanchard-Fields (2002) underline that biological, psychological and socio-cultural forces influence the human being.[199] These forces are, however, moderated through individual life events and experiences. This fact finally makes it so difficult, if not impossible, to predict their effect on an individual person.

In spite of the numerous theories which come to different conclusions it can be summarized that approaches are not always competing against each other, but that theories or models may be complementary, or, valid only in special cases (under certain premises). More recent theories emphasize the complexity and higher number of influence factors on the process of aging. Some changes occur universally (at different points in life) while other factors are individual and, therefore, likely to change for one person but not for another.

2.7 Economic Aspects of Aging

The following two sub-chapters deal with the economic aspects of aging. The financial situation of elderly persons is expected to have a prominent influence on many of the aforementioned aspects (e.g. financing health promoting measures, leisure time activities etc.). Vice versa biological, psychological, or sociological changes evidently impact the financial situation of older individuals (e.g. higher travel or communication expenses when children move out of the house, high expenses due to personal medical condition etc.). So, on the one hand the actual financial situation will be looked at in further detail in Ch. 2.7.1. On the other hand, the structure of expenses of senior households will be analyzed in Chapter 2.7.2.

2.7.1 Financial Situation of Old Consumers

The importance of the research of older persons' financial situations is beyond dispute. Studies found that the subjective well-being correlates with the average

197 See for example Birren, J. E./Birren, B. A. (1996), pp. 283.

198 Cf. Papalia, D. E. (2007), pp. 11.

199 Cf. Cavanaugh, J. C./Blanchard-Fields, F. (2002), pp. 7.

income[200] of a nation.[201] Although it seems that elderly compared to younger age groups are more satisfied with their financial situation,[202] this does not mean that a sound level of wealth becomes irrelevant when people age. To the contrary, Walker (2010) argues that at old age "the objective significance of sufficient earnings increases because the more health impairments that arise and the more fragile social support networks become, the more important is the availability of other supporting resources."[203]

At higher age the employment ratio decreases and therewith the financial means obtained through employed or self-employed work. As Figure 15 shows, a sharp drop of the employed persons share from 71% in the 55-59 age group to 33% in the group of the 60-64 old individuals can be observed. In the adjacent age group of 65-69 year old persons, the large majority is retired and only 2% are still working at that age. But, the data also unveils that people retire later on average. With 20% the working ratio of 60-64 year old persons was 13 percentage points lower back in 2002 compared to 2008.

It is evident that the average income households gain through employed or self-employed work drops due to the higher number of retired persons in the older age classes (s. Figure 16). While for example the average gross income of households, where the head of the household is between 35 and 44 years, is 3,603 €, "older" households merely earn an average of 371 € (head of household between 65 and 69 years).

200 Correlations have been found to be strong especially in low level income groups where small increments in income have a substantial impact on life satisfaction.

201 Cf. Diener, E./Oishi, S./Lucas, R. E. (2003), pp. 410. For the complex interplay of factors that determine the subjective well-being, see also Tesch-Römer, C./Wiest, M./Wurm, S. (2010), pp. 265.

202 Cf. Noll, H.-H./Schöb, A. (2002), pp. 254.

203 Walker, A. (2010), p. 578.

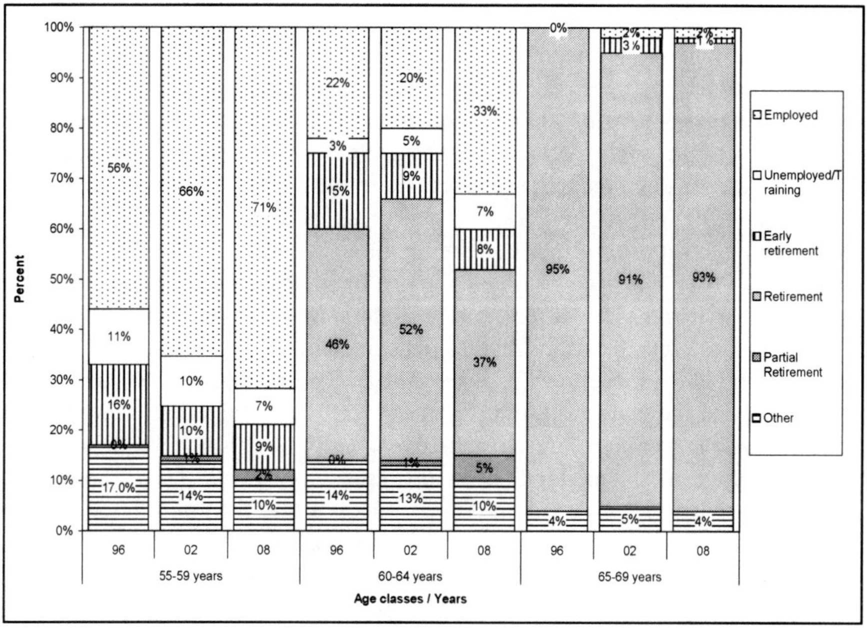

Figure 15: Employment ratios of different age classes in 1996, 2002, and 2008
(Source: **Naumann, D./Romeu-Gordo, L.** (2010), p. 125.)

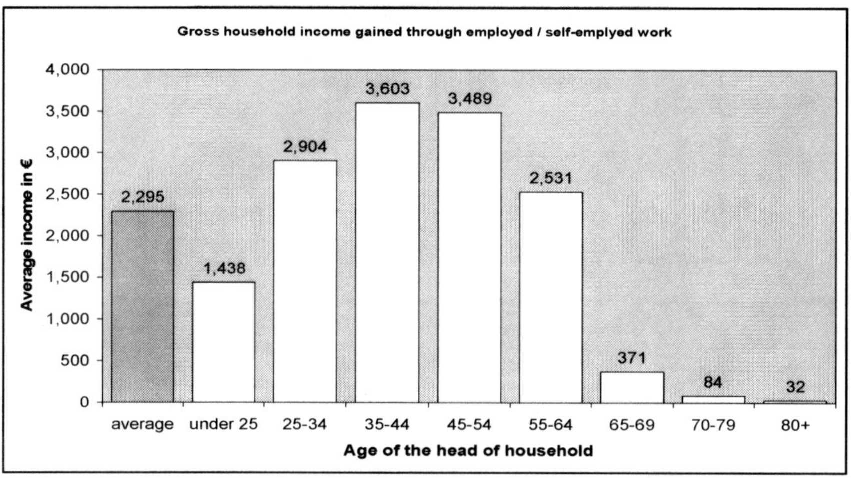

Figure 16: Gross household income gained through employed/self-employed work for different age classes
(Source: Own calculations according to **Statistisches Bundesamt** (2010c), p. 36.)

The drop of direct income is, however, compensated by other monetary sources when people retire. The income earned by employed or self-employed work is mainly substituted by pensions. In Germany the pension system is build on three pillars: the federal pension fund, the corporate pension plan / supplementary pension plan of the public service, and the private pension plan.[204] When including these transfer payments, households where the head of the household is between 65 and 69 years of age, have a total gross income of 2,820 €. For households with the head of the household being in the age class of 70-79 a total gross income of 2,707 € has been recorded.[205] As a breakdown of the gross income of the over 65 year old German citizens in 2004 shows, the large majority of the financial means stem from the federal pension fund. Only 10% stem from private pension plans (s. Figure 17). Even though the gross income of households of the different age classes introduces the overall financial situation of the elderly compared to younger households, several important influence factors have not been looked at so far. First, taxes on regular payments (salaries/pensions) need to be considered. As taxation on pensions is lower than on salaries, the net household income develops positively for elderly households and reduces the gap to younger households (s. Figure 18). Second, as elderly households are, in average, smaller than younger households, the net income is distributed between fewer persons – that means that the net income *per person* is even higher and differences compared to younger households are further attenuated. Due to children leaving the elderly households, even more money can be used for individual consumption purposes.[206]

---

204 Cf. Heinze, R. G./Naegele, G./Schneiders, K. (2011), pp. 46; Naegele, G. (2000), pp. 394.

205 Cf. Statistisches Bundesamt (2010c), p. 36.

206 The shift of the expense structures will be discussed in more detail in the next chapter (s. Ch. 2.7.2, pp. 62).

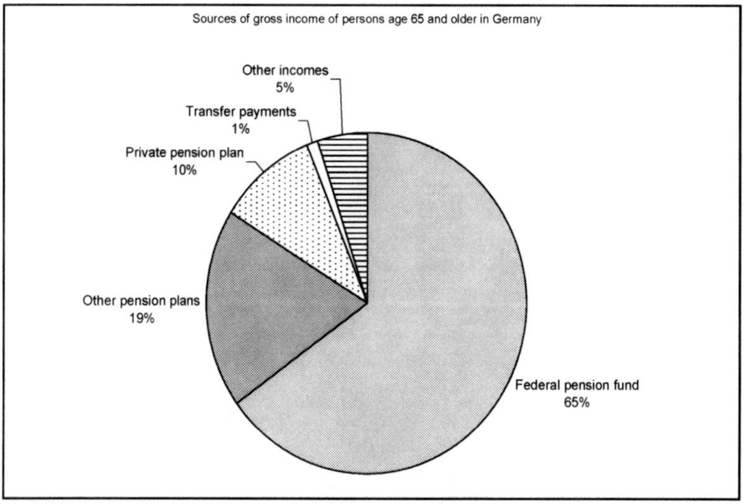

Figure 17: Gross income sources of persons aged 65 years and older in Germany (Source: **BMAS** (2008), p. 98.)

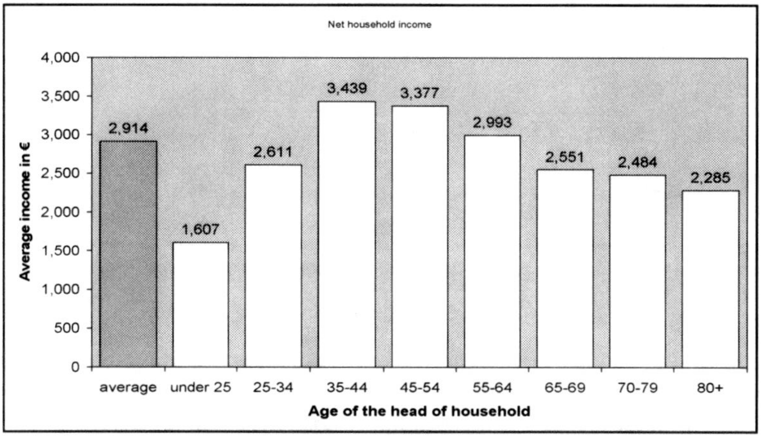

Figure 18: Net household income of different age classes in Germany (Source: Own calculations according to **Statistisches Bundesamt** (2010c), p. 36.)

Finally, the development of assets should be analyzed in the following. It can be observed that the gross total assets steadily increase from the age class of under 25 year old (19,200 €) to the 55-64 age class (194,100 €), the gross total assets is lower for the three oldest household types (s. Figure 19). Here again, an age related fact needs to be considered. At higher age it is more likely that mortgages have already been fully paid so that debts and the related financial burdens of the elderly households are reduced. This is expressed in the total net assets.[207] Here again the differences of the wealthiest household class (head of household is between 55-64 years of age) compared to the three oldest household types is attenuated.

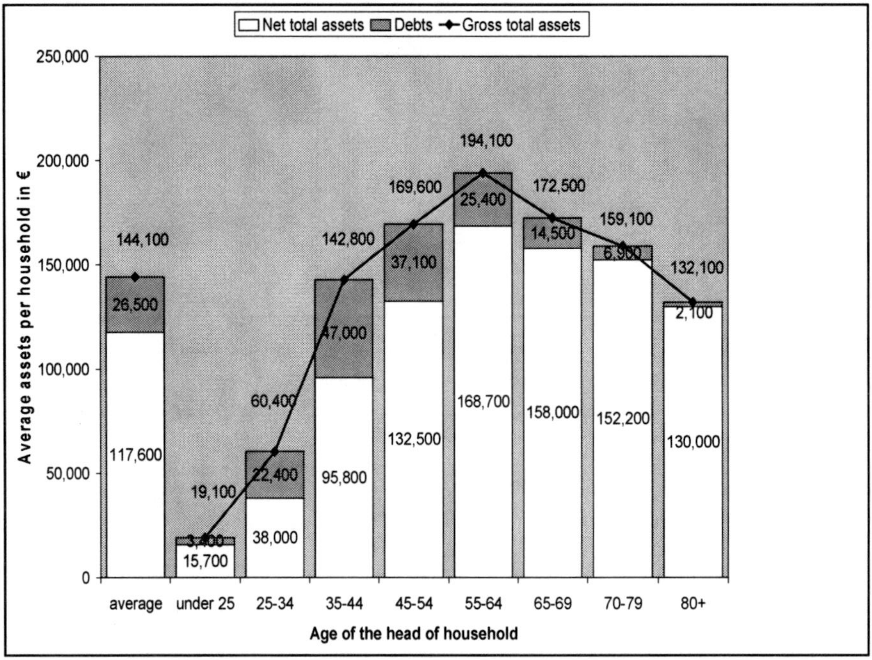

Figure 19: Average assets per household of different age classes in Germany (Source: **Statistisches Bundesamt** (2010d), p. 30.)

The picture that could be drawn from the overall financial situation of aged persons has been positive so far. However the data that has been interpreted referred to average numbers which are masking the diversity of individual conditions. Poverty at old age is a serious issue which is oftentimes neglected when talking about the wealthy "master consumers." In fact, 12.8% of Germans between 66 and 75 years and 16.2% older than 76 years of age are at risk[208] of

207 The total net assets are obtained by subtracting the total amount of debts from the gross total assets.

208 The threshold income of people at risk of poverty has been defined by the OECD (Organisation for Economic Co-operation and Development) as those household incomes that are below the value of 60% of the median equivalent income of all households. Cf.

poverty (s. Table 8). The situation of older people (aged 65 and older) has largely improved over the years and has stayed more or less on a constant level throughout the last ten years.[209] Several younger age classes are even more affected by poverty risk than elderly age classes. At younger age, though, it is not uncommon that people are affected due to temporary unemployment, whereas poverty at a later age is hardly changeable.[210] Moreover, the structure of old age poverty has not changed: it is predominantly single women, especially divorced and unmarried women who are affected by poverty.[211]

Age class	Percentage of people per age class who are at risk of poverty (2008)
17 and younger	16.4
18-25	24.4
26-40	14.0
41-50	11.0
51-65	12.7
66-75	12.8
76+	16.2
average	14.5

Table 8: Percentage of people per age class who are at risk of poverty
(Source: **Goebel, J./Grabka, M. M.** (2011), p. 15 on the basis of SOEP v26; DIW Berlin 2011.)

A positive trend that can be remarked is the higher employment rate of women, due to a change in life style, increased need of work force in the service sector, and better career opportunities.[212] A look at the net monetary assets shows how important the segment of older consumers is. In 2008 nearly 60% of the total net monetary assets belonged to households where the head of the household was 55 years or older (s. Figure 20). Future challenges will arise due to the fact that more non-retired persons were or still are unemployed compared to previous generations and so were not able to comparably pay into the pension fund. The rising number of part-time, or minimum wage workers create similar problems leading to higher numbers of elderly who, after their official retirement, have to accept jobs to increase their pension, which is – and that needs to be kept in

---

Tesch-Römer, C./Wurm, S. (2009), p. 114. In 2008 those households in Germany with an income lower than 935 € were accounted to the households being at risk of poverty. For a definition of the equivalent income see for example Tesch-Römer, C./Andrick, R. (2011), p. 43.

209 Cf. Goebel, J./Grabka, M. M. (2011), pp. 6.

210 Cf. Tesch-Römer, C./Wurm, S. (2009), p. 114.

211 Cf. Naegele, G. (2000), p. 400.

212 Cf. Bäcker, G. (2008), p. 364; Behrend, C. (2000), p. 404.

mind – not always possible, especially not for persons with handicaps or serious diseases. Apart from that, it can be expected that the expenses for aged persons will continuously increase (e.g. for medical care) which may again intensify the phenomenon of old age poverty in the next years. As the public provisions will be successively reduced, corporate and private pension plans will gain in importance in the future.[213]

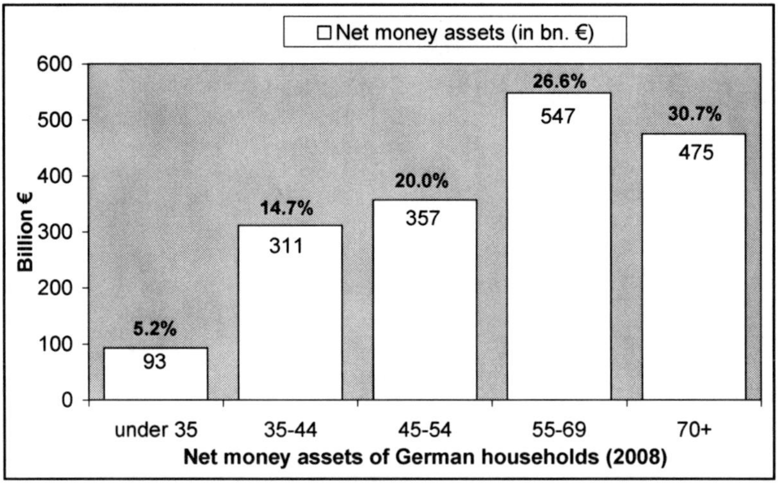

Figure 20: Estimated total net money assets of German households
(Source: Own Calculations according to **Statistisches Bundesamt** (2010d), p. 30.)

2.7.2 Expenses of Old Consumers

In the following sub-chapter it will be analyzed how older households spend their money compared to younger ones. The data will also allow concluding which industry sectors can expect growth due to the demographic changes. While very general aspects of household expenses will be discussed in this chapter, further information is provided about specific products when the final product category is selected.

The German income and expense study "EVS" (Einkommens- und Verbrauchsstichprobe) splits the monthly household expenses into eleven categories (s. Figure 21). As shown in Table 9 with nearly 2,500 €, expenses for private consumption peak where the head of the household is 35-45 and 45-55 years old. Up to that age expenses increase, and past that age expenses drop to an

213 Cf. Motel-Klingebiel, A./Simonson, J./Romeu Gordo, L. (2011), p. 11.

amount as low as about 1,850 € (households with head of the household 80 years and older). This, however, needs to be put into perspective as the average household size decreases with age.

As depicted in the overview of expenses, the order of importance for the four largest monthly expense areas of German households in 2008 (living; transportation; food; recreational activities) are identical in the considered age classes (35-44; 65-69; 70-79).[214] While no big differences exist in the living expenses and expenses for recreational activities, it can be seen that the older age classes spend less money per month for transportation (272 € in the 65-69 age class; 226 € in the 70-79 age class) than the average (328 €) and especially the considered younger age class (395 € in the 35-44 age class). The same holds true for food expenses as well as for several other smaller categories (household devices, accommodation/gastronomy, clothing, and education) where older households spend less money than younger ones. One further category which gains importance at older households are expenses in the healthcare category. Here, households with the head of the household being 65-69 (131 €) and 70-79 (154 €) spend considerably more money than the German average household (94 €) – this corresponds to an increase of 39% and 64% in expenses compared to the average household. Due to these mentioned tendencies, Heinze/Naegele/Schneiders (2011) identify the following "winner" and "loser" sectors[215]:[216]

Sectors that will benefit from the demographically induced increase of the expenditure ratio:

- Living
- Health care
- Recreational activities
- Transportation
- Communication
- Travelling
- Accommodation/gastronomy

214 The noted age is based on the age of the head of the household (the person who is mainly responsible for the household income).

215 Some of the sectors mentioned by these researchers are sub-categories of the sectors listed in Figure 21.

216 Cf. Heinze, R. G./Naegele, G./Schneiders, K. (2011), pp. 73.

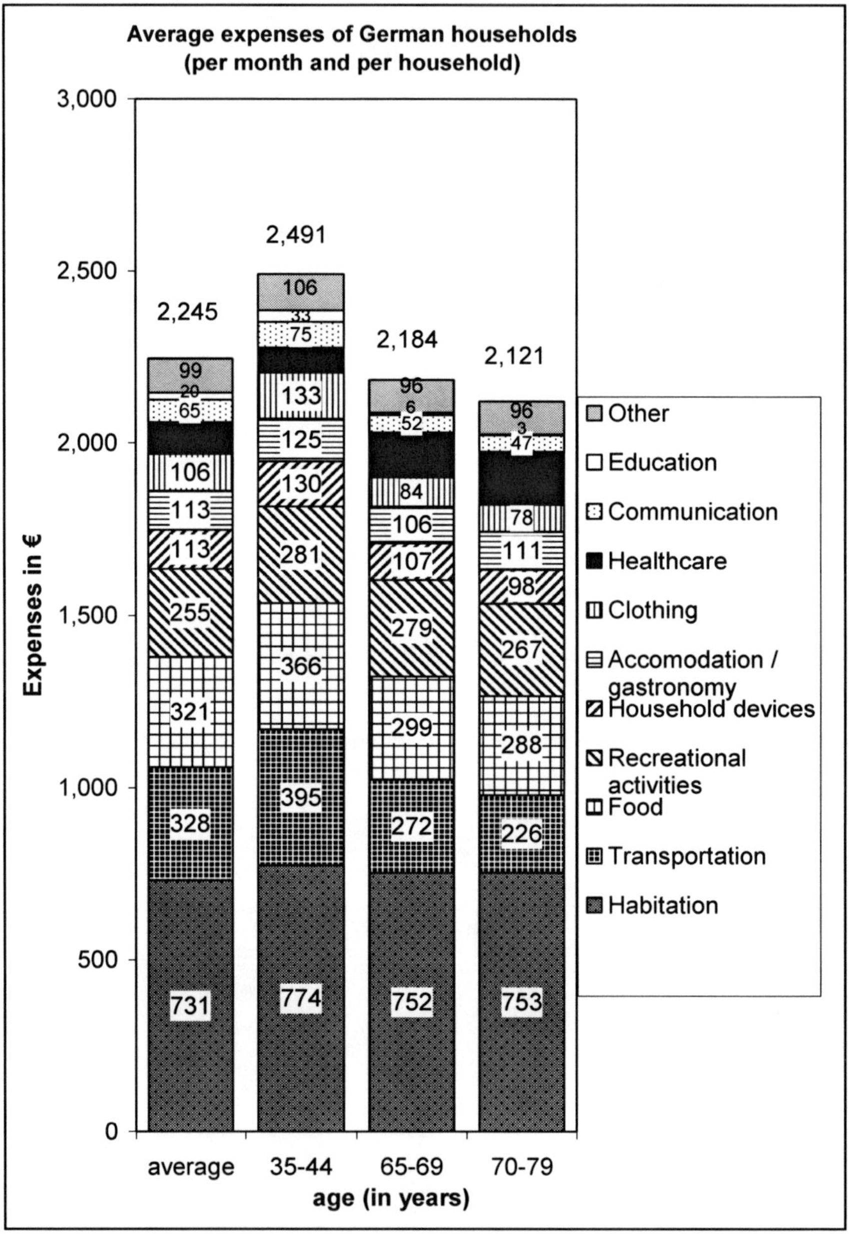

Figure 21: Average expenses of German households per consumption category in €
(Source: **Statistisches Bundesamt** (2010b), p. 43.)

Age class	Average	Under 25	25-34	35-44	45-54	55-64	65-69	70-79	80 and older
Expenses (in €)	2,245	1,378	1,993	2,491	2,490	2,309	2,184	2,121	1,841

Table 9: Average total expenses for private consumption per household in Germany (Source: **Statistisches Bundesamt** (2010b), p. 43.)

Sectors that will be disadvantaged by the demographically induced relative decrease of expenses are:

- Food
- Clothing
- Consumer durables[217]
- Energy and household maintenance

It can be concluded that the elderly households dispose over a sound financial situation. While researchers point to the fact that today's retirees are not in some sort of shopping spree (as oftentimes reported in the news or magazines), it can be observed that the propensity to save money decreased in the last years starting with the age of about 55 and increased again with individuals 70 years and older.[218] This has been explained with the changes in the society's values that has stronger consumptive interests in this age bracket. Another aspect is that due to the increased childlessness in academic households – it lacks potential inheritors. The saving ratio increases again starting with the age of about 70 due to safety motives that are related to expected health costs.

2.8 Conclusion

As this chapter has shown, the science of gerontology covers a broad range of aspects. In order to get a better understanding of "aging," it was attempted to give an overview of the most important findings of the different research areas. It is crucial to understand aging as a dynamic, multidimensional, and differential process.

Both, negative and positive aspects of aging have been discussed. The biological aspects of aging predominantly deal with functional decreases of organs when people age – an increase in multimorbidity at old age was diagnosed. Also deficits in learning and memory were found and critical live events as well as daily

217 Different to non-durables (like food or one-way packaging), durable goods are goods that may be used several times (like cars, furniture, or electronic applications). Cf. Runkel, M. (2004), p. 11.

218 Cf. Heinze, R. G./Naegele, G./Schneiders, K. (2011), pp. 74.

hassles are factors that are more prevalent at old age. They are, in turn, able to negatively affect the elderly individuals' well-being. Different to that, especially psychological and social research can report positive findings and the potentials at old age. Likewise, the outlook for the financial situation of German retirees appears positive. While age poverty should not be neglected, the overall financial situation is positive, particularly when compared to previous generations. The demographic shift in combination with the positive financial situation of elderlies on the one side and existing deficits in consumer knowledge and adequate market offers considering the needs of old individuals on the other side makes these consumers an interesting research focus.

3 Innovation Theory

As was mentioned in the introductory chapter, the dissertation focuses on the behavior of older *and* younger consumers regarding *innovations*. Influence factors on their behavior are to be identified and potential differences are to be revealed. The analysis of the gerontological research results from the last chapter led to the conclusion that many changes occur in later life. In order to accurately assess the decision process of the young and elderly concerning innovations and find the relevant influence factors, it is necessary to carefully study the general theoretical knowledge of innovations first.

Therefore, the following chapter commences with defining innovations (Ch. 3.1). After that, a brief overview on the research areas of diffusion (Ch. 3.2) and adoption (Ch. 3.3) is presented, followed by a closer look at the innovation decision process and its different stages (Ch. 3.4). Finally, the determinants of innovation adoption are studied (Ch. 3.5) which are subdivided into product-specific, personal, and situational influence factors.

3.1 Innovation Definition

The importance for companies to launch innovative products has already been noted earlier. However, it is not only for companies that innovations represent a frequently thematized issue. Innovations are an object of interest in numerous other areas. For example, economic research engages in analyzing the macroeconomic effects of an industry's or nation's innovation activities and the necessary conditions that foster innovation activities. Political science questions which objectives concerning innovation activities should be pursued by politicians and in which manner political influence should be exerted. Aside from that, a scientific-technical, as well as a sociological perspective can be identified. While the former deals with fundamental research in the area of new technological developments, the latter examines innovations from the viewpoint of society and tries to identify important social factors that are either beneficial or obstructive to the innovation and its diffusion among members of the society.[219]

Moreover, when asking the question "what is new?" (object-oriented perspective) not only products and services from the viewpoint of end-consumers are at the heart of the research interest. From a business' or company's perspective, an innovation can also be a process, or technical knowledge (technical innovations), structures, cultures, or systems (organizational innovations). Innovations

219 Cf. Hauschildt, J. (2004), p. 32.

can include the renewal of industry structures, market structures, market scope, business rules, ways of competition (business related innovations), as well as social technologies, political innovations, or new life-styles (social innovations).[220] From a company's perspective, it is also of major interest to research the success factors of innovations, the measurement of a firm's innovativeness, strategies to outperform the competition, the problem of invention, the management of innovations including their marketing, the changes of customer needs, the integration of consumers into the innovation process, or creativity techniques.[221]

This short enumeration of what constitutes an innovation already demonstrates that the term "innovation" is used in various different disciplines and contexts. Accordingly, a great variety of definitions[222] exists making a precise definition of "innovation" indispensable to avoid misunderstandings. As there is no generally accepted definition that incorporates all facets of innovations, it is reasonable to postulate distinctive definitions depending on the research objective.[223]

In spite of the absence of a universal definition, there are some characteristics that are valid across most innovations such as the complexity,[224] risks or insecurities about the innovation's properties or utility.[225] The most striking characteristic, however, is the novelty aspect. It is widely accepted that the aspect of "being new" is the most important and constituting attribute of an innovation.[226]

220 Cf. Zahn, E./Weidler, A. (1995), pp. 362; Zapf, W. (1989), pp. 175. For alternative systematization approaches see Disselkamp, M. (2005), pp. 20; Pepels, W. (2003), p. 5; Trommsdorff, V./Schneider, P. (1990), pp. 3; Dietz, J.-W. (1989), pp. 63; Thom, N. (1980), pp. 32.

221 Cf. Reichart, S.V. (2002); Pepels, W. (1999); Haedrich, G./Tomczak, T. (1996), pp. 171; Kotzbauer, N. (1992); Trommsdorff, V./Schneider, P. (1990); Cooper, R.G. (1979).

222 For an overview of scientific definitions of the innovation term see exemplarily Hauschildt, J. (2004), pp. 4; Garcia, R./Calantone R. (2002); Dietz, J.-W. (1989), pp. 40.

223 Cf. Pohl, A. (2004), p. 80; Frederickson, H. G./Rowe, L. A./Boise, W. B. (1974), pp. 284.

224 Complexity in this context does not necessarily refer to the fact that an innovation is "complex to use,“ but also to complexity in the sense of highly sophisticated technologies that are employed, which may however remain invisible to the end-user of the innovation.

225 Cf. Hoeffler, S. (2002), p. 267; Reichart, S. V. (2002), pp. 17; Sabisch, H. (1991), pp. 10.

226 Cf. Gabler (2004), p. 1496; Zahn, E. (1991), p. 120; Meißner, W. (1989), p. 16; Thom, N. (1980), p. 23.

This novelty attribute is generated by a special combination of "purpose" and "means" that has, until then, not existed before.[227] Schumpeter had already referred to this attribute in the year 1947, when he defined the process of innovating as "doing of new things or doing things that are already being done in a new way."[228] He added the notion that, when talking about innovations, the examined objects must not unavoidably be something completely new. In the same way, Disselkamp (2005) noticed that an innovation can be something already existing that becomes successful only at a later point of time well after its invention. Additionally, he even accounted objects that have only been changed gradually and not radically to innovations.[229] This points out that not only the fact of being new is of importance, but rather different *degrees* of the novelty – in the following also called "innovation degree," "innovation height" or "innovation extent" – will be taken into consideration.

Depending on the degree of novelty, innovations can be classified into incremental vs. radical, continuous vs. discontinuous, or improvement vs. fundamental innovations.[230] Incremental innovations are product or process improvements that lead to a higher quality or increased productivity. They can be found very frequently, whereas radical innovations are normally more spectacular as they satisfy needs for the first time. For the latter, no comparable solution has existed on the market before.[231] The cited dichotomies can be used synonymously. However, they are not able to precisely describe the diversity of existing forms of innovations. In principle, they can merely be understood as endpoints of a continuum that represents the different innovation degrees.[232] Especially companies need a more precise classification. Therefore, scoring models, ordinal scales, multidimensional approaches, and scales that measure the consequences arising from "innovating" were developed and implemented.[233]

The methods enumerated above cannot be simply used to analyze the degree of novelty from a consumer's perspective. The reason for that lies in the fact that the degree of novelty cannot be easily assessed objectively – it is the subjectively perceived extent of newness of an innovation that is decisive for the individual: "It matters little, so far as human behaviour is concerned, whether or not an idea is 'objectively' new. The perceived newness of the idea for the individ-

---

227 Cf. Hauschildt, J. (2004), p. 7.
228 Schumpeter, J. A. (1947), p. 151.
229 Cf. Disselkamp, M. (2005), pp. 16.
230 An overview of further term pairs is given by Hauschildt, J. (2004), pp. 14.
231 Cf. Brockhoff, K (2002), p. 28; Zahn, E. (1991), p. 121.
232 Cf. Dewar, R. D./Dutton, J. E. (1986), p. 1423; Robertson, T. S. (1971), p. 7.
233 Cf. Hauschildt, J. (2004), pp. 14.

ual determines his or her reaction to it. If an idea seems new to the individual, it is an innovation."[234] Personality and context variables are responsible for the subjective perception of things. According to Kelly's personality theory all objects are compared, evaluated, or distinguished within a framework consisting of personal constructs.[235]

The finding that individuals may differently perceive the same objects indicate that besides an object-oriented dimension there is a subject-oriented dimension of the innovation term which expresses for *whom* something is new and how new the innovation is perceived to be. This subject-oriented dimension can be subdivided in a supply and a demand side. The latter, again, can be classified depending on whether the demandside consists of an end-consumer or an organization – and therefore whether the situation involves an individual or group decision.[236] Comparably, Kotzbauer (1992) suggests a classification by opposing the perspective of the consumer and producer by differentiating between a micro and macro level (s. Table 10).[237]

Perspective	**Reference level**	
	Micro level	**Macro level**
Producer	(I) Company novelty	(II) Technical novelty
Consumer	(III) (Perceived) Individual novelty	(IV) (Perceived) Market novelty

Table 10: Classification of innovation perspectives
(Source: **Kotzbauer, N.** (1992), p. 10.)

Since end-consumers represent the central research unit of this study (innovation view (III)), it is essential to pay attention to some particularities that result from the subjective component of the consumers' innovation perception. For instance, it can be the case that a product which is classified as an innovation according to objective methods of measurement is not perceived as such by the consumer. The opposite case would be that a consumer perceives a product as innovative, although no changes of the objective product properties have been performed[238] – in this context researchers speak of a "pseudo-innovation."[239] In the same way, subjectivity also has an effect on the judgment of the innovation degree. It can

234 Rogers, E. M. (2003), p. 12.
235 Cf. Kelly, G. A. (1963), pp. 105.
236 Cf. Pohl, A. (2004), pp. 83.
237 Cf. Kotzbauer, N. (1992), pp. 9.
238 This can be a result of communication strategies.
239 Cf. Brockhoff, K. (2002), p. 28; Helm, R. (2001), pp. 49; Haedrich, G./Tomczak, T. (1996), p. 151; Kieser, A. (1974), pp. 3.

be assumed that two different individuals perceive the same product as new to a different extent. Besides the mere end-consumer perspective, issues regarding certain market segments are considered (e.g. younger vs. older consumers), too. Therefore, not only the individual evaluation whether a product is perceived as new, but also the superordinated market view (IV) can be of major significance.[240] This view helps adjusting the companies' marketing instruments effectively.

The hitherto mentioned classifications of innovations have all been of static nature. Considering a time component adds a dynamic aspect to the innovation concept. As consumers become increasingly accustomed with the attributes of an innovation, they are decreasingly perceived as new over the course of time.[241] Hauschildt (2004) transfers the dynamic view onto the company-level and describes the innovation process at an enterprise. He identified the following steps: (1) idea/initiative; (2) discovery; (3) research; (4) development; (5) invention; (6) start of innovation use; and (7) ongoing innovation use.[242]

As the vast number of innovation related publications showed, there are various perspectives from which innovations can be looked at. Apart from an object-related view which tries to clarify what is new and how new something is, the subject-related view addresses the question for whom something is new. The dynamic aspect will be discussed in more depth in the next chapters of diffusion and adoption. According to the proposed innovation categorization, this study deals with innovations that are concrete market offers in the form of *products* and simultaneously *market innovations* (IV). While product characteristics should be judged as perceived by the end-consumers, the categorization of products on a novelty scale should not be performed by the same individuals. This method of innovation assessment is preferred as it enables the researcher to compare consumers throughout different segments as for example young and old individuals.

3.2 Innovation Diffusion Research

Rogers (1962) defined diffusion as "the process by which an innovation spreads. The diffusion process is the spread of a new idea from its source of invention or

240 See also the discussion in Niedbal, M. (2005), p. 92.

241 Cf. Böcker, F. (1991), pp. 374; Aulinger, R. (1970), p. 31.

242 Cf. Hauschildt, J. (2004), pp. 24; similar see Weis, H.C. (2004), p. 250.

creation to its ultimate users or adopters."[243] The diffusion research takes an aggregated perspective – in other words the sum of all single adoption decisions.[244]

The concept of diffusion is closely related to the concept of the product life cycle.[245] Diffusion research analyzes rate, pattern and potential population penetration of a novelty.[246] The ideal-typical, cumulative pattern of diffusion takes the form of an S-curve (s. Figure 22). It shows the cumulative proportion of single adoption decisions over time in relation to a certain market potential (M) within a social system or market segment. The product life cycle, by contrast, is based on absolute sales levels or cash flow over time.[247]

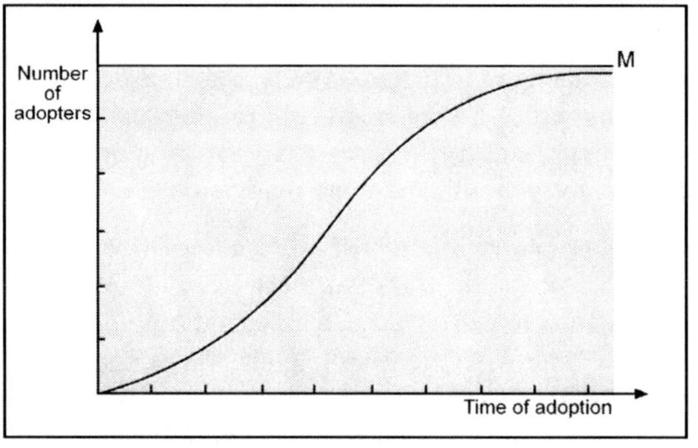

Figure 22: Generalized cumulative diffusion pattern
(Source: **Robertson, T. S.** (1971), p. 31.)

The central elements which are regarded within diffusion theory are: the innovation itself, communication channels, a social system, and time.[248] The innovation is spread through communication within a social system from one individual to another. Due to social relationships between individuals, knowledge about the innovation is communicated from the person who possesses information about the innovation to a person who does not possess any information about the innovation yet.[249] This new knowledge can then lead to an adoption of the novelty.

243 Rogers, E. M. (1962), p. 13.
244 Cf. Helm, R. (2001), p. 112.
245 Cf. Robertson, T. S. (1971), p. 30.
246 Cf. Gatignon, H./Robertson, T. S. (1985), pp. 858.
247 Cf. Sääksvuori, A./Immonen, A. (2005), pp. 181; Cunningham, M. T. (1969); Levitt, T. (1965).
248 Cf. Katz, E. (1961), pp. 72.
249 Cf. Rogers, E. M. (1962), pp. 13.

As the communication of new ideas and the adoption of innovations take time and take place in several stages, the diffusion can be described as a sequential process. Depending on the time needed to adopt the innovation, the individuals can be classified into different categories. For example, persons adopting an innovation relatively earlier compared to the average member of the social system show a higher degree of innovativeness, and are therefore referred to as "innovators." Adoption at a later point in time stands for a lower degree of innovativeness. Arranged in decreasing levels of innovativeness, those groups of individuals are referred to as early adopters, early majority, late majority, and laggards who are the last persons to adopt the innovation (s. Figure 23).[250]

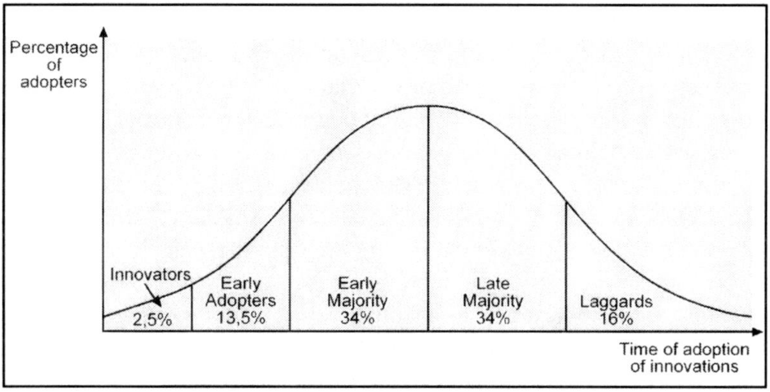

Figure 23: Adopter categories of innovations
(Source: **Rogers, E. M.** (1962), pp. 162.)

Persons belonging to the different categories have been found to possess peculiar characteristics. People with higher education, cosmopolitanism and social status, positive attitude towards change, strong use of mass media, and high knowledge in the innovation field are more likely to be innovators, while later adopters are described as more skeptical and traditional.[251] Rogers furthermore found that noncumulative diffusion curves, which are taking the different adopter categories into consideration, are almost normally distributed as shown in Figure 23. The bell shaped curve starts with innovators who represent 2.5 percent and ends with laggards who are accounting for 16 percent of the total

250 Cf. Ibid., pp. 17 and pp. 148.

251 Cf. Ostlund, L. E. (1972); Rogers, E. M./Shoemaker, F. F. (1971), pp. 184; and the cited references in Dickerson, M. D./Gentry, J. W. (1983), pp. 226.

adopters. While innovators lie two standard deviations to the left of the mean time of adoption, laggards lie one standard deviation to the right.[252]

Different to Rogers' qualitative view of diffusion, Bass (1969) developed the first quantitative model to describe and explain the diffusion process.[253] With this model, he analyzed the diffusion patterns and forecasted future sales of complete product groups. Bass' underlying assumption is that market penetration of the innovation determines its diffusion process.[254] Contrary to Rogers, it is not the social affiliation that is responsible for the differing points in time of the adoption, but innovators who influence later adopters and pressures of the social system. Bass assumes that the propensity for an initial purchase of a new consumer product is linearly related to the number of buyers who have already bought this new product.[255] In the description of his model, the author assumes the existence of just two categories of individuals:[256] innovators and imitators.[257] The former are among the first who buy the innovation and can be characterized as persons who are interested in novelties of a certain product category. They actively search for information and make their purchase decisions independently from other consumers.[258] Opposed to innovators, imitators[259] are influenced by other members of the social system during their purchase decision.[260] They only decide to adopt an innovation if technical, social, or economic risks of the buying situation can be reduced to an adequate level. Among others, risk reduction stems from the degree to which an innovation has already spread, and, associated with that, from the availability of positive experiences from other adopters.[261]

---

252 Cf. Rogers, E. M./Shoemaker, F. F. (1971), p. 182.

253 Cf. Bass, F. M. (1969).

254 Cf. Ibid.

255 Cf. Ibid., p. 217.

256 Though Bass notes the existence of two classes of persons (innovators and imitators), he does not integrate them into his mathematical model, but solely considers adopters and non-adopters. Cf. Tanny, S. M./Derzko, N. A. (1988), pp. 226.

257 For an in depth discussion of different "innovator" and "imitator" conceptualizations see Pechtl, H. (1991), pp. 10.

258 Cf. Schmalen, H./Xander, H. (2002), p. 443.

259 Imitators consist of the aggregated groups of early adopters, early majority, late majority, and laggards. Cf. Bass, F. M. (1969), p. 216.

260 Cf. Ibid.

261 Cf. Schmalen, H./Pechtl, H. (1992), pp. 185.

As Bass merely considers a monocausal determinant in his basic model,[262] it has been subject to different modifications and extensions.[263] Besides the focus on the endogenous variables of personal communication, experience, and social pressure, authors of subsequent studies added exogenous variables as for example marketing instruments, such as pricing, advertising, and allowed repurchases or influences of interventions by competitors.[264]

Other than the adoption research that will be discussed in the next chapter, diffusion research in particular focuses on the diffusion processes (descriptive and explicative function), quantitative sales forecasts of an innovation (prognostic function), and on drawing up strategies for an optimal use of marketing mix instruments (normative function).[265] Despite the fact that the main emphasis in this dissertation is on the individual level, diffusion research results are helpful as many overlaps between these research areas exist.

3.3 Innovation Adoption Research

Rogers (1962) described the adoption as a "mental process through which an individual passes from first hearing about an innovation to final adoption."[266] A person who has decided to adopt a new product is labeled as an adopter while an individual who is considering to purchase an innovation, but has not yet ultimately decided to adopt it, can be referred to as a potential adopter.[267]

This definition of adoption already points at the major difference compared to diffusion theory: Adoption theory focuses on the disaggregated level. Since the adopting unit is a single individual, neither the spread of innovations within a society or market segment nor market potentials play a crucial role here. Moreover, adoption research is aimed at finding explications for individual behavior, whereas diffusion theory lays stress on describing diffusion patterns, forecasting

262 Schmalen/Xander (2002), Jugel (1991), Mahajan/Muller/Bass (1990), and Mahajan/Peterson (1985) specify additional presumptions and limitations of early diffusion models. Cf. Schmalen, H./Xander, H. (2002), p. 450; Jugel, S. (1991), pp. 25; Mahajan, V./Muller, E./Bass, F. M. (1990), pp 11; Mahajan, V./Peterson, R. A. (1985), pp. 10.

263 Cf. Bass, F./Jain, D./Krishnan, T. (2000), p. 99.

264 Cf. Schmalen, H./Xander, H. (2002), p. 444; Kalish, S. (1985); Mahajan, V./Muller, E./Kerin, R. A. (1984); Robinson, B./Lakhani, C. (1975). For a further overview see Mahajan, V./Muller, E./Wind, Y. (2000); Mahajan, V./Muller, E./Bass, F. M. (1990), pp 20; Gatignon, H. A./Robertson, T. S. (1986), p. 40.

265 Cf. Schmalen, H./Xander, H. (2002), p. 445; Mahajan, V./Muller, E./Bass, F. M. (1990), pp. 15; Böcker, F./Gierl, H. (1987), p. 687; Gierl, H. (1987), pp. 24.

266 Rogers, E. M. (1962), p. 76.

267 Cf. Pohl, A. (1996), p 46.

the demand for a new product, and giving normative statements. Furthermore, Gatignon/Robertson (1986) argue that "the behavioural assumptions underlying consumer diffusion models are typically simple and do not provide a behavioral explanation for the rate or pattern of diffusion."[268] They cite Mate (1981) who found that most models simplified and employed assumptions that for instance people who are not aware of the novelty show the same probability to learn about the innovation through advertising and personal influence or that individuals are homogeneous in their disposition to be informed or in their capability to persuade others to try novelties. Further it is not uncommon that studies presume that people who adopted the new product do not lose the interest to provide information to persons who are uninformed. Additionally the ideas of opinion leadership or the existence of social barriers of communication are oftentimes neglected.[269] Despite these distinctions, both research areas are strongly interrelated, as the diffusion consists of the single adoption decisions made over time.

Following the above mentioned innovation definition, the adoption comprises the first time purchase of a *newproduct* through an individual person. As it is the individual on which this study focuses, the adoption represents intraindividual decisions compared to interindividual, collective decisions which occur in organizations.[270] The decision to adopt is a central step in the process, though it is by far not the only one. Adoption is not a single-unit act, but must rather be seen as a sequence of different steps with intermediary results.[271] Hence, Robertson (1971) recommended to use the term "adoption *decision* process" instead of just "adoption process" as the latter expression can be misleading – it suggests that only the positive course of the process is of interest.[272] For the same reason Rogers/Shoemaker (1971) introduced the terminology of the "innovation decision process."[273] This expression embraces a broader view, as it explicitly leaves open whether the process results in an adoption or non-adoption of the innovation. Besides, it should also include the behavior that sets in after the decision to adopt.[274] All three terms should, however, be used synonymously in the following. When talking about the "adoption process" the possibility to refuse an innovation should always be included.

---

268 Gatignon, H. A./Robertson, T. S. (1986), p. 38.

269 Cf. Mate, K. V. (1981) cited in Gatignon, H. A./Robertson, T. S. (1986), p. 38.

270 Cf. Lutschewitz, H./Kutschker, M. (1977), pp. 1.

271 Cf. Beal, G. M./Rogers, E. M./Bohlen, J. M. (1957).

272 Cf. Robertson, T. S. (1971), pp. 58.

273 Cf. Rogers, E. M./Shoemaker, F. (1971), pp. 99 and 101.

274 Cf. Ibid.

3.4 Innovation Decision Process

As noted in the previous chapter, the adoption or rejection of a novelty occurs in several stages and can thus be described as a process. Rogers (1962) credited Ryan/Gross (1943) as the first authors to have recognized this.[275] An in depth analysis of different decision models, the related assumptions, influencing variables and outcomes, as well as the understanding of the innovation decision process used in this study are discussed in the forthcoming sub-chapters. After a discussion of different innovation decision process models (Ch. 3.4.1), the stages of awareness (Ch. 3.4.2), interest (Ch. 3.4.3), evaluation (Ch. 3.4.4), and adoption/rejection (Ch. 3.4.5) are outlined in more detail.

3.4.1 Different Models of the Innovation Decision Process

In principle, the innovation decision process can be viewed as a specific form of the regular buying decision process – namely for the specific case that the analyzed object is an innovation.[276] Related to this premise of newness there are some other factors that need to be taken into consideration as they might differ from the regular purchase decision process. These include the non-existence of established buying patterns, or the higher likelihood of uncertainty with regard to the product properties and performance. Rogers (2003) characterized the innovation-decision process as being "essentially an information-seeking and information-processing activity in which an individual is motivated to reduce uncertainty about the advantages and disadvantages of an innovation."[277]

The history of adoption process models can be traced back to the work of Wilkening (1952) who was also among the first authors to analyze this process in the context of an individual's decision.[278] However, it was Rogers – originating from the area of sociological research – who carried forward and incorporated the ideas and research results achieved in disciplines such as rural anthropology, sociology, geography, and industrial economics. With his work, which has since been broadly used in the ensuing years, he provided a solid theoretical basis to introduce the concept of adoption process models into the field of consumer behavior research. Since then, many studies have refined and extended his concept, and so, a large number of varying models has evolved to describe

275 Cf. Ryan, B./Gross, N. C. (1943) cited in Rogers, E. M. (1962), p. 79.

276 Cf. Meffert, H. (1976), p. 93; Rogers, E. M./Shoemaker. F. (1971), p. 99.

277 Rogers, E. M. (2003), p. 172.

278 Cf. Wilkening, E. A. (1952) cited in Rogers, E. M. (1962), p. 80.

the innovation decision process. These differ in the number of stages as well as in the underlying presumptions.

In the first edition of his "Diffusion of Innovations" book in 1962, Rogers defined five steps an individual runs through in the adoption process. These are: (1) awareness, (2) interest, (3) evaluation, (4) trial, and (5) adoption.[279] Together with Shoemaker, he later revised his own model which then encompassed only four stages where the authors distinguished between (1) knowledge, (2) persuasion, (3) decision, and (4) confirmation.[280] Apart from the process and its consequences, they supplementary incorporated antecedents and other influencing factors in their model. The authors explained this measure with new knowledge obtained from other adoption research studies, the theories of attitude change, learning processes, and decision making.[281] In the third edition of his book, Rogers undertook the so far last change in his model concerning structure of the adoption process. He went back to the original number of five stages by adding the step of "implementation" at the second to last position between the stages of "decision" and "confirmation."[282]

As demonstrated, a huge number of different classifications of stages in the research of adoption processes exists. Even though there is no commonly accepted selection and definition of stages, several characteristics are likely to be found in the majority of adoption process models: They possess a comparable structure insofar as the processes start with the perception of an innovation and end with the purchase and use of the new product or with rejection. The steps in-between this initiating and concluding stage are, in contrast, very heterogeneous. A comparable schematic can especially be observed in conventional innovation decision models. They use some form of the "hierarchy of effects" model which was introduced by Lavidge/Steiner (1961) and named by Palda (1966).[283] Originating from investigations concerning the effectiveness of advertising, the "hierarchy of effects" approach supports the idea of consumers moving through a series of steps before making a purchase decision and spending time *at every step* before moving on.[284] This idea is conceptually quite similar to the traditional AIDA model of attention, interest, desire, and action, and has also been adapted by many researchers.[285]

---

279 Cf. Rogers, E. M. (1962), pp. 81.

280 Cf. Rogers, E. M./Shoemaker. F. (1971), pp. 101.

281 Cf. Ibid., p. 103.

282 Cf. Rogers, E. M. (1983), pp. 163.

283 Cf. Palda, K. S. (1966); Lavidge, R. J./Steiner, G. A. (1961).

284 Cf. Clow, K. E./Baack, D. (2005), pp. 90.

285 Cf. Barry, T. E. (1987), pp. 251.

Despite the development of a large variety of adoption process models and their broad reach, they have not solely triggered positive reactions in the research community. It can for instance be remarked critically that many of the proposed models are only effective under certain conditions. Most studies imply that a high-involvement[286] situation is given, resulting in cognitive steering of the process by the adopter. This is particularly the case for models using the "hierarchy of effects" approach to explain the innovation decision process.[287] Applying this approach, however, is not valid for adoption decisions in situations of low involvement.[288] The amount of cognitive processing is necessary to consider as it determines the flow of the adoption process.[289] Low involvement adoption processes significantly differ from those of high involvement as for example the purchase or trial can take place before the individual has formed an attitude towards the innovation. Therefore, the sequence of process steps, which would more accurately correspond to low involvement adoption decisions, would display the structure of (1) awareness, (2) trial, (3) attitude formation, and (4) adoption.[290] Other criticism revolved around the fact that it cannot be expected that individuals pass through each stage of the process prior to a purchase in a stair-step fashion and that the process may not be as rational as commonly assumed.[291] In general, older models can be criticized for their inflexible structure. The sequence of stages seems too rigid to adequately reflect the consumer's novelty decisions process, as neither the leaving out of stages nor feedback loops, or the premature withdrawal from the process are considered.[292]

Due to the vast number of adoption process models which use their own definitions and number of stages, every single one of those propositions has been the target of criticism. Yet, it would be too time-consuming and expedient to discuss each model and the comments which have been made. Just one further statement by Antil (1988) may be picked up here.[293] He mentions that the focus in the research of the consumer's new product decision process so far has almost completely been on the pre-trial and/or pre-adoption stages. The author therefore

286 For an exact conceptualization of involvement see Hohl, N. A./Naskrent, J. (2009).

287 Due to the importance of learning in hierarchy-of-effects models, Ray (1973) attributes them to "learning theory" approaches. Cf. Ray, M. L. (1973), p. 150.

288 Low involvement models have for example been advanced by Kassarjian, H. H. (1981); Robertson, T. S. (1976); and Krugman, H. E. (1965).

289 Cf. Gatignon, H./Robertson, T. S. (1991), p. 325.

290 Cf. Gatignon, H./Robertson, T. S. (1985), p. 854.

291 Cf. Campbell, R. R. (1966).

292 Cf. Robertson, T. S. (1971), pp. 58; Rogers, E./Shoemaker, F. (1971), p. 101.

293 Cf. Antil, J. H. (1988), p. 6.

supplementary integrated the variables of direct product experience and product evaluation into his model and places them as important mediators between the trial and adoption stage. This remark will be considered in the succeeding description of the single decision steps.

Although some authors denounce stages as dividing the adoption process in a somewhat arbitrary way and even question their existence, it will be referred to sequence-models here. Subdividing the process into several, typical stages is based on the presumption that the consumers who find themselves in a specific stage will show a similar behavior.[294] It is assumed that the underlying decision process is based on a high involvement situation. Though it is not alluded to at every stage, it is important to be aware that the potential adopter does not always pass through all of the hereafter proposed process stages in this ideal-typical way. It is possible that stages might be skipped, that a person returns to a stage at some other moment in time, or that the process is abandoned unexpectedly for some reason.

3.4.2 Awareness

The innovation decision process starts with the first notice of a novelty. It is absolutely possible that the individual does not know a single detail about the new product or that he or she has never seen the innovation before – the person just knows about the existence of the innovation and is, so far, not motivated to seek additional information about it.[295] Reasons that are likely to stimulate the start of an innovation purchase decision process are, e.g., dissatisfaction with the actual product, depleted stocks, different desires or needs, marketing influences, and new available product add-ons.[296] The reasons can be divided in two rough categories. A potential adopter can become aware of the innovation's existence either coincidentally or by active search. Rogers (1962) found that most researchers conceive awareness of an innovation as a "random or non-purposive occurrence" which may create a problem because of emerging dissonance on the side of the consumer (innovation-oriented decision).[297] Besides that, there is the possibility of a more explorative behavior which leads to awareness. The active search process is usually given in situations where a person is looking for a solu-

294 Cf. Helm, R. (2001), p. 108; Beal/Rogers/Bohlen (1957) note that consumers may use different information sources depending on the stage they are in. Cf. Beal, G. M./Rogers, E. M./Bohlen, J. M. (1957), p. 166.

295 Cf. Rogers, E. M. (1962), pp. 81.

296 Cf. Statt, D. A. (1997), pp. 231.

297 Cf. Campbell, R. R. (1966), p. 460; Rogers, E. (1962), p. 82.

tion for a specific problem and, in the course of searching, gets in contact with the innovation. This means, that in some cases, a need or problem-recognition might precede this first stage of awareness of a potential option or solution (problem-oriented decision).[298] However, the pressure of a problem is not always necessary to start the information search process. Persons might moreover display curiosity or interest in a certain product category and become aware of the innovation in this context.[299]

3.4.3 Interest

After having become aware of the innovation there has to be a sufficiently large level of interest to proceed to the next steps of the adoption process. To attract interest, the new product has to seem adequate enough to solve the person's initial problem, or improve a principally satisfactory situation.[300] That implies a state of passive acceptance of the innovation. If this is not the case, the individual will show passive resistance. The negative scenario prevails in situations where there is no general need for the innovative problem solution, low involvement, or absence of cognitive processing.[301] As a result, there will usually be no further interest in the new product. There is no intention to actively seek for information and therefore the innovation decision process will be abandoned – although the person is aware of the innovation. Besides the non-existence of the need for a special problem solution, habits pose another reason for passive resistance or inertia. Especially people who "strive for consistency and status quo rather than to continuously search for, and embrace new behaviours"[302] show a general resistance towards innovations. Ram/Sheth (1989) therefore attributed the resistance to the potential adopters' unwillingness to change a satisfactory status quo as well as to arising conflicts with their belief structure.[303] Sheth (1981) further hypothesized that the resistance to change and the innovations associated with change will be greater, the stronger a person's habits are.[304]

If the consumer is interested, he or she will favor the innovation in a global way. The novelty will from then on belong to the consideration set of possible alternatives for the problem solution. However, at this process stage no efforts have

---

298 Cf. Schiffman, L. G./Kanuk, L. L. (2004), p. 531; Campbell, R. R. (1966), p. 460.

299 Cf. Bloch, P. H./Sherrell, D. L./Ridgway, N. M. (1986), pp. 120.

300 Cf. Pechtl, H. (1991), p. 65.

301 Cf. Nabih, M. I./Bloem, S. G./Poiesz, T. B. (1997), p. 191.

302 Sheth, J. N. (1981), p. 275.

303 Cf. Ram, S./Sheth, J. N. (1989), p. 6.

304 Cf. Sheth, J. N. (1981), p. 276.

been made to more explicitly judge whether or not the innovation represents an adequate solution in one's own situation.[305] Yet, having accepted the innovation idea does not necessarily mean to automatically progress to the next stage in the adoption process. When the actual product is still fully functional for example (depending on the selected product category), it is also possible that the person just remembers the innovation for a future problem solving situation, but does not engage in advancing the decision process.[306] Otherwise, the positive course of the interest sequence leads to the next step where the potential adopter starts the evaluation of the innovation.

3.4.4 Evaluation

Now, that the potential adopter is aware of the new product and shows a general interest in the innovation he or she engages in activities to take the next step in the decision process. During the evaluation sequence the person intends to structure the decision situation and actively seeks valuable information to help him or her make a good decision.[307] The individual applies what Rogers (1962) called a "mental trial," which helps to find out whether the innovation fits the consumer's expectations.[308] During the evaluation phase, the person more intensely deals with the innovation than during the interest sequence. The potential adopter finds out more about the innovation and for example learns about the principal functioning, application areas, advantages and disadvantages of the new product.[309] In this context Rogers/Shoemaker (1971) differ between "how-to knowledge" which is the knowledge about how to utilize the innovation correctly, and "principles knowledge" which is the knowledge about the innovation's underlying technical-functional principles.[310] Apart from information gathering in order to generally understand the innovation and its technological features, the person needs to compare the characteristics of competing innovations which are satisfying the same need.

Not only the "mental trial" but also the trial as a real act is integrated in this stage. Contrary to many researchers' conceptualizations of the decision process, trial should not be interpreted as a separate step here. As Gatignon/Robertson (1991) remark, it is important for sales forecasts to divide the innovation acqui-

305 Cf. Rogers, E. M. (1962), pp. 82.
306 Cf. Schmalen, H./Pechtl, H. (1996), p. 818.
307 Cf. Brose, P. (1982), pp. 185.
308 Cf. Rogers, E. M. (1962), pp. 83.
309 Cf. Abraham, S. C./Hayward, G. (1984), p. 225.
310 Cf. Rogers, E. M./Shoemaker, F. F. (1971), pp. 106.

sitions made by the consumers into trial and repeat purchases.[311] When considering durable goods as study object, trial should be referred to as a "validity test or 'dry run'"[312] representing another possibility for the potential buyer to gain further information about the product.[313] Ergo, it helps to better evaluate the novelty *before* the actual purchase and is therefore integrated in this evaluative process stage. Rogers (1962) defined trial as an opportunity to "use[s] the innovation on a small scale in order to determine its utility" and underlines that the trial's main function is "to demonstrate its usefulness for possible complete adoption."[314]

During this more elaborated period of information search and analysis, the potential adopter creates an active acceptance or resistance towards the considered innovation compared to the passive view of acceptance and resistance in the stage before (interest stage). The active notion of acceptance and resistance is due to the just explained more extensive handling of the situation and explicit assessment of multiple attributes of the innovation. For example, active resistance is reported to be found when consumers feel that the innovation is too risky.[315] Both, acceptance and resistance are interpreted in this study as momentary states in which the potential adopter judges the innovation as favorable or unfavorable. No final decision has been made until then, so that resisting a new product at one point in time can still lead to acceptance at another point in time during the evaluation process.[316] This can occur for example in situations where the individual obtains stronger persuasive arguments, new incentives, or revises his earlier negative appraisal of the novelty.[317]

At the end of this sequence of evaluating the product, there is the development of a favorable or unfavorable attitude towards the considered innovation. This attitude can be defined as a consistent disposition in a certain situation to react towards an object more or less strongly in a positive or negative way.[318] This attitude ultimately leads to the final *decision* of this step as well as of the overall

---

311 Cf. Gatignon, H./Robertson, T. S. (1991), p. 325.

312 Rogers, E. M. (1962), p. 84.

313 This is especially the case for the product category of consumer durables.

314 Rogers, E. M. (1962), p. 84.

315 Cf. Ram, S./Sheth, J. N. (1989), pp. 6.

316 Cf. Bagozzi, R. P./Lee, K.-H. (1999), p. 219; Ram, S. (1987), pp. 208. Contrary to this, if further processing was unsuccessful, resistance will lead to the final decision to reject the innovation. The case of rejection is explained more precisely in the following subchapter.

317 Cf. Bagozzi, R. P./Lee, K.-H. (1999), p. 222.

318 See for example Trommsdorff, V. (2004), p. 159.

adoption process: whether or not to purchase the innovative product. The latter decision of non-adoption can be further subdivided into the decision to reject the innovation and the decision to postpone adoption.

3.4.5 Adoption or Rejection

The precedent stages including the creation of an active and passive acceptance or resistance as well as the decision to adopt, reject or postpone are mental processes and have an intentional character. The final step of the innovation decision process, now, relates to the *behavioral* stage.[319] As a result, one possible outcome of the last stage in the innovation decision process is the physical act of the innovation adoption/purchase. But, it is also possible that the last stage results in a postponement of the adoption or a final rejection.

Adoption is often synonymously used as the purchase of the considered innovation. However, it is a major shortcoming to assume that the innovation purchase stands for the end of the adoption process and that buying that product equals adoption. This is because adoption is a broader term which encompasses more than just the ownership of an innovation. In this context, Robertson (1971) pointed out that the "purchase of a product is a necessary but not a sufficient condition for adoption, since adoption includes a sense of product commitment, involving acceptance and use over time. Not every purchase results in adoption, nor is every purchase evidence of a consumer's decision to *adopt* a product or brand."[320] Rogers (1962) also integrated this idea in his definition of adoption which he described as "a decision to continue full scale use of an innovation."[321] A very similar definition is used by Robertson (1971) as he referred to adoption as "the acceptance and continued use"[322] of an innovation. The author does not merely talk about a use-*decision*, but the proper use of the product.[323]

Now, it is difficult to say at what moment one can talk about "continued use." This strongly depends on the nature of the considered product class – whether the product is a durable or non-durable good. Especially in the case of durables

---

319 Cf. Nabih, M. I./Bloem, S. G./Poiesz, T. B. (1997), p. 191; Rogers, E. M. (1983), p. 174. The only exception represents the "trial" which is not categorized to the last stage in the innovation decision process, though it may also imply behavioral aspects.

320 Robertson, T. S. (1971), p. 57.

321 Rogers, E. M. (1962), p. 17.

322 Robertson, T. S. (1971), p. 56.

323 Later, Rogers (1983) responded to this criticism by integrating a stage of "implementation" in his adjusted adoption model, where the new idea is put into practice. Cf. Rogers, E. M. (1983), pp. 174.

an adopter underlines the adoption by its "continued full scale use" after purchase. The usage intensity of durables can be measured by width, conceived as the number of applications or different uses, and by depth, conceived as the amount of usage.[324] When regarding non-durables, this conceptualization becomes obsolete because of the relatively smaller number of applications and shorter using period. Adoption, then, occurs through "repeat purchase decisions" of the same novelty.[325]

If the novelty has been purchased, but no implementation occurred, meaning that the product is either not regularly used or not repeatedly purchased after the initial purchase, researchers talk about discontinuance. The discontinuance phenomenon refers to findings made in dissonance theory which was established by Festinger (1957).[326] The rational is that if the individual obtains information or has experiences with the innovation after the purchase which do not meet the expectations, the individual may perceive dissonances.[327] In this situation the consumer's options encompass the reduction of dissonance by information search.[328] Alternatively, he or she avoids negative information or discontinues using the innovation.

Innovation rejection builds the counterpart to adoption. Gatignon/Robertson (1991) described rejectors as persons who "have processed the information that they required to make a decision and have concluded not to adopt."[329] The notions of resistance and the final *decision* not to adopt differ compared to the rejection. With rejection is meant that the novelty does not embody a possible solution of the initial need or problem at that moment and in future.[330] Hence, the process is terminated and no further efforts in information search are made. However, in accordance with more flexible adoption process models, rejection of a certain innovation conception may lead to some kind of feedback loop or the start of a new decision process where *other* innovation conceptions are taken into consideration.

Apart from innovation adoption and rejection, there is a third possibility of how the innovation decision process can end. That is, if the adoption is postponed.

---

324 Cf. Gatignon, H./Robertson, T. S. (1991), p. 325.

325 Cf. Gatignon, H./Robertson, T. S. (1986), p. 49; Robertson, T. S. (1971), p. 57.

326 Cf. Festinger, L. (1957).

327 Rogers/Shoemaker (1971) integrate this aspect in their confirmation stage. Cf. Rogers, E. M./Shoemaker, F. F. (1971), pp. 113.

328 Cf. Ehrlich, D. et al. (1957), pp. 99.

329 Gatignon, H./Robertson, T. S. (1991), p. 325.

330 Cf. Weiber, R./Pohl, A. (1995), p. 417.

Postponement can occur in the case of indecision, anticipation of new technologies, or due to temporal unavailability of the innovation.[331] In the situations mentioned earlier, the potential adopter is unwilling to commit to a problem solving innovation. The potential adopter remains undecided and keeps his decision open, for example because of uncertainty or task avoidance, what in turn leads to purchase delays.[332]

On the other side, leapfrogging[333] characterizes a special behavior of postponement which refers to technology anticipation. Leapfrogging can be described as the deliberate postponement of an innovation purchase because the person expects a novelty to be introduced to the market, which is generally characterized through a superior performance.[334] Under these circumstances, an instantaneous purchase would mean to opt for "a decision to saddle oneself with a soon-to-be-obsolete technology."[335] Weiss/John (1989) defined leapfrogging as follows: "Since improvements are expected to continue into the future, some potential buyers choose to bypass the current innovation, since a technology or product acquired now may soon be rendered obsolete by an even more advanced offering. This is leapfrogging behavior, the act of considering the adoption of an innovation, and then deciding to bypass it in favor of waiting to consider an improved future product."[336] In their study, Holak et al. (1987) empirically investigated the role of consumer expectations and indirectly found that individuals may delay their purchase if they expect benefits as for example price reductions or product improvements.[337] Greenleaf/Lehmann (1995) similarly found the expectation of product improvements, price reductions, substitute availability, and information search as delay reasons.[338] To sum up, it is important to understand that the term "leapfrogging" used here only considers the *intentional* decision in favor of a future technology as well as *voluntary*[339] decisions.[340] Furthermore, to

331 Nabih/Bloem/Poiesz (1997) additionally differ between postponement of adoption and postponement of trial. Cf. Nabih, M. I./Bloem, S. G./Poiesz, T. B. (1997), p.191.

332 Cf. Gatignon, H./Robertson, T. S. (1991), p. 325; Greenleaf, E. A./Lehmann, D. R. (1991).

333 Pohl (1996) gives an overview of different manifestation areas of leapfrogging. See Pohl, A. (1996), pp. 5 and the cited literature.

334 Cf. Rosenberg, N. (1976).

335 Rosenberg, N. (1976), pp. 534.

336 Weiss, A. M./John, G. (1989), p. 1 cited in Pohl, A. (1996), p. 10.

337 Cf. Holak, S. L./Lehmann, D. R./Sultan, F. (1987).

338 Cf. Greenleaf, E. A./Lehmann, D. R. (1995).

339 The case of forced leapfrogging, for example due to an individual's financial restrictions or due to unavailability of the innovation, should be assigned to the situational factors which are discussed later (s. Ch. 3.5.3, pp. 113).

leap frog does not mean that the individual necessarily adopts the future generation of the product, but that the decision process is resumed in future with rejection as a possible outcome.

The discussion of an optimal process classification has shown that a distinct separation of the different phases in the process is not possible in most cases due to flexible boundaries of the stages. It is further impossible to present a model that represents all kinds of adoption decisions. However, the presented sequence indicates the general new product decision process in the high-involvement context.[341] Furthermore, some links became visible in the discussion above, which are useful in the search for variables that may influence the adoption decision process. These will be discussed in the subsequent chapter.

3.5 Conceptualization of Adoption Determinants

After having analyzed the sequence of stages which an individual runs through during an innovation decision, it is now of interest which factors have an impact on the positive or negative course of the process – that means on the final adoption or non-adoption of innovations. When focusing on a single product these factors may also serve to explain the reasons for temporal differences in the adoption decision of different individuals which allows classifying them into categories such as innovators, early majority, etc.[342] To attain the goal of identifying variables which may be responsible for potential disparities in the adoption behavior of persons belonging to *different age groups*, it will first be referred to determinants from classical adoption theory. Later, those factors will be assessed under the light of current research results of marketing literature devoted to older consumers (s. Ch. 4.2, pp. 123).

Determinants that influence innovation decisions are generally classified into product-specific (Ch. 3.5.1), personal (Ch. 3.5.2), and situational factors (Ch. 3.5.3) – see also the overview given in Figure 24. Other authors include organizational or environmental determinants (for example to better explain the adoption of technological innovations in companies),[343] social system variables[344] or categorize determinants regarding competitive actions, or marketing mix in-

340 This is in line with Pohl, A. (1996), pp. 10.

341 The final form may, of course, vary depending on eventual specificities of the considered product category.

342 Cf. Rogers, E. M. (2003), pp. 16.

343 Cf. Pohl, A. (1996), pp. 68; Herbig, P. A./Palumbo, F. (1994), pp. 71.

344 Cf. Gatignon, H./Robertson, T. S. (1985), pp. 857; Rogers, E. M./Shoemaker, F. F. (1971), pp. 101.

struments.[345] Although many researchers categorize determinants into the three areas of "product," "person," or "environment," it is rarely possible to do so unambiguously. This is because the perception of a particular determinant is consistently influenced by all three areas[346].[347] Notwithstanding, each determinant possesses a dominant influencing area, so that the determinants will be systematized according to their major influencing area hereafter.

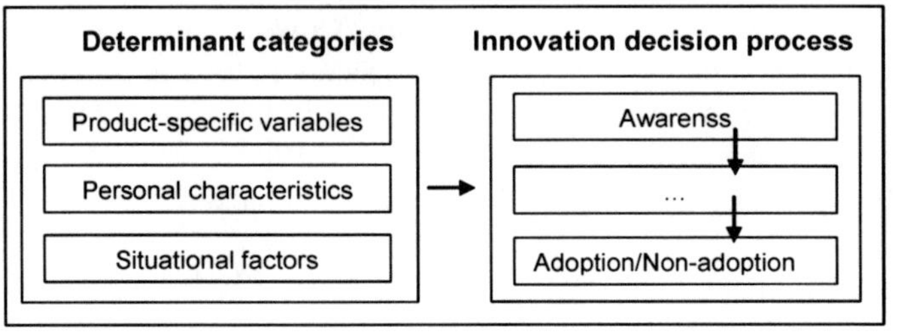

Figure 24: Influence of adoption determinants on the innovation decision process

3.5.1 Product-Specific Variables as Adoption Determinants

The product-specific determinants of adoption represent those *product* properties that have an influence on adopting an innovation. Traditionally these factors have been seen as crucial in the adoption and diffusion of innovations: Holak (1988) found that prior studies reveal a domination of innovation properties in predicting adoption behavior compared to psychological or demographic variables.[348] In the next sub-chapters, the six most commonly cited product-specific factors are picked up and explained. According to authors such as Rogers (1962), Fliegel/Kivlin (1966), Ostlund (1973), or Robertson (1971) these include (1) relative advantage, (2) compatibility, (3) complexity, (4) trialabil-

345 Cf. Mahajan, V./Sharma, S./Buzell, R. D. (1993); Gatignon, H./Robertson, T. S. (1991), pp. 317.

346 For example, the risk an individual is exposed to during the decision process depends on product-specific characteristics, such as the product's features, complexity, etc. But, personal characteristics such as the person's risk attitude also play a decisive role in this context, as well as prevalent situational factors such as time pressure, or financial restrictions.

347 Cf. Pohl, A. (1996), p. 58.

348 Cf. Holak, S. L. (1988), p. 53; see also LaBay, D. G./Kinnear, T. C. (1981) and Ostlund, L. E. (1974).

ity/divisibility, and (5) observability/communicability.[349] In addition to these five factors, Ostlund (1974) integrated a sixth dimension to the group of product-specific attributes:[350] the variable of "perceived risk" (6) originally introduced by Bauer (1960) in a more general, non-innovation context.[351]

Originally, these factors stem from analyses of the aggregated level of innovation diffusion. However, they have also been used by researchers who analyzed the *individual* adoption of an innovation.[352] Thus, the determinants discussed hereafter represent factors that are more or less extensively considered and analyzed by the individual during the innovation decision process. When discussing the issue of determinants, it is important to bear in mind that it is not the objective product properties that determine the (non)purchase. It is the *subjective perception* of characteristics which is responsible for the differing reactions of potential buyers.[353] Rogers (1983) explained it with these words: "It is the receivers' perceptions of the attributes of innovations, not the attributions as classified by experts or change agents that affect their rate of adoption."[354]

3.5.1.1 Perceived Relative Advantage

The relative advantage indicates in how far a novelty is seen as superior to those products that preceded the innovation.[355] However, the conceptualization of relative advantage as well as its measurement poses some difficulties. As the term comprises a large number of aspects, Tornatzky/Klein (1982) even labeled it as a "garbage pail characteristic."[356] In the following it is necessary to avoid applying an all too general view of relative advantage and, instead, specify the notion of the term.

Early diffusion and adoption models almost exclusively employed aspects of financial benefits to describe the utility of an innovation.[357] Ostlund (1974) for example expanded the common use of relative advantage by regarding time sav-

---

349 Cf. Ostlund, L. E. (1973), pp. 93; Robertson, T. S. (1971), pp. 191; Rogers, E. M./Shoemaker, F. F. (1971), pp. 137; Fliegel, F. C./Kivlin, J. E. (1966), pp. 237; Rogers, E. M. (1962), pp. 124.

350 Cf. Ostlund, L. E. (1974).

351 Cf. Bauer, R. A. (1960).

352 Cf. Holak, S. L. (1988), p. 52.

353 Cf. Weiber, R./Pohl, A. (1995), p. 412.

354 Rogers, E. M. (1983), p. 212.

355 Cf. Rogers, E. M. (1962), pp. 124.

356 Tornatzky, L. G./Klein, K. J. (1982), p. 34.

357 For an overview of such studies see Mahajan, V./Muller, E./Bass, F. M. (1990), p. 7.

ings and effort savings next to an innovation's monetary value.[358] While effort savings unequivocally relate positively to adoption, the same is normally the case for the time-saving aspect. A positive relationship, however, is not always prevailing as consumers may also desire innovations with an opposed, time-consuming effect. This is the case for hobby or recreational items (e.g. entertainment electronics) where time consumption does not have a negative consequence on adoption.[359] Economic savings predominantly also foster adoption, except for those products for which it is believed that price and quality are positively correlated. In this case the price takes the place of a quality indicator.[360] Additionally, Schmalen/Pechtl (1996) mention the technical superiority of a novel product as another facet of relative advantage.[361]

To subsume it can be stated that the relative advantage correlates positively with the likelihood of an individual to adopt the new product.[362] The different aspects of relative advantage are, in contrast, not uniform in their influence. They affect the relative advantage either positively or negatively. Later, in the empirical part of the dissertation, it will be necessary to consider these differing impacts. In addition, the focus lies only on the aspects of relative advantage that seem to be relevant for the innovations in question.

Perceived relative advantage $\xrightarrow{+}$ adoption

3.5.1.2 Perceived Compatibility

Rogers (1983) defined compatibility as the "degree to which an innovation is perceived as consistent with the existing values, past experiences and needs of potential adopters."[363] He differentiates between a novel product's compatibility or incompatibility as regards to (1) sociocultural values and beliefs, (2) previously introduced ideas, and (3) customer needs for innovation.

(1) Culturally acquired values can be an important obstacle in adopting innovations. The examples given by Rogers (1983) and Rogers/Shoemaker (1971) refer

---

358 Cf. Ostlund, L. E. (1974), p. 27.

359 Cf. Gatignon, H./Robertson, T. S. (1985), p. 858; Strober, M. H./Weinberg, C. B. (1977), pp. 141.

360 Cf. Kamakura, W. A./Balasubramanian, S. K. (1988), p. 4. Similarly prestige products are seen to have higher quality, but also greater exclusivity. Cf. Lichtenstein, D. R./Ridgway, N. M./Netemeyer, R. G. (1993), pp. 236.

361 Cf. Schmalen, H./Pechtl, H. (1992), p. 819.

362 Cf. Rogers, E. M. (1983), p. 218.

363 Rogers, E. M. (1983), p. 223.

to innovations that are incompatible to traditions of villagers, ethnic groups, or a whole countries' society.[364] In the reported cases, innovation adoption is low as they demand large changes in culturally acquired behavior and ways of thinking passed on from generation to generation. Therewith, the authors center more on societal values than personal values.[365] The studies which followed predominantly embodied the latter kind of values or consumer traits to measure innovativeness or predict new product adoption.[366]

(2) The compatibility of a novelty with a preceding product represents another aspect which has an influence on adoption. In his remarks intending to explain "compatibility with previously introduced ideas," Rogers (1983) uses aspects with respect to a person's knowledge, familiarity, and past experiences with a product or an old idea.[367] This threefold differentiation shows few connections to product related attributes. This issue, therefore, is abridged here and picked up again later in Chapter 3.5.2.2 (pp. 103) in the context of the personal determinants of adoption.

There is one further aspect of compatibility with previously introduced ideas the product plays a certain role for: Namely, cognitive or behavioral changes that are caused by an innovation are widely used as a way to categorize types of innovation. Numerous authors draw on knowledge and attitude changes or changes in consumption patterns followed by an innovation or necessary prior to adoption.[368] This approach allows positioning novelties on a continuum with the endpoints of incremental and radical innovations. If consumers cannot recur on existing knowledge to assess a new product, researchers talk about really new products or discontinuous innovations.[369]

(3) Apart from compatibility with culturally learned values and with product category knowledge, the link between the characteristics of the innovation and the needs on the side of the potential adopter comprises another aspect of compatibility. In this context, Rogers/Shoemaker (1971) introduce the idea of change agents trying to determine the needs of the consumers. Subsequent to their analysis, change agents are able to recommend innovations which can ful-

---

364 Cf. Ibid.; Rogers, E. M./Shoemaker, F. F. (1971), pp. 146.

365 Cf. Blackwell, R. D./Miniard, P. W./Engel, J. F. (2001), p. 317.

366 For an overview of studies see Holak, S. L. (1988), p. 51.

367 Cf. Rogers, E. M. (1983), pp. 224.

368 See for example Ziamou, P. L. (1999), p. 369; Veryzer, R. W. (1998), p. 138; Schmalen, H./Pechtl, H. (1992), p. 87; Weiber, R. (1992), p. 3; Robertson, T. S. (1971), p. 7 and pp. 227.

369 See for example Hoeffler, S. (2003), p. 406.

fill these needs. A better compatibility with felt needs results in an occurrence of faster innovation adoption.[370] Analogously, companies have to take the role of change agents and have to find out about the needs of the targeted consumers. They are required to conceptualize their new products in a way that they represent a benefit for potential customers and satisfy their needs.[371]

As it emerges from the literature research, innovation studies mainly refer to individual aspects of compatibility which are discussed later. The product oriented view is, in contrast, underrepresented. The very basic understanding of novelty-related compatibility – the technological compatibility with newer and older related devices and accessories – is rarely mentioned there. It can be summed-up that high levels of compatibility with values, previous ideas, and needs correspond to a higher speed or likelihood of adoption.[372] Here, again, it must be referred to the fact, that perceived compatibility as conceptualized by the cited authors does not depend on the product, but is more related to personal characteristics of a potential adopter.

Perceived compatibility $\xrightarrow{+}$ adoption

3.5.1.3 Perceived Complexity

In terms of an innovation's complexity, it can be referred to Rogers (1962) again who describes complexity as "the degree to which an innovation is relatively difficult to understand and use."[373] Any novel product may be classified on a complexity-simplicity continuum, meaning that some innovations are clear in their purpose and functioning to members of a social system, while others are not.[374]

The issue of complexity is closely related to the constructs of perceived risk and compatibility.[375] The more compatible a new product is with the preceding product, the easier it is for the individual to understand it – hence the novelty is perceived as less complex.[376] Opposed to that, a positive relationship between complexity and perceived risk exists. The more complex a novelty, the greater

370 Cf. Rogers, E. M./Shoemaker, F. F. (1971), p. 148.

371 See for example Herrmann, A./Huber, F. (2000), pp. 3.

372 Cf. Weiber, R. (1992), p. 5; Meffert, H. (1985), pp. 35; Rogers, E. M./Shoemaker, F. F. (1971), pp. 148.

373 Rogers, E. M. (1962), p. 130.

374 Cf. Ibid.

375 Cf. Bähr-Seppelfricke, U. (1999), p. 25; Meffert, H. (1985), p. 36.

376 Cf. Bähr-Seppelfricke, U. (1999), p. 25.

the perceived risk related to the usage of the innovation.[377] As complex products cannot be easily understood by the potential user, it will require an investment in time and effort to learn about the benefits of the innovation and its usage.[378] If a consumer passed the interest-stage of the innovation decision process, he or she is likely to engage in a more extensive information search regarding a complex product. Hence, the evaluation-stage, also including possible trial activities, will take longer compared to unsophisticated products.[379] In addition to a slowed adoption process, numerous studies have confirmed that the more complex a new product appears to be, the lower the propensity to adopt the innovation will be.[380]

While in classical adoption and diffusion research the issue of complexity falls somewhat short (it is usually referred to *product* characteristics alone), a second aspect of complexity should be added here to broaden the concept. It can be assumed that a purchase decision is not only perceived as complex because of expected difficulties to understand and use the novel product, but also because potential adopters are confused by the large number of choice options. In product categories with strong competition, consumers are confronted by a largely diversified assortment of product solutions. Competing standards, wide product assortments, mistakable product descriptions, non-uniform product labels, a proliferation of brands, and a large number of stores to choose from represent aspects that describe a situation of overchoice.[381] This perception of overchoice, in turn, can lead to confusion[382] on the side of the consumers and may finally result in suboptimal product choices, product purchase delays, or even withdrawal from the decision process, leading to non-adoption – sometimes just in order to avoid the frustrating purchase situation.[383]

It can be maintained that perceived overchoice causes the potential adopter to perceive the purchase decision as more complex. Hence, similar to product complexity, overchoice is very likely to increase time and efforts necessary for

---

377 Cf. Holak, S. L./Lehmann, D. R. (1990), p. 63.

378 Cf. Veryzer, R. W. (1998), p. 145.

379 Cf. Schulz, R. (1972), p. 46.

380 Cf. Rogers, E. M./Shoemaker, F. F. (1971), p. 154. See also the studies cited in Gatignon, H./Robertson, T. S. (1985), p. 853.

381 Cf. Schweizer, M./Kotouc, A. J./Wagner, T. (2006), p. 185; Gourville, J. T./Soman, D. (2005), p. 382; Sproles, G. B./Kendall, E. L. (1986), pp. 273.

382 For an overview of different notions of consumer confusion, see Schweizer, P. (2004), pp. 22.

383 Cf. Chernev, A. (2003); Walsh, G./Henning-Thurau, T. (2002), pp. 96; Mitchell, V.-W./Papavassiliou, V. (1999), p. 327.

the decision process. Therewith, it negatively impacts the likelihood of adoption.[384]

Perceived complexity $\xrightarrow{-}$ adoption

3.5.1.4 Perceived Trialability and Divisibility

The degree to which it is possible to try an innovation prior to fully adopting it describes its trialability. This means that consumers can experiment with the novelty on a limited basis without taking the full risk inherent to the complete adoption.[385] A novelty might be tried either on a small-scale (e.g. a promotional package), or, if it is not possible to divide the new product, for a certain period of time (e.g. renting a receiver for cable-TV/on demand movies that are desired to be purchased eventually). Trying the novelty in-store as it is done for TV sets, surround sound speakers, home video games and the like, or borrowing the new product from friends also represent possible ways to try the innovation even though divisibility from the supply side is not given.[386] It is this reduced investment of time and/or money that affects the perceived risk: If trialability is given, the amount of perceived risk will decrease.[387]

Rogers (1962) further presumes that individuals adopting an innovation relatively earlier may differently judge the importance of trialability compared to later adopters. Innovators tend to perceive this characteristic as more important because only few experiences with the innovation exist at the early stages of diffusion.[388] In contrast to this, later adopters benefit from experiences of precedent adopters who may take the role of serving as a psychological or vicarious trial.[389] In general it can be expected that the relationship between trialability of a novelty and rate of adoption is positive.[390]

Perceived trialability/divisibility $\xrightarrow{+}$ adoption

384 Cf. Settle, R. B./Golden, L. L. (1974), p. 37.

385 Cf. Rogers, E. M./Shoemaker, F. F. (1971), p. 155.

386 Cf. Rogers, E. M. (1995), pp. 246; Cf. Rogers, E. M. (1962), p. 131.

387 Cf. Holak, S. L./Lehmann, D. R. (1990), p. 63; Ostlund, L. E. (1973), pp. 100.

388 Cf. Rogers, E. M. (1962), pp. 131.

389 Cf. Rogers, E. M./Shoemaker, F. F. (1971), p. 155.

390 Cf. Ibid.

3.5.1.5 Perceived Observability and Communicability

A definition of observability is given by Rogers/Shoemaker (1971) outlining it as "degree to which the results of an innovation are visible to others."[391] In addition to visibility, the authors include the characteristic of "communicability," which is expressing whether an innovation can be easily described to other members of the social system or not.[392]

Social aspects may influence product choice. Regarding socially desired product innovations, it can be said that high levels of observability enhance the rate of the new product's adoption. But social pressures that force individuals to conform to the peer group behaviors may also be prevalent. This issue is not discussed here in more detail as it is focused on later in Chapter 3.5.2.2 (pp. 103) dealing with social pressures and its effects on adoption.

In the context of observability interdependences among some product-specific determinants exist. The most pertinent relationship of observability is hypothesized to be with variables of perceived risk and complexity. On the one hand it is assumed that, if characteristics of an innovation are difficult to observe or communicate, individuals will perceive them as more risky.[393] On the other hand highly complex novelties – especially in the early stages of diffusion – are difficult to explain and communicate to others. With his study, Ostlund (1973) was able to sustain these negative relationships for a number of product innovations.[394] Overall, Rogers/ Shoemaker (1971) found that the observability and the adherent communicability of an innovation are positively related to its rate of adoption.[395]

Perceived observability/communicability $\xrightarrow{+}$ adoption

3.5.1.6 Perceived Risk

In addition to the five product-specific factors introduced by Rogers (relative advantage, compatibility, complexity, trialability, and observability), the perceived risk also plays an important role in the innovation decision process. Ostlund (1974) is credited as the first author who integrated this factor as a variable

391 Rogers, E. M./Shoemaker, F. F. (1971), p. 155.

392 Cf. Ibid.

393 Cf. Holak, S. L./Lehmann, D. R. (1990), pp. 63.

394 Cf. Ostlund, L. E. (1973), pp. 100.

395 Cf. Rogers, E. M./Shoemaker, F. F. (1971), p. 156.

influencing *adoption*.[396] The integration of the variable and the following discussion are important as, according to Sheth (1981), the perceived risk represents a major determinant of resistance towards novel products.[397] As most of the mentioned product-specific factors, perceived risk is closely related to individual characteristics as well.[398] The consumers' personal risk tolerance determines whether the perceived risk of an innovation purchase is judged as too high or as still acceptable. Though the perception and consequences of risk also depend on the individual, the objective risk is inherent to the product and can best be understood in the context of the specific product or product category.[399]

The *general* concept of perceived risk, which has originally been introduced into consumer research by Bauer (1960), stresses the effect of possible negative consequences of product purchases. Central to his perspective is the presumption that all consumer driven actions are possibly accompanied by negative consequences, which are difficult to predict or even not predictable at all – hence, consumer behavior involves risks.[400] Again, perception is pivotal to the effect of risks, meaning that the subjective amount of perceived risk determines the consumers' responses. Subsequent to Bauer's suggestions, several researchers developed component approaches to further analyze the role and dimensions of risk in buying decisions. Cox (1967), Cunningham (1967b), and Bettman (1973) advanced models which decompose the risk into several elements. The latter two authors even factor in a functional relationship between the different components.[401]

Cox (1967) identifies two factors determining perceived risk: the amount of loss if consequences after the purchase are unfavorable, and the subjective feeling or level of certainty that the consequences will be unfavorable.[402] Cunningham (1967b) similarly operationalizes risk as uncertainty and negative purchase consequences, and shows that products can be ranked along a "riskiness continuum" depending on the number of interviewed consumers who either rate the products with high, medium, or low levels of risk. Furthermore, when determining the level of risk, he explicitly assumes a multiplicative combination of the two ma-

---

396 Cf. Ostlund, L. E. (1974).

397 Cf. Sheth, J. N. (1981), p. 276.

398 For this reason, the part dealing with "psychological risks" is discussed later in the chapter of personal characteristics that influence the adoption of innovations (s. Ch. 3.5.2.2, pp. 103).

399 Cf. Cunningham, S. M. (1967b), p. 108.

400 Cf. Bauer, R. A. (1960), p. 390.

401 For an overview of further contributions see Pohl, A. (1996), p. 122.

402 Cf. Cox, D. F. (1967), pp. 37.

jor components "uncertainty" and "consequence."[403] In a comparable way, Bettman (1973) distinguished between an uncertainty of purchase consequences and the subjective importance of these consequences. However, he allowed multiplicative as well as additive relationships between the two components of his model.[404] Different to the just cited authors, Schweiger/Mazanec/ Wiegele (1976) do not consider perceived risk and its components globally. Instead, they combine the uncertainty-level of *specific purchase consequences* with their respective importance in their measurement approach. Therewith, the authors expand the two component approach (uncertainty and importance) to a multi-dimensional view of perceived risk. The dimensions of risk encompass the single, concrete threats due to undesired consequences as perceived by the consumer.[405]

After the essential establishment of perceived risk in consumer behavior theory, the construct has received much attention and has been applied to a multitude of issues related to consumer behavior. Many conceptualizations have been proposed by researchers, and the nature of risk dimensions added to the concept, are manifold. Primarily, however, researchers have used the following set of dimensions to describe the construct:[406]

- Performance/functional risks (likelihood that the product will not work properly)
- Financial risks (chances of losing money)
- Physical/health risks (likelihood that the product may be injurious to someone's health)
- Psychological risks (chances that the product may cause displeasure)
- Social risks (negative judgment of peer group)
- Loss of time (due to product-related inconveniences)

This demonstrates that the concept of "risk" is multi-layered and overlaps with similar concepts are unavoidable.[407] In consumer behavior studies multiple in-

403 Cf. Cunningham, S. M. (1967b), pp. 84.

404 Cf. Bettman, J. R. (1973), pp. 185.

405 Cf. Schweiger, G./Mazanec, J./Wiegele, O. (1976), pp. 96.

406 Cf. Bänsch, A. (2002), p. 76; Solomon, M. R. (1999), p. 281; Taylor, J. W. (1974), pp. 54; Jacoby, J./Kaplan, L. B. (1972), pp. 382; Roselius, T. (1971), p. 58; Cunningham, S. M. (1967b), p. 83; for an example of concrete risk aspects see Mitchell/Boustani (1993), who analyzed the perceived risk involved in the purchase of breakfast cereals. Cf. Mitchell, V. W./Boustani, P. (1993), p. 20.

407 E.g. psychological risks and loss of time are taken together and discussed later (s. Ch. 3.5.2.2, pp. 103). Social risks that have been mentioned in the context of observability are outlined within the scope of social pressure in the same chapter.

fluence factors have been used to explain the phenomenon of perceived risk. The following factors are repeatedly used in studies, whereas the level of perceived risk is the higher...[408]

- ...the higher the degree of novelty
- ...the more complex a product/purchase decision
- ...the higher the price
- ...the lower the relative advantage associated to the product
- ...the more difficult a comparison (especially the comparison of the quality) between competing products of the same product class
- ...the less it is possible to try the product prior to purchase
- ...the higher the expected level of social deprecation of a wrong decision
- ...the lower the level of own or external experience of the product

The listed factors determining perceived risk are assumed to equally apply for purchase decisions of *innovations* – not only for purchase decisions of regular products. In principle, all of Rogers' five adoption determinants can be regarded as variables influencing perceived risk.[409] However, the variables mentioned here are only indicative of the large number of determinants that have been contemplated. To portray a more complete picture of factors influencing the level of *perceived risk*, consumer traits (as self-efficacy, interest and the like) and situational factors (e.g. time-pressure) should be mentioned here, too. This, however, is beyond the scope of this study.

Consumers attempt to minimize the extent of perceived risk or, at least, attempt to reduce it to a tolerable level.[410] Depending on the level of perceived risk, a consumer might use different strategies to reduce this level. Analogous to his component view of perceived risk, Cox (1967) differentiates between two ways of risk reduction: risk reduction by lowering the amount at stake, and risk reduction by lowering the amount of uncertainty concerning the consequences of the product purchase.[411] The former strategy includes modifying the goal of the product purchase: reducing the original level of aspiration and even avoiding setting goals or avoiding the purchase may be the result.[412] Both, product trial and the use of free samples, which imply an adoption on a limited scale, fall un-

---

408 Cf. Helm, R. (2001), p. 118; Schmalen, H./Pechtl, H. (1996), p. 821; Kotzbauer, N. (1992), pp. 38; Holak, S. L./Lehmann, D. R. (1990), p. 63; Panne, F. (1977), pp. 130; Schweiger, G./Mazanec, J./Wiegele, O. (1976), p. 94.

409 See also Mitchell, V. W./Boustani, P. (1993), p. 18.

410 Cf. Hansen, F. (1972), p. 439; Cox, D. F. (1967), p. 630.

411 Cf. Ibid., pp. 38.

412 Cf. Ibid., pp. 72.

der the approach of stake reduction. The latter strategy of reducing the amount of uncertainty has been suggested to be the more common strategy used by individuals.[413] Uncertainty reduction particularly comprises information search which is often named in the first place.[414] However, the relation between perceived risk and the amount of information search is not direct. This is due to the different risk reduction strategies that act like a mediator between these two variables. Still, it can be stated that the higher the perceived risk, the higher the motivation of an individual to search for additional information.[415] The numerous measures to acquire additional information and alternative strategies that fulfill the task of uncertainty reduction are listed next.[416]

a) Information related strategies of uncertainty reduction (through information seeking)

- Reliance on past experiences
- Reliance on the experience of others (e.g. through opinion-leaders or word-of-mouth (WOM) communication)
- Mass media information (e.g. TV commercials, leaflets etc.)
- In-store information (e.g. salespeople/shop assistants, displayed product information)
- Testing journals

b) Information substitution and other strategies of uncertainty reduction[417]

- Product/brand/producer-loyalty
- Purchase of brand/from producer with high level of reputation
- Retail store loyalty
- Product warranties
- "Money back" guarantees
- Endorsements by celebrities or experts
- Certified products/quality seals
- Purchase of high priced products (price as a quality indicator)

---

413 Cf. Ibid.

414 Cf. Schweiger, G./Mazanec, J./Wiegele, O. (1976), p. 99.

415 Cf. Kroeber-Riel, W./Weinberg, P. (2003), p. 251.

416 Cf. Bänsch, A. (2002), p. 77; Helm, R. (2001), pp. 118; Meffert, H. (2000), pp. 124; Mitchell, V.-W./Boustani, P. (1994), p. 63; Mitchell, V. W./Boustani, P. (1993), p. 21; Roselius, T. (1971), pp. 57; Arndt, J. (1968), pp. 333; Cox, D. F. (1967), pp. 54, pp. 65, and pp. 74.

417 Some of the listed strategies are performed by the consumers and some of the strategies (e.g. warranties or endorsements) are offered by the companies for the consumers in order to reduce potential uncertainties.

- Delegating buying responsibility to other competent persons
- Free sample or trial purchase
- Buy insurance policy

As a conclusion, it can be stated that perceived risk plays an important role in the consumers' innovation decision process. The large number of factors determining the level of perceived risk as well as their eventual consequences on the purchase procedure demonstrate the complexity of the construct. However, it can generally be said, that the higher the perceived risk, the less likely consumers will adopt the new product and the more persons will reject novelty acquisition.

Perceived risk $\xrightarrow{-}$ adoption

3.5.2 Personal Characteristic Variables as Adoption Determinants

Just as well as in diffusion research, the *personal* characteristics of consumers have been drawn on in the research of innovation adoption determinants to explain individual differences next to the product-specific characteristics.[418] In this context, personal characteristic variables are defined as those factors influencing adoption that are primarily determined by the person and only secondarily determined by the novel product.[419] Due to the large number of different personal characteristic variables proposed by the research community to influence adoption behavior, varying approaches of classifying these personal attributes exist.

Some authors divide the personal attributes into sociodemographic factors, psychographic factors, and observable consumption behaviors.[420] This is nearly identical to the classic innovation publication by Rogers/Shoemaker (1971) who have proposed a similar classification to explain differences between earlier and later adopters including communication behavior variables instead of the broader category of consumption related variables.[421] Fantapié Altobelli (1991) uses another approach and differentiates between intraindividual determinants (demographic variables, personality attributes, and social characteristics), interindividual determinants (interactions of an individual with its social environ-

418 Cf. Gatignon, H./Robertson, T. S. (1991), pp. 319.

419 Cf. Litfin, T. (2000), p. 35.

420 See for example ibid., pp. 36; Pohl, A. (1996), pp. 64; Kaas, K. P. (1973), pp. 25.

421 Cf. Rogers, E. M./Shoemaker, F. F. (1971), pp. 185.

ment), and external system determinants (mass communication and other marketing mix variables).[422]

To state more precisely which characteristics have, among others, been used in predicting innovative behavior, it is exemplarily referred to Ostlund (1974) who employed the following set of variables: respondent age, occupational status of husband, respondent education, family income, self-confidence in problem-solving and psychosocial matters, interest polymorphism, privilegedness, social integration, social mobility, cosmopolitanism, and venturesomeness.[423] In a subsequent study he discovered that omitting quite a few consumer characteristics[424] would just have a negligible impact on the predictive quality.[425]

Although there does not seem to be a consistent picture of the predictive value and the direction of the influence of personal characteristic variables,[426] Holak (1988) points out that in certain cases they may have a considerable impact on innovative behavior.[427] She further advises to include consumer traits in future studies as a mere focus on product-specific variables may lead to less than optimal results. This remark is acknowledged here by describing sociodemographic variables, several psychographic variables, and social integration.

3.5.2.1 Sociodemographic Variables

The sociodemographic variables can be subdivided into social status variables (income, education, occupation), family-life-cycle variables (age, gender, marital status, age and number of children), and geographic variables (region, place of residence – urban vs. suburban vs. rural area).[428] Out of these socio-economic variables, especially age, income, education, and occupation have regularly been used in diffusion and adoption studies to explain innovative behavior. While higher income, social status and education are all assumed to bear a positive effect on the likelihood of innovation adoption, the way how age affects the decision to adopt is less clear.

In the search for variables determining an individual's innovativeness, for example Robertson/Zielinski/Ward (1984), Dickerson/Gentry (1983), La-

---

422 Cf. Fantapié Altobelli, C. (1991), pp. 27.

423 Cf. Ostlund, L. E. (1974), p. 24.

424 The author reduced the set to socio-economic status, social mobility, venturesomeness, general self-confidence in problem-solving, and respondent age.

425 Cf. Ostlund, L. E. (1974), p. 27.

426 See also the review of Kassarjian, H. H. (1971).

427 Cf. Holak, S. L. (1988), pp. 65.

428 Cf. Freter, H. (2001b), pp. 1074.

Bay/Kinnear (1981) and Robertson (1971) found innovators or early adopters to be better educated, have a higher income, higher social status and tend to be younger.[429] The positive correlation between education and innovation behavior is explained by Bohlen et al. (1968) with the persons' high levels of available information processing capacities and the ability to more easily differentiate between essential and non-essential information. In consequence, the functions of the novelty can be more easily understood.[430] In case of higher income, it is supposed that disposing of more financial resources reduces the adopters' perceived risks of innovation purchases – the amount at stake is relatively lower compared to individuals with lower incomes. In this way, the potential loss plays a rather marginal role for high income adopters.[431] Additionally, there may be expensive innovations – especially in consumer durables – that can only be afforded by affluent consumers. For other consumers this may result in what Katona (1964) described as "divergence of buying-willingness and ability."[432] All three aspects – occupation, education, and income levels – are closely related to each other. For the ownership of home-computers, for example Dickerson/Gentry (1983) presume that with higher status occupations, persons will have more work-related experiences with computer and databases which reduces the perceived complexity and facilitates the understanding of the innovation as well as its adoption.[433]

Rogers/Shoemaker (1971) performed a meta-analysis and confirmed the above mentioned relationships. However, different to studies which propose that age is negatively related to innovation adoption, the larger part of studies they have analyzed (about 50%) found no correlation. Only about 20% indicate that innovative consumers are younger while 30% of the studies report that they tend to be older.[434] Midgley's (1987) meta analysis adds to the confusion as he, in turn, was not able to reconfirm Rogers' and Shoemaker's conclusion and finds that earlier adopters are more likely to be older than later adopters.[435] Authors assuming age to be negatively related to innovativeness claim that the reasons for this phenomenon are that younger persons have not yet developed rigid purchase

429 Cf. Robertson, T. S./Zielinski, J./Ward, S. (1984), pp. 389; Dickerson, M. D./Gentry, J. W. (1983), pp. 293; LaBay, D. G./Kinnear, T. C. (1981), pp. 272; Robertson, T. S. (1971), pp. 85.

430 Cf. Bohlen, J. M. et al. (1968), p. 355.

431 Cf. Dickerson, M. D./Gentry, J. W. (1983), p. 226.

432 Katona, G. (1964), pp. 28.

433 Cf. Dickerson, M. D./Gentry, J. W. (1983), p. 226.

434 Cf. Rogers, E. M./Shoemaker, F. F. (1971), pp. 185 (no definition of "younger" and "older" consumers is given by the authors).

435 Cf. Midgley, D. F. (1987), pp. 205.

behaviors, are more apt to deviate from society norms and are willing to take on more risks.[436] Other researchers assume the opposite to be true and disagree with the postulated decreasing interest in new things when people age. It is rather the personality – which is formed during childhood and adolescence and which does not significantly change in later years – that determines a person's actual consumer behavior.[437] Likewise, Szmigin/Carrigan (2001b; 2001c; 2000), who are devoting their research to older consumers, try to overcome traditional and stereotype views about elderlies by repeatedly stressing that their behavior is changing and that resistance to innovative products is not a phenomenon uniquely related to older persons.[438] Although the consequence and impact of age is discussed controversially, this variable is surveyed in almost every research study: it represents a standard questionnaire item not only in the domain of consumer behavior research. Due to its central role in this study, age is subject of a more detailed analysis in a chapter dealing specifically with the role of age in innovation adoption (s. Ch. 4.2.1, pp. 124).

3.5.2.2 Psychographic Variables

Psychographic analyses encompass the consumers' values, personalities, and lifestyles.[439] In innovation adoption research, variables from these three areas have been integrated more or less extensively to explain innovative behavior. In search of personality determinants of innovative behavior, Robertson/Myers (1969) for example analyzed a broad number of variables and included the California Psychological Inventory in their questionnaire.[440] The authors were able to identify relationships between these basic personality variables and the consumers' innovative behavior. However, they needed to conclude that these relationships were of "questionable statistical significance."[441] This statement is indicative of numerous other studies that obtained dissatisfactory results when

436 See for example Schulz, R. (1972), p. 49.

437 Cf. Laurinkari, J. (1988), p. 157.

438 Cf. Szmigin, I./Carrigan, M. (2001b), p. 27; Szmigin, I./Carrigan, M. (2001c), p. 117; Szmigin, I./Carrigan, M. (2000), p. 509.

439 Cf. Hoyer, W. D./MacInnis, D. J. (2004), pp. 416; Freter, H. (2008), pp. 135.

440 The inventory is composed of the following variables: Dominance, capacity for status, sociability, social presence, self-acceptance, sense of well-being, responsibility, socialization, self-control, tolerance, good impression, communality, achievement via conformance, achievement via independence, intellectual efficiency, psychological mindedness, flexibility, and femininity. Cf. Robertson, T. S./Myers, J. H. (1969).

441 Robertson, T. S./Myers, J. H. (1969), pp. 165.

personality variables were used to explain innovative behavior.[442] Even in predicting more general aspects of consumer behavior other than innovation adoption, weak relationships regarding personality or consumer traits are reported.[443] Despite the mixed findings some factors nonetheless seem promising in this research context. From the large number of suggested psychographic variables, the potential adopter's innovativeness, interest, experiences/knowledge, psychological risk, self-efficacy, and social pressure are discussed more thoroughly in the following.

Traditionally, innovativeness has been described as the "degree to which an individual is relatively earlier in adopting new ideas than the other members of his social system."[444] With this definition Rogers equates the time of adoption with innovativeness. Consequently, Rogers (1962) based the different adopter categories with innovators first and laggards as last group of persons adopting the novelty on this understanding of innovativeness.[445] However, the traditional interpretation of innovativeness has been found to operationalize the theoretical construct inadequately. Thus, Midgley/Dowling (1978) criticize that studies so far have predominantly been based on simplistic trait-behavior models which measure innovativeness by merely detecting either the observable act and time of adoption, or by surveying the ownership of novel products from a given list with different innovations.[446] In the same way, Venkatraman/Price (1990) objected that widely embraced conceptualizations dispose of a too strong focus on acquisitiveness.[447]

When comparing the different ways other studies operationalized innovativeness, Midgley/Dowling (1978) found that actually not the same concept is measured, but that different concepts of innovativeness prevail which are characterized by varying levels of generalizability. Accordingly, they propose to specify innovative behavior at three levels: the specific product level, the specific product category level, and across several product categories (with the latter approaching the idea of innovativeness as a general personality trait).[448] To avoid confusion the authors termed the (unobservable) construct of innovativeness "innate innovativeness" while, on the other hand, the visible behavior –

---

442 Cf. Jacoby, J. (1971), p. 244 and the cited literature.

443 Cf. Hoyer, W. D./MacInnis, D. J. (2004), p. 435; Blackwell, R. D./Miniard, P. W./Engel, J. F. (2001), p. 214.

444 Rogers, E. M. (1962), p. 19.

445 See Chapter 3.2, pp. 71.

446 Cf. Midgley, D. F./Dowling, G. R. (1978), pp. 230.

447 Cf. Venkatraman, M. P./Price, L. L. (1990), pp. 293.

448 Cf. Midgley, D. F./Dowling, G. R. (1978), pp. 231.

which corresponds to Rogers/Shoemakers' (1971) conceptualization – was named "actualized innovativeness."[449] At the same time the authors assume these two forms of innovativeness to be conceptually distinct, meaning that results from one level of innovativeness cannot simply be applied to the other – with actualized innovativeness being a complex function of innate innovativeness.[450] In her model, Hirschman (1980) enhanced the idea of Midgley/Dowling and further subdivides the actualized innovativeness into the subcategories of "vicarious innovativeness" (acquisition of information about new products or consumption situations) and "adoptive innovativeness" (acquisition and continued use of new products)[451].[452]

The propensity of an individual to favor new products and engage in the search for new experiences has regularly been used in adoption and diffusion research, however – as the discussion has already showed – often by using differing concepts. Today, it is acknowledged that consumer innovativeness can be conceptualized at different levels of abstraction and in a multidimensional way.[453] It is important to refer to its notion as a "latent underlying preference for new and different consumption experiences."[454] The reactions of an individual to the new and different are likely to vary on a continuum from a very positive attitude to a very negative attitude toward change.[455] It should also be noted explicitly what researchers have implicitly assumed for a long time. That is, that innovativeness represents a consumer trait which is possessed, at a higher or lower level, by *all* members of a society.[456] Comparable constructs like venturesomeness, open-mindedness, change orientation, need for change, and novelty seeking have been positively associated with the likelihood of innovation adoption.[457] On the other hand, opposed constructs like rigidity, dogmatism, resistance to change, tradi-

---

449 Midgley, D. F. (1977), p. 107.

450 Cf. Midgley, D. F./Dowling, G. R. (1978), pp. 234.

451 The adoptive innovativeness as a special form of actualized innovativeness is what corresponds to adoption as it was discussed in Chapter 3.3 (pp. 75), being the final step in the innovation decision process.

452 Cf. Hirschman, E. C. (1980), p. 285.

453 Cf. Tellis, G. J./Yin, E./Bell, S. (2004), pp. 5.

454 Venkatraman, M. P./Price, L. L. (1990), p. 294.

455 Cf. Goldsmith, R. E./Flynn, L. R./Goldsmith, E. B. (2003), p. 55; Robertson, T. S. (1971), p. 94.

456 Cf. Hirschman, E. C. (1980), p. 283; Midgley, D. F./Dowling, G. R. (1978), p. 229.

457 Cf. Wood, S./Swait, J. (2002), pp. 3; Manning, K. C./Bearden, W. O./Madden, T. J. (1995); Hirschman, E. C. (1980), pp. 284; Rogers, E. M. (1962), p. 304; Rokeach, M. (1960).

tional or habitual attitudes toward existing practice or behavior, as well as closed-mindedness are all linked to an individual's reluctance to innovate.[458]

Innovativeness $\xrightarrow{+}$ adoption

Interest

Another consumer-specific determinant which was found to impact innovation adoption is a person's interest in the product or product category. Holak (1988) describes product class interest as an "individual's inherent interest in a product category, whether it be photographic equipment, electronic gadgets, or kitchen items."[459] Midgley/Dowling (1978) also used product category interest in their model of innovativeness conceptualization. They defined it as an intervening variable between innovativeness as a personality trait and innovativeness interpreted as innovative behavior (innovativeness at the lowest degree of abstraction), arguing that interest is one step closer to the observable phenomenon.[460] It does not seem reasonable to consider a person's general interests to be a determinant with much explanatory power, but instead restrict the interest variable to a specified product or product category. Midgley/Dowling (1978) explain this with the fact that "most individuals do not have all-encompassing interests or activities, especially with respect to the purchase of goods or services."[461] Constraints regarding time and limited funds also prevent a person from being an innovative buyer across very many product categories.[462] Interest in a product category impacts several other areas of the purchasing process. The propensity to read articles about a product, for example, increases with an individual's interest for the product category. Higher information search intensity results in more knowledge of the interested consumer about the specific product category.[463] More generally, it has been found that it is especially the interest in a product category that influences the extensiveness of the decision process.[464]

To sum it up, it is assumed that a higher level of interest in a product category corresponds to a higher probability of adopting a specific novelty of this product

---

458 Cf. Ram, S./Sheth, J. N. (1989), pp. 8; Ram, S. (1987), p. 208; Sheth, J. N. (1981), pp. 275; Jacoby, J. (1971), p. 245; Robertson, T. S. (1971), p. 94; Vacchiano, R. B./Strauss, P. S./Hochman, L. (1969); Rokeach, M. (1960).

459 Holak, S. L. (1988), p. 53.

460 Cf. Midgley, D. F./Dowling, G. R. (1978), p. 236.

461 Ibid.

462 Cf. Wilkie, W. L. (1986), p. 122.

463 Cf. Wilkie, W. L. (1986), p. 122.

464 Cf. Kapferer, J.-N./Laurent, G. (1985), p. 292.

category, while people displaying only little product category interest are unlikely to purchase the innovation.[465]

Interest $\xrightarrow{+}$ adoption

Experiences/Knoweledge

Experiences also seem to play a role in influencing the innovation decision process. Gatignon/Robertson (1991) point out that heavy product category usage and experience represent one of the few consistent findings when analyzing the personal characteristics of innovators.[466] This assumption is based on results of studies such as the one by Taylor (1977) who identified a significant positive relationship between the intensity of product category usage and time of adoption of a novel product which has been introduced into that product category.[467] Through product (category) usage consumers gain experiences and become familiar with the novel item. In this process the adopter acquires knowledge which he or she can use for future purchase decisions. These outlined relationships are reflected in the definition of knowledge put forward by Alba/Hutchinson (1987) who suggest that knowledge is composed of two components, familiarity and expertise. With familiarity they mean "the number of product related experiences that have been accumulated by the consumer."[468] The authors also specify from which sources these experiences may originate from (e.g. advertising exposure, information search, interactions with salespersons, product usage, etc.). In addition, with expertise they consider the ability of an individual to successfully perform product related tasks in their categorization. The authors include cognitive structures (beliefs about product attributes) and cognitive processes (decision rules) as the two components of expertise.[469] Although there are differences between the concepts of familiarity and expertise, they seem not to be entirely independent from each other. This is, for example, because familiarity with a product also facilitates cognitive processes as the evaluation of new information about an innovation of the same or related product category.[470]

An alternative conceptualization of knowledge about innovations is given by Rogers/Shoemaker (1971) who distinguish between awareness-knowledge, how-

465 Cf. Holak, S. L. (1988), pp. 53; Wilkie, W. L. (1986), p. 122.
466 Cf. Gatignon, H./Robertson, T. S. (1991), p. 320.
467 Cf. Taylor, J. W. (1977).
468 Alba, J. W./Hutchinson, J. W. (1987), p. 411.
469 Cf. Alba, J. W./Hutchinson, J. W. (1987), p. 411.
470 Cf. Gatignon, H./Robertson, T. S. (1985), p. 861.

to knowledge, and principles knowledge.[471] While the latter is supposably only attributed to experts, as it describes the knowledge about the underlying functioning principles of the innovation, awareness-knowledge means the individual "knows" an innovation exists motivating potential adopters to acquire "how-to knowledge" which in turn comprises the information how to correctly use the novel product.[472]

When consumers are exposed to the existence of an innovation, they will be faced with new information. This is in line with the innovation definition which describes innovations as "something that has not existed before." Novelties are assessed by means of existing knowledge.[473] If the product is compatible, the existing knowledge base helps the potential adopter to comprehend the new product's attributes. This process is also referred to as learning by analogies.[474] As incremental innovations are modifications of older products or just entail a small number of new attributes, only little knowledge needs to be transferred. Opposed to that, in case of radical innovations, consumers are not able to refer to their present knowledge. As there is little overlap between the new, discontinuous product and existing ones, entirely new knowledge structures need to be created.[475] Furthermore, it is possible that existing knowledge becomes obsolete with an innovation introduction of that type.[476] This incompatibility to previous knowledge and the higher learning requirements to understand the functions, benefits and how to properly use the novel product are reasons for the relatively slower diffusion of actual new products compared to incrementally new products.[477]

Although the traditional belief proposes that innovators are characterized by considerable product category knowledge, more recent studies reveal that expert knowledge can also have a negative effect on adoption. Exceptions to the traditional view have been found for potential adopters who are particularly competent and able to adequately evaluate the innovation, but who come – by virtue of

---

471 Cf. Rogers, E. M./Shoemaker, F. F. (1971), p. 106.

472 Cf. Polanyi, M. (1985), p. 16; Rogers, E. M. (1983), pp. 167; Rogers, E. M./Shoemaker, F. F. (1971), pp. 106; Ryle, G. (1963), pp. 27.

473 Moreau/Lehmann/Markman (2001) specify that it is especially product category knowledge which helps to assess a novel product and, hence, represents a central factor in altering the adoption process. Cf. Moreau, C. P./Lehmann, D. R./Markman, A. B. (2001), p. 14.

474 Cf. Gregan-Paxton, J./John, D. R. (1997), pp. 267.

475 Cf. Ibid., p. 275.

476 Cf. Niedbal, M. (2005), pp. 104.

477 Cf. Gatignon, H./Robertson, T. S. (1985), p. 863.

their high level of knowledge – to the conclusion that the new product is unfavorable or, at least, not superior compared to the older product which is replaced by the innovation. In this case, the higher knowledge leads to rejection.[478] Hence, the relative advantage of an innovation mediates whether high levels of knowledge or expertise accelerate the novelty adoption. Additionally, the rejection of a novel product by knowledgeable consumers can be expected as outcome of the innovation decision process because – when comparing the existing, old solution and the innovative, new product – they become aware of more differences and things they do not understand.[479] This means, that experts can be limited by established knowledge patterns.[480] Still, it can be assumed for the large majority of individuals that compatibility of existing knowledge with the new product facilitates its adoption.

Knowledge compatibility $\xrightarrow{+}$ adoption

Psychological Risk

In the context of perceived risk the issue of psychological risks has already been touched on. There, the psychological risk has been seen as possibility that the product causes displeasure. This can, for example, be the case when the innovation requires the adopter to learn new things or necessitates changing his or her behavior after the purchase – that means especially in those cases where low compatibility with existing knowledge prevails. The same holds true for product-related inconveniences that evolve. But even prior to the actual product usage and resulting consequences, psychological risks can be perceived. Regarding the loss of time as an example, not only the need to spend time to learn the proper use of the innovation may be prevalent, but even before that, information search prior to the novelty purchase causes time investments. The fear of the necessary time investment as well as the inconvenience to deal with the details of this innovation purchase is, therefore, also accounted to the psychological risk. The perceived psychological risk is likely to be higher in situations where network externalities exist. Network externalities prevail when an innovation depends on previous adoption or the adoption by other individuals.[481] As these are far-reaching decisions that may also influence the purchase of related products in the future, they are assumed to evoke the perception of risks on the side of the individual.

---

478 Cf. Leonard-Barton, D. (1985), pp. 915; Hirschman, E. C. (1980), p. 289.

479 Cf. Statt, D. A. (1997), p. 234.

480 Cf. Sääksjärvi, M. (2003), p. 96.

481 Cf. Hirunyawipada, T./Paswan, A. K. (2006), p. 186.

Psychological risk $\xrightarrow{-}$ adoption

Self-Efficacy

Apart from the common determinants used in adoption and diffusion literature, self-efficacy is a less usual variable picked up to be further analyzed. Similar to self-confidence in problem solving, self-efficacy should be understood as the perceived ability to cope with day-to-day problems and as self-confidence exhibited in consumer decision making.[482] It entails the belief of someone's capability and expectation to overcome these problematic situations.[483] Self-efficacy can be seen as a factor that reduces the perceived risks of an individual in the purchase decision process. Thus, it can be assumed that with more self-efficacy, the chances to adopt an innovation increase.

Self-efficacy $\xrightarrow{+}$ adoption

Social Pressure

The social aspects of an innovation adoption have already briefly been mentioned in the chapter dealing with the observability of innovations and in the context of perceived risks. Rogers (1983) even integrated social aspects of innovations in his relative advantage discussion.[484] According to him innovations not only possess their core benefit but also an added, social status benefit. That means that social aspects contribute to the adoption decision due to the fact that ownership or consumption of the innovation may display the adopter's status, or can lead to rewards or avoidance of negative consequences on the part of the reference group.[485] Peer group influence is most notably present in situations of luxurious good purchase or purchase of products that are visible to others.[486] In the case of existing peer group pressure it can be assumed that taking over a "required" novelty occurs faster among members of the group than an innovation which is exposed to less social pressure.[487]

Social pressure $\xrightarrow{+}$ adoption

---

482 Cf. Schiffman, L. G./Sherman, E. (1991), p. 189; Ostlund, L. E. (1974), p. 24.

483 Cf. Städtler, T. (2003), pp. 971; Bandura, A. (1986), pp. 390.

484 Cf. Rogers, E. M. (1983), pp. 215.

485 Cf. Schuster, H.-W. (1994), pp. 112.

486 Cf. Bearden, W. O./Etzel, M. J. (1982), pp. 184.

487 Cf. Fisher, R. J./Price, L. L. (1992), pp. 482.

3.5.2.3 Social Integration

In search of factors determining whether a person adopts an innovation earlier or later than the majority of society, or not at all, attention also needs to be directed to the issue of social integration. Similar to the understanding of social networks in the gerontological part, social integration can be defined as extent of social interaction with other members of the community.[488] Three major perspectives can be found in research dealing with interpersonal relations. While research has almost exclusively focused on the normative perspective which discusses social influence (e.g. conformity in groups or family decision making) for a long time,[489] other studies added the informational perspective which sheds light on the communication behavior of potential adopters (e.g. to resolve informational uncertainties) and the aspect of relationships as sources of social support.[490] More specifically, the latter two aspects refer to the degree to which a consumer interacts with other individuals and engages in interpersonal communication which is usually referred to as WOM communication.[491] Voss (1984) estimates that even in the era of mass communication and mass advertising, as much as 80% of all purchase decisions are impacted by someone's direct recommendation.[492] Due to its strong persuasive character, WOM has been extensively discussed and is defined by Arndt (1967b) as "*oral, person-to-person* communication between a perceived *non-commercial communicator* and a *receiver* concerning a *brand*, a *product*, or a *service* offered for sale."[493] Compared to other interpersonal communication sources which also meet the criterion of face-to-face contact (like personal selling or shopping personnel), independence from the manufacturer makes WOM a reliable and credible source of information and therefore creates strong influence on the individual adoption decision.[494]

The non-commercial information sources can be further subdivided into communication with persons from primary formal and informal as well as secondary

---

488 Cf. Robertson, T. S. (1971), pp. 105 and pp. 201.

489 Cf. Ibid., pp. 201; Bourne, F. S. (1968).

490 Cf. Price, L. L./Feick, L. F. (1984), pp. 250; Katz, E. (1968), p. 350.

491 Other terms used synonymously to WOM comprise for example product-related conversations, personal recommendations, or informal communication. Cf. Lau, G. T./Ng, S. (2001), p. 164. Leonard-Barton (1985) further notes, that there exists an overlap between the phenomenon of opinion leadership and WOM communication. Cf. Leonard-Barton, D. (1985), p. 914.

492 Cf. Voss, P. Jr. (1984), p. M-10.

493 Arndt, J. (1967b), p. 190.

494 Cf. Lau, G. T./Ng, S. (2001), p. 164; Gatignon, H./Robertson, T. S. (1991), p. 331; Kaas, K. P. (1973), p. 54; Arndt, J. (1967b), p. 205.

formal and informal groups. Persons with whom the individual has regular, close and (in)formal contact can be accounted to the primary group. This group, above all, consists of family members, friends and neighbors. Different to that, secondary groups are relatively larger and entail colleagues, club members and other persons with whom a formal, more distant relationship is formed and contacts are less frequent and also less intensive.[495] Although only weak ties between the individual and members of the latter group types exist, they can still constitute an important source of WOM.[496]

The effect of social connectedness on innovation adoption has been analyzed by several researchers. In a study employing depth interviews, Robertson (1967a) found that innovative consumers are more integrated into their neighborhoods and interact with a larger number of people than non-innovative persons.[497] The author accounts the higher levels of personal communication which result from better social integration as reason why those persons adopt earlier. This relationship is confirmed by another study which researched product-related conversations (WOM communication) of an innovation among female residents of a large apartment complex. Arndt (1967a) found that those females who were well integrated into the social structures were also faster in adopting the novelty than isolated residents.[498] Furthermore, support for this hypothesis is delivered by Midgley (1987) who included social participation – wherewith he comprehends the integration with social communication networks – in his meta analysis.[499] In accordance with these results and in the context of communication behavior, Rogers/Shoemaker (1971) generalized that earlier adopters show higher social participation, are more integrated into the social system, and have greater exposure to interpersonal communication channels than later adopters.[500] Exposure to WOM and to social interaction in general, enhances innovation adoption by creating awareness for that innovation, receiving recommendations or support in the purchase decision process.

Social integration $\xrightarrow{+}$ adoption

495 Cf. Kroeber-Riel, W./Weinberg, P. (2003), pp. 439; Assael, H. (1992), pp. 403; Reimann, H. (1974), pp. 134 and the cited literature; Cooley, C. H. (1909), p. 23 cited in Strodtbeck, F. L. (1973), pp. 646.

496 Cf. Brown, J. J./Reingen, P. H. (1987), pp. 352.

497 Cf. Robertson, T. S. (1967), p. 28.

498 Cf. Arndt, J. (1967a), p. 293.

499 Cf. Midgley, D. F. (1987), p. 206.

500 Cf. Rogers, E. M./Shoemaker, F. F. (1971), pp. 188. Similarly see Robertson, T. S. (1971), p. 94.

3.5.3 Situational Variables as Adoption Determinants

Next to product-specific determinants and personal characteristic variables, there are situational variables that may impact the outcome of the innovation decision process. Although situational factors have predominantly been researched in the context of regular buying behavior, they are supposably effective to the same extent when the item to be purchased is of innovative nature. The study of situational factors has emerged from results of empirical studies which approved that even if consumers resembled each other in several important determinants of the purchase behavior, many of them still showed a different behavior depending on the specific purchase situation. Likewise, different persons may behave in the same way when faced with similar situations.[501]

Given the poor conceptual investigations, it was Belk (1974) who engaged in the definition of situational variables which subsequently experienced widespread recognition. According to the author, consumer situations comprise "all those factors particular to a time and place of observation which do not follow from a knowledge of personal (intra-individual) and stimulus (choice alternative) attributes, and which have a demonstrable and systematic effect on current behaviour."[502] The author further developed a categorization which employs five groups to differentiate between the sources of situational effects:[503]

1) Physical surroundings (e.g. location, sound, odors, weather)
2) Social surroundings (e.g. presence of other people, interpersonal interaction)
3) Temporal perspective (e.g. time of day/season, time relative to past or future events, time constraints)
4) Task definition (e.g. quantities to be purchased, purchase of gift, differing buyer and user roles)
5) Antecedent states (e.g. momentary moods like anxiety and pleasantness, or momentary conditions like cash on hand, fatigue, and illness)

While in Belk's (1975) taxonomy, financial constraints are limited to "cash on hand," this aspect needs to be broadened. As for larger purchase items (durable goods), more financial means are required, the temporary (overall) economic situation of a person should also be considered.[504] Furthermore, a look at prod-

501 Cf. Kakkar, P./Lutz, R. J. (1981), pp. 204; Miller, K. E./Ginter, J. L. (1979).

502 Belk, R. W. (1974), p. 157.

503 Cf. Belk, R. W. (1975), p. 159. Likewise, see Solomon, M. R. (1999), pp. 303; Pepels, W. (1995), pp. 95.

504 Cf. Wilkie, W. L. (1986), pp. 275.

uct depletion and -failure seems promising in explaining situations that directly stimulate purchases. While in the first situation of depletion (re-)purchases are made due to the using up of a product through consumption, the latter situation of product failure includes the breakdown of a product which causes the need for repair, purchase of a used unit, or purchase of a new product to replace the current one.[505]

Assael (1992) uses a different, more marketing-oriented approach in differentiating between types of situations.[506] He distinguishes between:

1) Consumption situations (e.g. at home vs. at workplace; alone vs. with friends)
2) Purchase situations (e.g. product availability; long lines at check-out counters; time pressure)
3) Communication situations (e.g. contact with radio commercial in car vs. at home; magazine read inside vs. outside the home)

Moreover, in order to determine the importance of situational influences, Sinha (1994) proposed to classify situations based on the level of familiarity.[507] While "common situations" include routine and habitual scenarios, "ad-hoc situations" are experienced less frequent and are fairly unfamiliar to the individual.

Due to the huge diversity of situational variables, not only their complete coverage and unequivocal categorization poses difficulties, but also the analysis of their effect on consumer behavior. Taking one important reason for deviations from intended purchasing behavior which is related to the in-store environment as an example, out of stock situations either result in looking for another store with the item in stock, purchase postponement, or the purchase of an alternative product or brand.[508] Concerning the influence of situational aspects on innovation adoption behavior, situational influences may enhance or slow down the innovation decision process and lead to adoption or rejection of the new product. For example, searching for more information, waiting for social approval or waiting for one out of several competing standards to establish take time and, hence, have a temporal extension of the decision process as a consequence. A potential adopter may withdraw from his or her decision to purchase a novelty, or discontinue its use. Apart from that, the reverse effect also seems possible

505 Cf. Ibid., pp. 276.

506 Cf. Assael, H. (1992), pp. 527. See also the classification in Hansen, F. (1972), p. 53.

507 Cf. Sinha, I. (1994), pp. 479.

508 Cf. Hoyer, W. D./MacInnis, D. J. (2004), p. 150; Assael, H. (1992), p. 528.

when the particular situation influences the individual to repeal his reluctance to adopt and purchases the novel product.

Some concrete consequences of situational variables on adoption are also offered, e.g., by Pohl (1996) who states that the more urgent it is for an individual to purchase an innovative product,[509] the faster the adoption process will be run through and the more likely the individual is to adopt the innovation.[510] Concerning time limitations it has been stated that consumers use different information search and decision strategies, which allow reducing the required time to come to a final decision.[511] In these time pressure situations, individuals adapt the extensiveness of information search: The amount of additional information acquisition declines with increasing levels of time pressure.[512] However, if consumers under time pressure are able to decide between a new and a known brand or product, the propensity to reject the new alternative increases because additional product information needs to be searched first.[513] In contrast, when urgency of product purchase is low, situational time limitations may lead to a postponement of the adoption decision. Finally, it should be noted that the power of situational influences seems to depend on the product category which is examined.[514]

Situational variables $\xrightarrow{+/-}$ adoption

509 Urgency for a product purchase can result either from product depletion or product failure as mentioned before, but also from the emergence of new needs.

510 Cf. Pohl, A. (1996), p. 67. See also Katona, G./Mueller, E. (1955), p. 32.

511 Cf. Kuß, A./Tomczak, T. (2004), pp. 221.

512 Cf. Bettman, J. R. (1979), p. 127.

513 Cf. Kuß, A./Tomczak, T. (2004), p. 222.

514 Cf. Belk, R. W. (1975), p. 160.

4 Older Consumers and Consumer Electronic Products

In the previous chapter numerous determinants have been identified which are believed to generally influence the innovation decision process and its outcome. However, prior to testing this large number of eventually influencing variables empirically, it is indispensable to restrict the focus exclusively on promising determinants and on a single product category. Therefore, in the following Chapter 4.1 it will be explained why consumer electronic products (CEP) represent a promising research area. After that, it will be assessed which of the previously explained determinants (s. Ch. 4.2) are relevant in explaining potential *differences* of older and younger individuals regarding ownership of *CEP*.

4.1 Product Category Selection: Consumer Electronic Products

One goal of the study is to identify the relevant explanatory variables of innovation adoption of younger and older consumers and determine their strength of influence. As it was mentioned in the introductory part of the first chapter, the innovative products are of interest, which do not represent a specific benefit exclusively for elderly consumers. Objects that have traditionally been associated with old individuals include products such as walking frames, denture cleaners, stair lifts or similar products that are deficit-oriented as they mainly deal with the physical changes of older consumers. The exemplarily enumerated products are adapted to the elderly consumers' specific biological needs. Although in some cases improvements are supposably still needful to guarantee a problem-free use of these products, the benefits and their use are clear and involved companies generally possess accurate knowledge how to adequately design and profitably market them.

However, by taking a look at the market it can easily be observed that the overwhelming majority of (innovative) products not specifically targeting older consumers lack some essential adaptations. To gain insights into the innovation adoption and its determinants, these "regular" novelties will be focused on in this study. To further delimit the vast range of products neglecting the needs of older buyers, either consumer durables (e.g. cars, TVs, furniture), non-durables (e.g. food items, cosmetics, detergents), or services (e.g. traveling, banking, healthcare) can be researched. Tangible products are at the center of analysis, whereas services are discarded. Non-durables are eliminated from deeper analysis, too, as for example Hupp (1999) was not able to identify any significant differences between the information and decision behavior of younger and older consumers for this category of products.[515] For *innovative* non-durables similar results can be expected. Additionally, as low-ticket consumer goods are generally purchased under conditions of low-involvement, it is likely that purchases are habitualized and decisions are often made without any explicit evaluation of the product in question. This would further complicate surveying consumers regarding those variables which are determining their purchase behavior.[516]

Therefore, the search will be directed to consumer durables, for which it has been found that high levels of engagement in the choice process prevail and that running through all steps of the innovation decision process is very likely.[517] Different to non-durables, for these kind of products the determinants in Chapter 3.5 (pp. 87) are thought to have a considerable influence on the outcome of the adoption process. Due to the extensive decision process, respondents are expected to be more able to evaluate the details of their decisions in a survey.[518]

Consumer electronic products (CEP) seem to meet the necessary criteria. Characterized by a high rate of (not seldom discontinuous) innovations which occurs along with important technology changes, purchase decisions of products in this market segment involve a considerable amount of risk (e.g. because of high prices and the perceived uncertainty in the purchase decision). They are typically complex and require substantial levels of knowledge. Individuals who do not possess such prior knowledge are required to engage in learning efforts in order to be able to adequately assess the innovation prior to purchase as well as to operate the product correctly after purchase.

In a next step it needs to be decided whether a single product or a group of innovative products of the CEP-category should be focused on in the study. Therefore, a pre-study was set up to find out whether a single innovative CEP represents a promising research object. The initial idea of surveying the adoption determinants of a single CEP innovation (digital video recorders [DVRs]) was

---

515 Cf. Hupp, O. (1999), pp. 150.

516 Among others, one of the reasons why purchase decisions of innovative non-durables are not processed extensively is because they are typically of incremental nature. Cf. Hawkins, D. I./Best, R. J./Coney, K. A. (2004), p. 248; Trommsdorff, V. (2004), pp. 58; Meffert, H. (2000), pp. 102; Gatignon, H./Robertson, T. S. (1991), pp. 325.

517 Cf. Hansen, F. (1972), pp. 314; Robertson, T. S. (1971), p. 62; Lavidge, R. J./Steiner, G. A. (1961), p. 60.

518 Cf. Sääksjärvi, M. (2003), p. 91; Meffert, H. (2000), pp. 112; Gatignon, H./Robertson, T. S. (1985), p. 854.

dropped soon after the pre-study. The study was designed to capture the ownership and use frequency of different generations of video recording and playing devices (like videocassette recorders, DVD-Player, DVD-Recorder, Hard Disk Drive Video Recorders, and the like). Furthermore, the general opinion of older and younger respondents regarding the specific DVR features was aimed to be assessed. The respondent feedback promptly showed the difficulties of the single product focus. Many elderly consumers refused participation due to non-ownership of DVRs or fear that they could not answer questions regarding this unknown innovation properly.[519] Most functionalities of DVRs were understood and even seen as beneficial. Still, most of the time, they failed to explain why no adoption had occurred or why a purchase would be unlikely in the future. It seemed that DVRs and their functionalities were not relevant enough for the older consumer segment.

This experience portended to the sensitivity of the results when interrogating persons about a *single* product. It also conforms to the findings from the analysis of adoption determinants. As mentioned before several studies explicitly point out the relevance of a product category focus for their particular context. Despite the assumption that the predictive power of adoption determinants is higher for a single product, there is the drawback that results cannot easily be transferred to other innovations. Instead, it was found that a product category focus improves predictions.[520] Moreover, product category ownership is more resistant to situational influences than the focus on a single innovation. The fact that a consumer might not have finished the adoption process of a concrete innovation is balanced by other products of that category which this person possesses. In case of technological innovations, Rogers (2003) argues that considering technology clusters is necessary because the assumption of new technological products being independent from other technological products would represent an oversimplification.[521] As the product category focus will allow transferring the results of the survey to a greater extent – if not on a global level, than at least to the products of the same product category – the focus on product category ownership is pursued here as well.

519 Despite the attempt to convince the potential respondents and pointing out that no specialist knowledge would be necessary to answer the questions, the participation rate could not be improved.

520 Cf. Hirunyawipada, T./Paswan, A. K. (2006); Schiffman, L. G./Kanuk, L. L. (2004), p. 538; Roehrich, G./Valette-Florence, P./Ferrandi, J.-M. (2002); Goldsmith, R. E./Goldsmith, E. B. (1996); Robertson. T. S. (1971), pp. 111; King, C. W./Summers, J. O. (1970).

521 Cf. Rogers, E. M. (2003), p. 15.

Following the classification of the German association for CEP, the "Gesellschaft für Unterhaltungs- und Kommunikationselektronik" (gfu),[522] used in their industry statistics, the CEP segment can be split into several areas which comprise products that are predominantly used privately. Besides home entertainment electronics (TVs, DVD-Players and -recoders, digital cameras, etc.) and telecommunication (phones, mobile phones, fax machines, etc.), the areas of video game software and consoles as well as information technology products (desktop PCs, notebooks, TFT-displays, etc.) are also included.[523]Figure 25 illustrates the recent market trend of the CEP market in Germany and the relative changes in sales of its different product areas.[524] The German CEP market plays an important role in Europe, being the second largest market after the United Kingdom (s. Figure 26).[525]

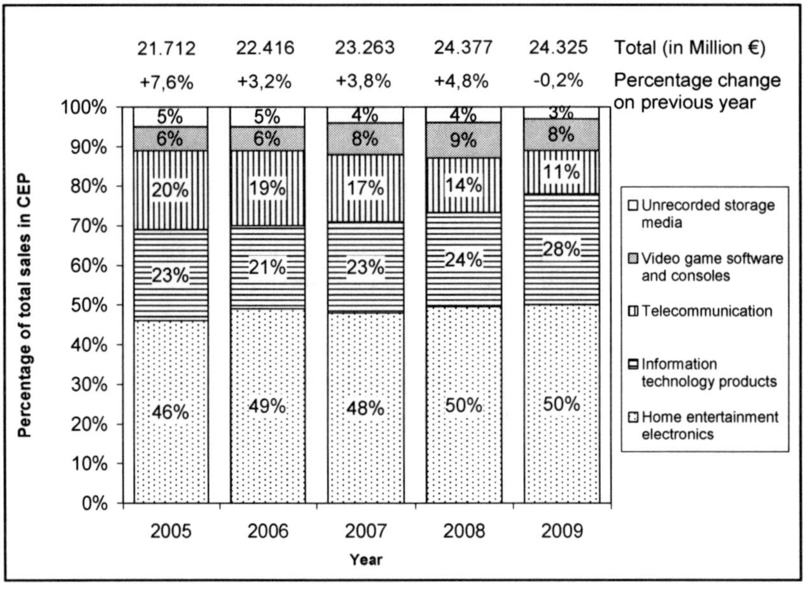

Figure 25: Sales development of different CEP categories
(Source: **gfu/GfK** (2009).)

522 The German gfu can be compared to the U.S.-American CEA (Consumer Electronics Association) which both promote and advance the members' interests in their country.

523 See for example Gfu/BVT/GfK (2006).

524 Cf. gfu/GfK (2009), p. 4.

525 Cf. GfK (2008), p. 2; GfK (2006), p. 8.

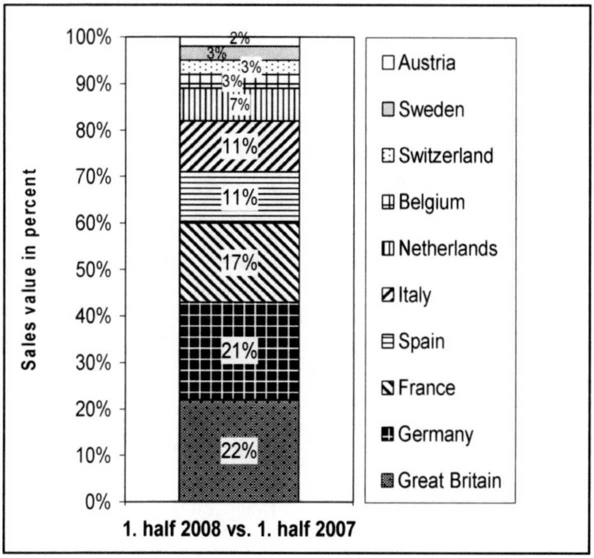

Figure 26: Relative CEP sales values of European countries (Source: **GfK** (2008), p. 2.)

Analysis of specific CEP ownership

Regarding household ownership of entertainment electronics, TVs (99.3%), radios (88.1%), hi-fi systems (80.9%), video-recorders (72.2%) and DVD-players (47%) rank among the top five. While ownership of TVs and radios reaches nearly 100% in all age classes (14-29 years; 30-49 years; 50 years and older), fewer old consumers possess a hi-fi system (69.4%) or a video-recorder (63.8%). However, the situation for the more innovative DVD-Player technology is different. A sharp drop in relative ownership and use can be observed as only 24.1% of persons over 50 years of age possess this home entertainment device – this is more than 40 percentage points less than for the other two age classes.[526] Other studies report higher numbers in DVD ownership of German households.[527] Those surveys find that between approximately 49% and 69% of the households possess a DVD-Player (s. Table 11). Nevertheless, an age effect

526 Cf. VuMA (2007), pp. 7.

527 This is partly due to the fact that some studies calculate the ownership average of several years (e.g. as in the VuMA (2007) study for the years between 2004 and 2006). In contrast to that, surveys which are only considering one (recent) year come to higher ownership results. Sometimes neither DVD-players or -recorders are separated, nor is a distinction between stand alone devices and DVD computer drives made - which is another cause for differences in the survey results.

is visible. More old consumers possess a DVD-Player compared to previous years, but the gap between younger and older consumers is of roughly the same extent. This situation is worth to be assessed in more detail. Furthermore, it is interesting to find out whether similar tendencies can be observed for related products. For this reason, CEP ownership (CEPO) will constitute the researched product area of the dissertation.

Name of the study[528]	**Year(s)**	**Share of households with DVD-Players**	**Share of households with DVD-Recorders**	**Share of house-holds with VCRs**	**Misc.**
VA (2007)	2005-2007	49,1%	13,1%	58,5%	age classes
Statistisches Bundesamt (2007a)	2002-2006	59,3%*	(see left)*	-	age classes
Statistisches Bundesamt (2007b)	2006	see above	see above	69,4%	
Ma2004 radio II	2004	37%	10%	67%	
Ma2005 radio II	2005	51%	17%	68%	
Ma2006 radio II	2004-2006	55,7%	20,4%	64,6%	2004-2006 +5,3% HDD
Ma2007 radio II	2007	57,3%	23,9	58,5%	7,6% HDD
Comm Netw. 10 (FOCUS)	2006	60%	10%	83%	+5% HDD
Comm Netw. 11 (FOCUS)	2007	68%	13%	82%	+6% HDD
GfK Panel	2002-2007	66%	6%	72%	

528 The corresponding sources are: FOCUS Magazin Verlag GmbH (2007); GfK Panel Services Deutschland (2007) cited in BVV (2007), p. 3; Media Analyse 2007 Radio II (2007) cited in Mediendaten Südwest (2007); Statistisches Bundesamt (2007a); Statistisches Bundesamt (2007b), p. 19; VA (2007); FOCUS Magazin Verlag GmbH (2006); GfK Panel Services Deutschland (2006) cited in BVV (2007), p. 3; Media Analyse 2006 Radio II (2006) cited in van Eimeren, B./Frees, B. (2006), p. 566; Ma2005 radio II cited in Gerhards, M./Klingler, W. (2006), p. 77; ARD/ZDF-Langzeitstudie Massenkommunikation (2005) cited in Stadik, M. (2005), p. 6; Ma2004 radio II cited in Gerhards, M./Klingler, W. (2006), p. 77.

GfK Panel	2002-2007	69%	9%	67%	
ARD/ZDF	2005	63%	15%	75%	+5% HDD
* = DVD-Players and DVD-Recorders					

Table 11: Overview of studies dealing with the ownership of CEP

In order to get a first insight into the CEP market, some general aspects of innovative CEP will be analyzed. Hence, the first research question (RQ 1) deals with age differences regarding several important CEP characteristics. It is, for example, interesting to find out whether differences in the perception of the technical or functional superiority of CEP exist. The compatibility of these technical devices, button sizes and labels of CEP, and the conformance of innovative CEP with the needs of the consumers are other interesting aspects that are addressed.

4.2 Assessing the Relevance of Determinants in the Context of Older and Younger Consumers Regarding Innovations

With help of the gerontological and innovation research findings from Chapters 2 (pp. 17) and 3 (pp. 67), combined with the specifications of the chosen product category and knowledge about elderly consumers (Ch. 4, pp. 117), the relevance of the determinants will be assessed and hypotheses will be proposed. This corresponds to one of the major research fields in consumer behavior: Finding out why a consumer purchases or owns a product.[529] Later on, it will be analyzed how well the theoretically derived determinants predict the ownership of CEP (RQ 2) including their order of importance (RQ 3). If not all of the determinants have a significant influence, it will be tempting to know which alternative models can be proposed for the different age groups (RQ 4).

This chapter is structured corresponding to Chapter 3. Only two changes have been made. First, due to the prominent status of "age" in this dissertation, this variable will be discussed separately. Secondly, innovativeness will not be discussed further as an *independent variable* in the context of a consumer trait which is influencing CEP adoption. Instead, it is assumed that the respondents' innovativeness is, at least in a fairly big way, expressed in the ownership of innovative CEP.

529 Cf. Freter, H. (1974), p. 15.

4.2.1 Role of "Age" in Innovation Adoption

In this chapter, the relevant literature is reviewed which deals with the relationship between age and innovativeness. Wherever possible it is also referred to age related research results dealing with the issue of innovative CEPO.

As briefly eluded before (s. Ch. 3.5.2.1, pp. 101), the impact of age on adoption behavior has been and still is discussed controversially. While some studies report a negative relationship between consumer age and innovativeness, others contradict and point out that older consumers are also innovative, while again others found little or no influence of age and deemed this variable irrelevant.

Studies supporting the thesis that older consumers are less innovative have, for example, been conducted by Gilly/Zeithaml (1985) who found that for innovative products or services like scanner-equipped grocery stores, automated bank teller machines (ATMs), and telephone calling services, trial and adoption were less frequent among elderly consumers.[530] Among others, similar results were attaint regarding purchases over the internet and the adoption of home computers, CEP (TVs, hi-fi systems, VCRs) solar energy systems, dishwashers, and vitroceramic-hobs.[531] But also products which were specially targeted at older consumers (like electronic pillboxes and auto-dialer telephones) were likely to be refused.[532] Furthermore, Lunsford/Burnett (1992) ascertain that innovation refusal is higher compared to the non-elderly even if adoption is limited to the acceptance of just the product idea.[533] In line with these authors, Kerschner/Chelsvig (1981) summarize that the older the consumer, the more negative his or her view towards technology and the lower his or her use of various technologies is going to be.[534] The strongly decreasing ownership of technological devices as it was exemplarily portrayed for DVRs earlier support this statement. Altogether, these findings sustain the prevailing image of older individu-

530 Cf. Gilly, M. C./Zeithaml, V. A. (1985).

531 Cf. Akhter, S. H. (2003), pp. 323; Martinez, E./Polo, Y./Flavian, C. (1998); Dickerson, M. D./Gentry, J. W. (1983); LaBay, D. G./Kinnear, T. C. (1981); see also Brünner, B. O. (1997), p. 195; Lewis, H. G. (1997), pp. 119; Kölzer, B. (1995), pp. 184; Federsel-Lieb, C. (1992), pp. 65; Pfaff, D. (1990), pp. 246 and p. 258.

532 Cf. Bowe, F. (1988), pp. 34.

533 Cf. Lunsford, D. A./Burnett, M. (1992), p. 54.

534 Cf. Kerschner, P. A./Chelsvig, K. A. (1981) citied in Gilly, M. C./Zeithaml, V. A. (1985), p. 353.

als in society which traditionally assumes the elderly having a decreasing interest in novel products and showing a general tendency to resist innovations.

Contrary to that, for instance Moschis (1992; 1994) and Brünner (1997) take the view that elderlies cannot be characterized as generally non-innovative. Only in some cases or for some products mature consumers might be a little less innovative than average consumers, but the authors emphasize that they do not generally resist new ideas.[535] For example, despite Gilly/Zeithaml (1985) discovered low acceptance for three out of four innovations, they received opposite results for the adoption of the electronic funds transfer (EFT) among elderlies.[536] Concerning this novelty, elderly individuals were more innovative than their younger counterparts. This is also why Schiffman/Sherman (1991) identify elderly consumers as "selectively innovative."[537] Other research studies came to similar results, supporting the thesis that elderly consumers are also innovative consumers.[538]

Aside from research that comes to opposed results (older consumers are *not* innovative vs. they *are* innovative), other studies suggest that young and old respond similarly as regards to innovations. For leisure and tourism service innovations this hypothesis is supported by Szmigin/Carrigan (2001c) who state that no age differences exist between older innovators and non-innovators, or by Steenkamp/Gielens (2003) who found no age effect on the trial probability of new consumer packaged goods.[539] In an earlier study, Prisuta/Kirner (1985) came to similar results.[540] They discovered that age was not a significant factor in explaining the attitude toward new technologies. Also an analysis of adoption barriers in the purchase of innovations by older persons can be interpreted to support this thesis. Despite Lunsford/Burnett (1992) put forward several barriers which possibly result in the elderlies' resistance to adoption, their identified barriers are not very different to those barriers which generally apply for consumers regardless of their age.[541] In this context, Szmigin/Carrigan (2000) subsume that "resisting innovation is not something particular to older consumers."[542]

535 Cf. Brünner, B. O. (1997), p. 196; Moschis, G. P. (1994), pp. 107; Moschis, G. P. (1992), pp. 273.

536 Cf. Gilly, M. C./Zeithaml, V. A. (1985).

537 Cf. Schiffman, L. G./Sherman, E. (1991), p. 190.

538 See for example McDonald, H./Corkindale, D./Sharp, B. (2003), p. 90; Hock, E.-M./Bader, B. (2001), pp. 43; Silvers, C. (1997), pp. 306; Midgley, D. F. (1987), pp. 205.

539 Cf. Steenkamp, J.-B./Gielens, K. (2003), pp. 378; Szmigin, I./Carrigan, M. (2001c).

540 Cf. Prisuta, R./Kriner, R. (1985) cited in Moschis, G. P. (1994), p. 108.

541 Cf. Lunsford, D. A./Burnett, M. (1992).

542 Szmigin, I./Carrigan, M. (2000), p. 509.

In spite of these mixed results, a certain resistance against innovations seems to dominate the older consumers' behavior. And even if recent studies claim that older consumers today are more innovative compared to previous generations of elderlies, it still can be assumed that great age differences exist in adoption, at least in the case of technological innovations like new CEP. Therefore, it is hypothesized that:

H_1: The ownership of CEP (CEPO) is negatively related to age.
H_2: The ownership of *innovative* CEP (CEPO_i) is negatively related to age.
H_3: The *weighted* ownership of CEP (CEPO_w) is negatively related to age.[543]

While the hypotheses mentioned above deal with the general relationship between age and different types of CEPO, the following hypotheses concretize the explicit direction of relationship. Thus, it is assumed that:

H_4: The older an individual the lower the ownership of CEP (CEPO).
H_5: The older an individual the lower the ownership of innovative CEP (CEPO_i).
H_6: The older an individual the lower the weighted ownership of CEP (CEPO_w).

While the previous discussion implicitly dealt with chronological age, which refers to the number of years that have passed since a person's birth, it is necessary to point out that age is a multidimensional concept. While in consumer behavior research it is the chronological age which represents the most frequently used demographic variable, in gerontology the "age" variable has been, among others, differentiated into biological, psychological, and social age for a long time.[544] Consequently, Moschis/Mathur (1993) criticize marketing research which is neglecting the fact that people "age differently and at different rates."[545]

Since the early 1980s the concept of age has also been broadened in the domain of marketing.[546] Apart from simply using chronological age, different other age measures have been employed to explain or predict consumer behavior. Studies

---

543 The weighted index "CEPO_w" will be explained later in Chapter 5.2.4.1, pp. 170.

544 Cf. Hoyer, W. J./Rybash, J. M./Roodin, P. A. (1999), pp. 13; Barak, B./Schiffman, L. G. (1981), p. 602.

545 Moschis, G. P./Mathur, A. (1993), p. 42.

546 Cf. Wei, S.-C. (2005), p. 633 and the cited literature; Barak, B./Gould, S. (1985); Barak, B./Schiffman, L. G. (1981).

about the subjective age of individuals have emerged,[547] referring to an age construct which measures a person's perceived age. Barak/Gould (1985) even list five different non-chronological kinds of age, including:[548]

- Cognitive age (see below for operationalization)
- Ideal age (the age an individual considers to be ideal)
- Youth age (discrepancy between a respondent's chronological and cognitive age)
- Discrepancy age (discrepancy between a respondent's cognitive and ideal age)
- Disparity age (discrepancy between a respondent's chronological and ideal age)

Kastenbaum et al. (1972) operationalize cognitive/perceived age by self-reports, distinguishing between:[549]

- Feel-age
- Look-age
- Do-age
- Interest-age

According to Schiffman/Sherman (1991), the main reason why this separation of cognitive age from chronological age is necessary is because age can be described as "state of mind" and that therefore people frequently perceive their age differently from their actual birth age.[550] More precisely, Underhill/Cadwell (1983) discovered that younger individuals perceived themselves older whereas older individuals perceived themselves to be younger than their actual age.[551] In total they found that 63% of men and women feel younger (on average 13 years) than their chronological age, with an increasing difference of perceived and actual age the older the respondent was. Still, the cognitive age of older individuals should be higher than that of younger individuals in almost all cases.[552] But, interestingly, it was found that "the cognitive age concept is applicable to people of all ages, as a few studies have shown that even young and middle-aged adults

547 See for example Moschis, G. P./Mathur, A. (2006); Mathur, A./Moschis, G. P. (2005); Gwinner, K. P./Stephens, N. (2001); Wilkes, R. E. (1992).

548 Cf. Barak, B./Gould, S. (1985), p. 53.

549 Cf. Kastenbaum, R. et al. (1972), pp. 197.

550 Cf. Schiffman, L. G./Sherman, E. (1991), p. 188; see also Barak, B./Schiffman, L. G. (1981), p. 602.

551 Cf. Underhill, L./Cadwell, F. (1983).

552 Cf. Ibid., pp. 19.

have cognitive ages different from their chronological ages."[553] Analogous to the findings that chronologically older consumers are less innovative, studies confirmed that this also holds true for the subjective age – at least regarding technological innovations. E.g. Eastman/Iyer (2005), in the context of internet, report that seniors with a younger cognitive age will use the internet more than those seniors who are older in cognitive age.[554] As some researchers claim that an individual's behavior generally depends more on cognitive than on chronological age,[555] also the subjective age measure will be drawn on for explaining CEP innovation adoption and eventual differences in the adoption determinants.

H_7: Cognitive age is a better predictor of the CEP adoption determinants than the chronological age.

As demonstrated, existing studies show ambiguous results when discussing the effect of age on innovativeness. Three different positions were identified which assign chronological age either to have a positive, negative, or no impact on innovativeness. Although with cognitive age it seems to be possible to increase the predictive power of this demographic variable, it is very likely that relying on age alone is insufficient to explain differences between age classes in the complex issue of innovation adoption behavior. This study expands on this idea of age as an *indirect* influence variable by more precisely analyzing the adoption determinants discussed in Chapter 3.5 (pp. 87). Thus, in the next sub-chapters product-specific, personal characteristic, and situational variables will be examined in terms of their relevance to explain novel CEPO. The idea behind it is, that age per se does not explain differences but that with age, biological, psychological, and sociological changes occur that are manifested in the adoption determinants. Therefore, age constitutes an important variable and it will be interesting to see whether, and if yes to what extent, chronological and cognitive age has an influence not only directly on CEPO, but also how it may have an influence *indirectly* on CEPO by affecting the adoption determinants.

4.2.2 Assessing the Relevance of Product-Specific Variables

The product-specific determinants are assessed with the goal to identify those variables that might influence novel CEPO. At the same time, this study refers

---

553 Stephens, N. (1991), p. 38.

554 Cf. Eastman, J. K./Iyer, R. (2005).

555 See for example Wei, S.-C. (2005), p. 638; Szmigin, I./Carrigan, M. (2001a); Auken, S. v./Barry, T. E./Anderson, R. L. (1993); Barak, B./Schiffman, L. G. (1981), p. 602.

to age related knowledge – wherever available – in order to integrate possible age-specific influences that the determinants are exposed to. The considered product-related variables include the relative advantage, compatibility, complexity (product complexity and in-store complexity), trialability, observability, and risk (performance risk and financial risk).

4.2.2.1 Perceived Relative Advantage

Rogers (1983) identifies relative advantage as one of the best predictors of an innovation's rate of adoption.[556] For home entertainment innovations[557] as well as in the context of age related marketing research it is repeatedly pointed out that relative advantage or product-related benefits play an essential role for consumers in deciding in favor or against adopting an innovation. Superficial or trivial product variations do not have the power to attract the attention of consumers and persuade them to purchase one of those devices. Instead, real needs must be addressed and be satisfied by the novelties – otherwise mature individuals show no interest in their adoption.[558] In a first step, independent from age influences, it is hypothesized that:

H_{20}: The higher the perceived relative advantage of innovative CEP the more innovative CEP are possessed (rel $\xrightarrow{+}$ CEPO_w).[559]

However, previous studies have recognized that older consumers may try new products for different reasons than younger ones and that the perceived relative advantage may differ significantly between those age groups.[560] As an example, Hock/Bader (2001) ascertained that many senior consumers own mobile phones in case of an eventual emergency while on the other hand younger mobile phone user view it as fashion item and as possibility to stay connected with friends.[561] The latter also favor innovations that save time or which represent the latest technology whereas these aspects are not valued by older persons to the same extent – they also refuse novelties which do not demonstrate any clear benefit

---

556 Cf. Rogers, E. M. (1983), p. 217.

557 Cf. van Eimeren, B./Frees, B. (2006), pp. 570.

558 See for example Gunter, B. (1998), p. 35; Schiffman, L. G./Sherman, E. (1991), p. 190; Gilly, M. C./Zeithaml, V. A. (1985), p. 356.

559 "More innovative" refers to individuals possessing a larger number of (innovative) CEP and/or CEP that score high on the innovativeness index. Both situations are realized through higher CEPO_w scores. This index will be explained later in Chapter 5.2.4.1, pp. 170.

560 Cf. Leventhal, R. (1997), p. 279.

561 Cf. Hock, E.-M./Bader, B. (2001), p. 44.

for them.[562] These differences between young and old are important to be taken into consideration in conceptualizing new products. With this, prompt resistance as a reaction to wrong needs that have been appealed to can be avoided. Doubts about the elderlies' general appreciation of the CEP advantages are justified.

H_8: The older an individual the lower the perceived relative advantage of innovative CEP.

4.2.2.2 Perceived Compatibility

The conceptualization of compatibility in Chapter 3.5.1.2 (pp. 90) revealed that compatibility with sociocultural values and beliefs, previously introduced ideas, and client needs for the innovation depends more on personal than product-specific characteristics which will, thus, be discussed later in the age related context of personal variables (s. Ch. 4.2.3.2, pp. 139). Here, the technological compatibility is at the center of attention.

Dhebar (1995) for example lists several examples of discontinuous innovations and also specifies the nature of incompatibility compared to the innovations' precursors.[563] Besides expenses related directly to the innovation, changes in technology can necessitate potential adopters to invest in other, compatible products, because the older versions of products have become obsolete.[564] Taking the large number of aged VCR-owners as an example, older consumers are more likely to be faced with considerable system switching costs due to the necessity to purchase new equipment. In the case of DVR, adoption means that consumers become aware during the adoption decision process that they cannot use their videocassette device any more as it is incompatible with new technology, building up another barrier to adoption.

Finally, another aspect is referred to which is usually disregarded in general theory dealing with the variable of compatibility, but becomes relevant when older consumers are of dominating interest: A novelty may have a product design which is incompatible to the elderlies' physiological needs. Different to general innovation research, age related research picks up this issue, because older con-

562 Cf. Melenhorst, A.-S./Rogers, W. A./Caylor, E. C. (2001), pp. 244; Lunsford, D. A./Burnett, M. (1992), pp. 55.

563 Cf. Dhebar, A. (1995), p. 138.

564 Cf. Dhebar, A. (1995), pp. 139.

sumers frequently criticize the lacking ease of use[565] of technological innovations.[566] Physiological changes which affect the performance of the sensory organs, especially deficits in sight faculty and touch sensitivity, often result in difficulties to operate new technological products. This is because engineers, when replacing the preceding generation of products, continuously develop smaller devices. This miniaturization tendency automatically results in the necessity to reduce the size of buttons and in markings that are difficult to read. Therewith, the likelihood increases that problems to operate the new technological item arise. Innovative devices are too often designed for an "average user" which tends to be well-educated, motivated, and younger.[567] This approach of product usability testing neglects physiological needs of older users.

As most of the just reported compatibility aspects are evidenced in other variables like the perceived financial risk or complexity[568] the technological compatibility will not be seen as a determinant of CEPO and no concrete hypotheses will be put forward. Still, it will be incorporated in the survey to give respondents the chance to express their opinion about some concrete aspects of technological compatibility and their consequences.

4.2.2.3 Perceived Product Complexity

The perceived complexity during the innovation decision process was, one the one side, defined as the degree to which a product is difficult to understand or use and, on the other side, as the extent to which the in-store assortment complexity (overchoice) potentially generates confusion and impedes an optimal product choice (s. the next sub-chapter).

The term "understanding" refers to potential adopters comprehending the functions the innovation has to offer. Hence, a close relationship to compatibility persists because experiences with preceding CEP or knowledge about related products like PCs, mobile phones, receivers and the like may help potential users to understand the principle functions of novel CEP as well. Thus, if the perceived compatibility with the CEP is high, then perceived complexity will be lower compared to those individuals who were rarely in touch with CEP. Al-

565 Here, ease of use relates to aspects of the DVRs' physiological appearance (hardware component) and not to perceived complexity during usage by reason of unclear or confusing menu structures etc. (software component).

566 Cf. Hock, E.-M./Bader, B. (2001), p. 34; Lunsford, D. A./Burnett, M. (1992), pp. 55; Schewe, C. D. (1988), pp. 67.

567 Cf. Carmichael, A. (2002), p. 217.

568 Cf. Moreau, C. P./Lehmann, D. R./Markman, A. B. (2001), p. 14.

though it can be assumed that the principle features are relatively easy to understand, some other aspects might not be understood as easily.

The second component of product complexity consists of the perceived difficulty of how to use the innovation. Here, the ease of use is placed at the center of interest again, however, not in the sense of physiological abilities (small sized buttons etc.), but in the sense of psychological abilities which are necessary to perform well in the interaction with the device.[569] For example, the questions how complex actual or potential innovative CEP-user perceive the tasks which are necessary to make use of the devices' functions.

The aspect of complexity will also be analyzed from a gerontological perspective. Consumer behavior theory already postulates that the likelihood of product purchases generally decreases with higher complexity. Research results discussed earlier in this book point to the particular relevance of complexity in older consumers' purchase decisions. Understanding and correctly using complex CEP requires time and efforts. Time does not appear to be the restricting factor for older consumers as with retirement elderly consumers have more leisure time. In consequence, they would also have more time for such activities as learning about the new functions of a novel device – it even might be an advantage over younger, employed individuals who have less time they can invest to understand or use a recently purchased CEP innovation. In contrast to that, gerontological studies revealed that some cognitive resources are likely to decrease when people age which require higher efforts by older adults to come to the same results as their younger counterparts. For example, older adults differ in their ability to *learn* compared to younger individuals.[570] The large amounts of information that need to be processed pose difficulties to elderly persons.[571] Ziefle/Bay (2005) provide further support for negative effects of complexity on elderly consumers.[572] They found that the older adults' performance to use mobile phones was significantly lower compared to those of younger persons, especially regarding the more complex tasks the authors tested.

When looking at technological innovations, CEP included, the large number of product features seems important. For long, it has been postulated that a growing number of features provides greater functionality and hence that those products are more beneficial to consumers. Opposed to that, it has been reported more

569 Cf. Dix, A. et al. (2004), pp. 277.

570 Cf. Meyer-Hentschel, H. (1990), pp. 110; Ross, I. (1982), pp. 31; Phillips, L. W./ Sternthal, B. (1977), pp. 447.

571 Cf. John, D. R./Cole, C. A. (1986), p. 302.

572 Cf. Ziefle, M./Bay, S. (2005), pp. 382.

recently that the accumulation of features may also have negative effects.[573] With "feature fatigue" Thompson/Hamilton/ Rust (2005) denominate the tendency – especially in electronics and information technology – that consumers are overwhelmed by products which are too complex because of their increased capabilities.[574] It is likely for the same reason, that older persons rated the ease of use of mobile phones lower than younger consumers in an experimental study by Ziefle/Bay (2005).[575] As a consequence, it is hypothesized that:

H₉: The older an individual the higher the perceived product complexity of innovative CEP.

H₂₁: The higher the perceived complexity of innovative CEP the fewer/less innovative CEP are possessed (cx_prod $\xrightarrow{-}$ CEPO_w).[576]

4.2.2.4 Perceived In-Store Complexity

Apart from product related issues it has become apparent that assortment size also plays a role in complexity perception of the purchase decision. Besides an increasing number of competitors of CEP-manufacturers who offer a multitude of different product models, the size of home entertainment retailers in Germany (like Saturn, MediaMarkt, Expert, ProMarkt, MediMax,...) also contribute to larger choice sets with their expansion of selling space.[577] The products they offer differ in technical specifications, the number of provided features, quality, design, ease of use, and price so that with the larger shopping spaces the assortment size is expected to be perceived as large and confusing, especially by individuals with few knowledge about the CEP novelties.[578] Analogous to the elderly consumers' perception of the innovative CEP complexity, assortment complexity is very likely to be perceived as high. It has been reported that, in general, elderly are more prone to confusion than younger consumers.[579] Thus, it is hypothesized that:

573 Cf. Mukherjee, A./Hoyer, W. D. (2001).

574 Cf. Thompson, D. V./Hamilton, R. W./Rust, R. T. (2005), pp. 431.

575 Cf. Ziefle, M./Bay, S. (2005), p. 382.

576 "Fewer/less innovative CEP" refers to individuals possessing a smaller number of (innovative) CEP and/or CEP that score low on the innovativeness index. Both situations are realized through lower CEPO_w scores. This index will be explained later in Chapter 5.2.4.1, pp. 170.

577 Cf. Cleve, J. (2011), pp. 165; Wengler, T. (2005), pp. 167.

578 Cf. Lunsford, D. A./Burnett, M. (1992), pp. 60; Ross, I. (1982), p. 37.

579 Cf. Mitchell, V.-W./Papavassiliou, V. (1999), p. 323.

H_{10}: The older an individual the higher the perceived in-store assortment complexity of CEP.

H_{22}: The higher the perceived in-store assortment complexity of CEP the fewer/less innovative CEP are possessed (cx_is $\xrightarrow{-}$ CEPO_w).

4.2.2.5 Perceived Trialability and Divisibility

The trialability for CEP depends on the specific device and on the selected retail channel. Large home entertainment retailers, as well as smaller and more specialized stores, generally offer the possibility to try out new CEP (operate/check ease of use) even if it is on a limited scale. It is rather the complex products that are more likely to be tried in the circle of family members or friends who might possess such devices. It can be assumed that trialability has no direct influence on adoption, but more likely takes the role of a mediator between for example perceived complexity or perceived risk and the adoption. As its influence on adoption is assumed to be weak and as trialability of CEP is given in most cases, this variable is excluded from further consideration.

4.2.2.6 Perceived Observability and Communicability

As well as trialability, the perceived observability of an innovation can be omitted in the ensuing discussion. Different to ideological innovations, practices or ideas which are not visible, most advantages of CEP are observable or can be at least communicated. Of course new functions of innovative devices and their advantages may be difficult to explain in some cases, but they are generally by far easier to understand than social innovations. There the success becomes visible much later after the point of adoption or even only indirectly. It is also likely that in the case of CEP the determinants divisibility and trialability, as well as observability and communicability are no distinct influence factors of CEPO, but that they are implicitly integrated in the determinants such as the judgment of the relative advantage or the perceived risk of CEP which they have been reported to be intercorrelated with. Therefore, this variable is excluded from further discussion.

4.2.2.7 Perceived Risk

The purchase of innovative products may be accompanied by negative consequences. Hence, the potential adopter's decision involves risks which he or she becomes aware of and needs to take into consideration when evaluating the novelty. Several dimensions of perceived risk have been identified (s. Ch. 3.5.1.6, pp. 95). However, as Dowling (1986) remarks, not all of these dimensions are necessarily relevant for each research setting, but depend on the type of product, respondent, and purchase situation.[580] In the following analysis, the physical risk is excluded as it is not expected to be of any relevance in the context of CEP. The loss of time risk and network externalities will be discussed as part of the psychological risk. Thus, a deeper analysis will be realized for:

- Performance risks
- Financial risks
- Psychological risks, network externalities and loss of time (s. Ch. 4.2.3.2, pp. 139)
- Social risks (ibid.)

Perceived Performance Risk

Performance risks represent a major product related dimension of risk in the purchase process of high-tech products. Due to the little diffusion of highly innovative CEP, deeper knowledge about the product is generally lacking and few experiences of other consumers exist. While younger consumers might generally be better informed about new technologies, they possess the advantage compared to mature individuals that they can draw on previous knowledge about other devices as it has already been pointed out during the compatibility discussion. Elderly, in contrast, might be uncertain about whether the product meets their expectations and whether it will perform as desired. This is consistent with findings that showed the preference of quality and service over price among older consumers as strategies to reduce perceived functional risks.[581] Intuitively, it is clear that concerns regarding performance and quality are the more prevalent, the more novel aspects the product features. Particularly for highly innovative product solutions, consumers might fear that the technology is not fully developed. Potential product failures or performance problems of first generation products can urge people to wait with purchases, a phenomenon which corresponds to leapfrogging behavior, or to complete non-adoption.

H_{11}: The older an individual the higher the perceived performance risk of innovative CEP.

H_{23}: The higher the perceived performance risk of innovative CEP the fewer/less innovative CEP are possessed (r_perf $\xrightarrow{-}$ CEPO_w).

Perceived Financial Risk

580 Cf. Dowling, G. R. (1986), p. 195.

581 Cf. Simcock, P./Sudbury, L./Wright, G. (2006), p. 357.

The high pace of post-innovative improvements or product successors causes expectations on the side of the consumers that a new and improved product will be launched in the foreseeable future. This, in turn, might create the expectation that prices of CEP will fall soon or that the product will be outdated shortly. The anticipation of economic savings in the future and the launch of a new generation of that product may result in postponing the novelty purchase, and, under these circumstances, have a negative effect on the initial adoption rate. In this case, the purchase of the innovation is perceived as unfavorable or too risky. This phenomenon is also known as "leapfrogging."[582] The result of the leapfrogging behavior is that (at a given point in time) people possess fewer products that are perceived as financially risky.

H_{24}: The higher the perceived financial risk of innovative CEP the fewer/less innovative CEP are possessed (r_fin $\xrightarrow{-}$ CEPO_w).

Previous studies oftentimes proceeded on the assumption that mature consumers purchase fewer novelties because of financial anxieties.[583] It is often reported that older consumers perceive high levels of financial risks when they are in purchase situations of rather expensive or new products which then would also negatively influence the adoption of CEP.[584] This can eventually be attributed to personality traits. Older individuals are frequently said to be thrifty, critical in evaluating innovations, risk averse, more cautious than younger individuals, avoid spending money frivolously or self-indulgently and are careful shoppers.[585] However, it is by far not clear whether these consumer traits and behaviors generally apply for all elderly shoppers and whether they are still valid for the current generation of older consumers. More recently, researchers proclaim that traditional views of elderlies need to be re-evaluated. Instead, they observe a growing number of aged persons who are active and self-confident consumers who spend considerable amounts of their money, seek novel experiences and are labeled by Schiffman/Sherman (1991) as the "new-age elderly."[586] Furthermore, as far as the *objective* financial situation is concerned, most elderlies are well-off and can be described as wealthier than previous generations.[587] Still it is as-

582 See for example Pohl, A. (1996).

583 See for example Wallach, M./Kogan, N. (1961), p. 32.

584 See the litature cited in the following.

585 Cf. Gwinner, K. P./Stephens, N. (2001), pp. 1036; Federsel-Lieb, C. (1992), p. 65; Lunsford, D. A./Burnett, M. (1992), pp. 55; Fox, M. C./Roscoe Jr., A. M./Feigenbaum, A. M. (1984), p. 565.

586 Cf. Schiffman, L. G./Sherman, E. (1991); Tongren, H. N. (1988), p. 140.

587 See also the discussion in the context of socio-economic variables (Ch. 4.2.3.1, pp. 137), where the "perceived financial situation" is further elicited.

sumed that for the majority of older consumers the purchase of innovative CEP is perceived to be tied to considerable amounts of financial risks.

H_{12}: The older an individual the higher the perceived financial risk of innovative CEP.

4.2.3 Assessing the Relevance of Personal Characteristic Variables

So far determinants were discussed which are dominated by the influence of product characteristics. But, throughout the discussion, it has been mentioned several times, that for certain determinants the portion of personal influence is larger. These variables are picked up subsequently. They are subdivided in sociodemographic variables (s. Ch. 4.2.3.1), psychographic variables (s. Ch. 4.2.3.2), and social assistance (s. Ch. 4.2.3.3).

4.2.3.1 Sociodemographic Variables

The sociodemographic variables have been stated to be problematic in explaining or predicting consumer behavior. This is because of their oftentimes criticized low predictive power of actual behavior. Still, they are often used in studies as they are easily accessible. While the age variable has already been analyzed separately in Chapter 4.2.1 (pp. 124), income, education, and gender remain to be discussed in light of gerontological research, age related marketing research and CEP. However, as these socio-economic variables do not play a dominating role here, but rather serve as control variables and survey sample descriptors, no concrete hypotheses are put forward regarding their effect on CEPO except for the financial situation of consumers.

Moschis (1992) assumes that lower education levels of elderlies compared to younger individuals are responsible for differences in new product acceptance.[588] In the case of complex, *technological* innovations this could, contingently, also be the case. Akhter (2003) found support for his hypothesis, that the likelihood to purchase over the internet increases with education.[589] The author puts forward that the reasons therefore are greater evaluation propensities and the fact that better-educated consumers are more informed about the internet. However, due to the wide spread of home electronic appliances it is more realistic that experiences with such devices or cohort effects are more influential than

588 Cf. Moschis, G. P. (1992b), p. 275.

589 Cf. Akhter, S. H. (2003), pp. 324.

a potential adopter's education. Only the indirect effect of education – which implies that lower education is accompanied by lower incomes – might seem to be able to impact the adoption decision of CEP. Lower-income earners are less likely to be in the position to afford rather expensive CEP innovations. Opposed to the general beliefs about the financial situation of elderly households it has been rectified in Chapter 2.7.1 (pp. 55) that they have a sound financial status. Although it should not be neglected that old-age poverty exists, the average financial situation has noticeably improved. This can be explicated with better financial provisions compared to previous generations of aged individuals. Even in comparison with their younger counterparts the disposable financial resources of elderly are substantial because of the differing household situations and life stage positions. While income of younger, employed persons is usually higher than pensions of retirees, older age groups have fewer fixed expenses as children have generally moved out, mortgages and retirement provisions have been paid. While younger consumers might be strongly interested in novel CEP technologies, they probably do not dispose of the financial means like middle-aged or old consumers to purchase the rather expensive CEP innovations. Corresponding to that, it has been observed that younger adopters can be found for innovations that do not request large financial investments.[590] Though a difference between the objective and subjective (perceived) financial situation of older persons might exist and some authors have claimed that the perceived financial situation will deteriorate with age,[591] it is assumed that:

H_{13}: The older an individual the better the perceived overall financial situation.

H_{25}: The better the perceived overall financial situation the more innovative CEP are possessed (fin_situ $\xrightarrow{+}$ CEPO_w).

There is no reason to assume gender differences when looking at the general proneness to show innovative behavior.[592] For the adoption of novel technologies, though, Akhter (2003) could track down some disparities. Using the example of internet shopping, results showed that men were more prone to purchase over the internet than women.[593] More general, Haimann (2005) states that men show more interest, and are more likely to buy the latest technological products compared to women, who in contrast seem to buy a novel product rather due to

590 Cf. Dickerson, M. D./Gentry, J. W. (1983), pp. 226.

591 E.g. Topritzhofer, E. (1969), p. 66.

592 For example Rogers (2003) does not mention gender as one of the socio-economic characteristics to influence early adoption of novelties. Cf. Rogers, E. M. (2003), pp. 288.

593 Cf. Akhter, S. H. (2003), pp. 323.

malfunction of the actual device.[594] As the main focus of this study is not on gender-differences regarding CEPO, no concrete hypotheses are postulated. Nonetheless, as with CEP a product category has been chosen where gender differences might prevail, its general role will be assessed.

RQ 8: Are any significant gender differences observable?

4.2.3.2 Psychographic Variables

The structure of the following chapter has been adapted to the findings of the determinant conceptualization. The issue of age related differences of innovativeness has already been discussed in the context of the "age" variable. Furthermore, in this study innovativeness is not considered directly but indirectly through the CEPO. Therefore it will not be discussed as an adoption determinant. Besides interest, knowledge compatibility, psychological risk, self-efficacy, and social pressure will be discussed more concisely.

Interest

Prior research results indicated that it is reasonable to analyze *product category*-interest rather than an individual's general interests (too wide) or interests regarding one specific product (too narrow) in order to predict a person's behavior. Support for a significant role of interest in innovation adoption is given by Holak (1988) who researched possible influencing factors which determine the acceptance of innovative durables across several product categories. The author found that acceptance of home entertainment goods is best expressed by a model which includes product class interest.[595] Hence the interest in innovative CEP or related technological appliances of younger and older individuals is under consideration in the following.

Current data from a German periodical consumer analysis supports the hypothesis that older consumers tend to be less interested in technological devices.[596] The numbers show that about 54% of the 30-39 year-old persons agreed or strongly agreed with the statement that they are interested in innovative technology devices, while only less than half of this percentage among the 70-79 year old persons (26%) answered similarly. Even more significant were the results of another analysis which focused more specifically on CEP. Here, as few as approximately 10% of the 70-79 year-old persons agreed or strongly agreed that

594 Haimann (2005) cites an interview with a strategy consultant who has dealt with elderly consumers. Cf. Haimann, R. (2005), p. 41.

595 Cf. Holak, S. L. (1988), pp. 58 and pp. 64.

596 Cf. VA (2007).

computers and other electronic devices "are fun/enjoyable." In turn, more than 60% of the 30-39 year age group answered similarly.

The same tendency applies for the concrete novelty of DVRs, how a recent study by IBM Global Business Services discovered which analyzed the interest of DVR non-owners.[597] While the highest interest rate can be found in the 20-29 age class with 64%, those rates are continuously decreasing for all following age classes with the oldest age group showing the lowest percentage of respondents interested in DVR-technology (29%). The different shares of interested consumers per age class are depicted in Table 12.

Age in years	19 and younger	20-29	30-39	40-49	50-59	60-69
Share of DVR non-owners who are interested in DVR	60%	64%	47%	43%	37%	29%

Table 12: Percentage of DVR non-owners who are interested in DVR (Source: **IBM Global Business Services** (2007), p. 13.)

Although innovative solutions are marketed which the elderly would also benefit from, their interest in new technologies seems to be of rather limited scope. While one part of the elderly consumers is up to date regarding new technological developments, the majority seems to avoid those devices and shows much less interest in CEP compared to younger individuals.

H_{14}: The older an individual the lower the interest in CEP.

H_{26}: The higher the interest in CEP the more innovative CEP are possessed (int $\xrightarrow{+}$ CEPO_w).

Knowledge Compatibility

Compatibility has been conceptualized on three levels:[598] compatibility with culturally learned values, with previously introduced ideas/existing knowledge, and with needs of the potential adopter. This broad definition makes it impossible to include all of its aspects. Moreover, Moreau/Lehmann/Markman (2001) give a substantial reason why compatibility can be left out from explicit consideration.[599] They argue that compatibility is evidenced in complexity because innovations that are perceived as non-compatible entail higher levels of perceived complexity. Still, the influence of compatibility should not be completely left

597 Cf. IBM Global Business Services (2007), p. 13.

598 See Chapter 3.5.1.2, pp. 90.

599 Cf. Moreau, C. P./Lehmann, D. R./Markman, A. B. (2001), p. 14.

out of further processing because the focus is exclusively on *knowledge* compatibility.

Many authors mention that a major advantage of elderly people is their large shopping and consumption experience which makes them competent consumers.[600] In product choice situations they can refer to multitudinous past experiences which help them distinguish between products and identify what they really need and which product solutions are inadequate.[601] The consequences of shopping related experiences are reported to be manifested in the tendency of the elderly to highly esteem quality, rely on brand names, and to focus less on price.[602] Although it seems that aged consumers know exactly what they want and what they can expect from a product, there are doubts whether this general supposition also applies for innovative CEP. For innovations of this product category it can be assumed that older adults can refer to fewer compatible knowledge and experiences than younger individuals. As a consequence, they are likely to be confronted with more information which is entirely new to them when evaluating CEP innovations. On the other hand, very old consumers may be less able to process this new information. These aspects in combination with the general tendency of mature persons to overestimate the amount necessary to learn, it is more likely that they resist the new technology.[603] Conversely, it has been acknowledged that, in general, extended expertise necessitates less cognitive efforts to understand features of novelties. That facilitates an adequate evaluation of the new product which, in turn, results in (earlier) adoption.[604] Thus, it is hypothesized that:

H_{15}: The older an individual the lower the knowledge compatibility of older and innovative CEP.

H_{27}: The higher the knowledge compatibility of older CEP with innovative CEP the more innovative CEP are possessed (compa $\xrightarrow{+}$ CEPO_w).

Psychological Risk

The assessment of CEP knowledge in Chapter 4.2.2.2 (pp. 130) has shown that elderly individuals supposably possess less compatible knowledge as regards to CEP innovations. From this follows that older consumers are likely to be forced

600 See the references in Hupp, O. (1999), pp. 103.

601 Cf. Kirsch, J. (2003), p. 190; Härtl-Kasulke, C. (1998), pp. 41; Schiffman, L. G./Sherman, E. (1991), p. 190.

602 Cf. Krieb, C./Reidl, A. (2001), pp. 45.

603 Cf. Gassmann, O./Reepmeyer, G. (2006), p. 106.

604 Cf. Gatignon, H./Robertson, T. S. (1985), p. 861.

to invest time and effort in extensive information research prior to an innovative CEP purchase. Despite the fact that the elderly dispose of more spare time compared to employed individuals to acquire necessary product information prior to the purchase or learn eventually unknown product features, it can be doubted that they generally *like* to spend their time for this activity. It can be assumed that they also dislike the risk of losing time by getting the product adjusted, repaired or replaced if it malfunctions, or by trying to understand how to operate the CEP.[605] Even if it is not the wasted time per se which causes displeasure, the expected frustration of the consumer in operating the novelty may have a relevant effect on adoption. Furthermore, Gassmann/Reepmeyer (2006) found that it is especially the older persons who miss out on the advantages of a novel technology because they misjudge the efforts which are required to understand it. The elderly often estimate the necessary efforts as too high compared to the advantages the person obtains in return.[606]

Another aspect of the psychological risk which is important in the context of CEP is network externalities. If, for example, several manufacturers offer different, non-compatible standards of an innovation, there is the risk of choosing the "wrong" technology which does not establish itself in the long term. The consumer, who has opted for the system which does not prevail over the competing technology and who has bought the corresponding technical appliances, is usually confronted with considerable costs if he or she wants to change to the technology with the established standard.[607] In this situation of high switching costs, Greenstein (1993) or Farrell/Shapiro (1989) talk about "lock-in-effects."[608] Due to the consequences which are related to the decision of a certain technology option, this decision is likely to be associated with risks.

$\mathbf{H_{16}}$: The older an individual the higher the perceived psychological risk of innovative CEP.

$\mathbf{H_{28}}$: The higher the perceived psychological risk of innovative CEP the fewer/less innovative CEP are possessed (r_psy $\xrightarrow{-}$ CEPO_w).

Self-efficacy

How effective the potential dangers are on the individual consumer depends on his self-esteem or self-perception. A literature review by Simcock/Sudbury/Wright (2006) aiming to find age related differences in the per-

605 Cf. Roselius, T. (1971), p. 58.

606 Cf. Gassmann, O./Reepmeyer, G. (2006), p. 106.

607 Cf. Schmalen, H./Pechtl, H. (1996), p. 821.

608 Cf. Greenstein, S. M. (1993), pp. 19; Farrell, J./Shapiro, C. (1989), p. 51.

ception of self, revealed, unfortunately, diverging positions and inconsistent results.[609]

Regarding the product category focus, it was found that self-efficacy has been researched in the domain of computers.[610] Results may be transferred to the category of CEP as potential adopters might also expect problems with understanding its features or with correctly operating the innovative device. Creating the analogy to Bandura et al. (1977) who states that self-efficacy influences behavioral decisions, it can be assumed in the CEP context that self-efficacy influences the adoption decision.[611] Even when users of technological devices have had negative experiences in the past, it depends on their self-efficacy whether they perceive themselves capable to manage potential problems in the future. Thus, self-efficacy influences whether negative experiences become effective and whether individuals tend to refrain from adoption or, alternatively, whether they can cope with eventual difficulties and nevertheless "dare" to purchase the item.

In spite of recent research findings which take the diversity of older persons and changes compared to prior generations into consideration, it is expected that differences in self-efficacy between them and younger individuals exist.[612] Difficulties with CEP items in the past and the elderly's low perceptions of self-efficacy may lead to reluctance to use the product, or to avoid buying similar technological products in future.[613] Hence it is hypothesized that:

H_{17}: The older an individual the lower the self-efficacy in dealing with innovative CEP.

H_{29}: The higher the perceived self-efficacy in dealing with innovative CEP the more innovative CEP are possessed (s_eff $\xrightarrow{+}$ CEPO_w).

Social Pressure

Jacoby/Kaplan (1972) – and Kaplan/Szybillo/Jacoby (1974) who validated the study – analyzed the perceived risk of twelve products stemming from different product categories, including one consumer electronic product (Color TV).[614] The authors found that social risks account for a considerable amount of perceived risks attached to TV purchases. More specifically it is assumed for the

609 Cf. Simcock, P./Sudbury, L./Wright, G. (2006), p. 360.

610 See the cited literature in Compeau, D. R./Higgins, C. A. (1995), pp. 189.

611 Cf. Bandura, A./Adams, N. E./Beyer, J. (1977), pp. 126.

612 Cf. Porter, C. E./Donthu, N. (2006), p. 1001.

613 Cf. Ibid., pp. 1001.

614 Cf. Kaplan, L. B./Szybillo, G. J./Jacoby, J. (1974); Jacoby, J./Kaplan, L. B. (1972).

category of CEP that certain social pressures may exist in the form that non-adopters of CEP innovations feel the pressure to keep up with the reference group and to likewise purchase similar items. One reason why this effect of social pressure can be observed is that TVs as well as most CEP are "observable" products. As regards to older consumers it appears that they are less concerned with perceived social risks.[615] The same is assumed for the social pressure to innovate. Thus, it is hypothesized that:

H_{18}: The older an individual the lower the perceived social pressure of possessing innovative CEP.

H_{30}: The higher the perceived social pressure of possessing innovative CEP the more innovative CEP are possessed (soc_pres $\xrightarrow{+}$ CEPO_w).

4.2.3.3 Social Assistance

Three major perspectives of social integration have been identified in the previous discussion: social integration from the normative, informational, and the social support perspective.[616] Especially the informational and social support aspects seem highly promising in the context of CEP and elderly persons. In the purchasing situation of a novel, complex or expensive product – as it is the case with many CEP – the level of risk perceived by the potential adopter is relatively high. To reduce risks, one common strategy is to search for additional information.[617] It has been stated before that WOM communication through social interaction is an important source of information used by the undecided individual in these situations. Research question five (RQ 5) will cover the informational aspect and identify the most important information sources used by younger, middle-aged, and old consumers regarding the purchase of innovative CEP.

Research studies dealing with the adoption of technologies found that elderly more likely refer to family member sources than younger persons. Other sources of interpersonal communication that have been considered in these studies were friends and neighbors. They were preferred by younger information seekers compared to older information seekers. But, TV and newspapers largely surpassed interpersonal communication as information sources for both, elderly and

---

615 Cf. Simcock, P./Sudbury, L./Wright, G. (2006), pp. 359.

616 See Chapter 3.5.2.3, pp. 111.

617 Cf. Statt, D. A. (1997), p. 236; Fitzgerald Bone, P. (1992), pp. 582; Schweiger, G./Mazanec, J./Wiegele, O. (1976), p. 99; Cox, D. F. (1967), pp. 38; Cunningham, S. M. (1967a), pp. 265.

non-elderly individuals.[618] The domination of impersonal information sources and the low preference of WOM are surprising. One explanation for the preference of mass media (especially TV, radio and newspapers) might be their frequent and intense use by older persons, thus providing them with many product related information. Print media additionally allow the older reader to adjust the pace of information presentation for better comprehension by, for example, re-reading difficult passages.[619] Besides the mentioned advantages of impersonal information sources, the elderly cannot draw on so many personal contacts as with age the structure and size of social networks change. Life-cycle research elucidates this process which is mainly characterized by the loss of formal roles in later adulthood. Life stages which are included in such models and which have a bearing on the social integration are e.g. that children move out of the family's home, retirement, death of spouse or relatives and the like. However, apart from the mere focus on network size, the *quality of relationships*is important and determines how valuable family members, friends, colleagues and so forth are when it comes to a CEP purchase. For that reason it is refrained from the term "social integration" and instead the term "social assistance" will be used whenever qualitative aspects are included. Therewith, in this study, the understanding of social assistance is broader than that of social integration, including quantitative as well as qualitative aspects.

Also, personal and impersonal information sources are credited with different functions. Even when studies report of the elderly's extensive use of TV, print media, and radio, the mass media predominantly merely create knowledge about the existence of a newly introduced product (awareness function). Contrary to that, direct communication with others is essential for forming an opinion and deciding to adopt or reject an innovation (persuasion function).[620] Opposed to publications which proclaim the importance of mass media for older age classes, other researchers could not confirm the rejection of WOM. Instead, the source preference seems to depend on the product type. For expensive appliances, the spouse, children, grandchildren, and friends, but also retail personnel are largely more important as information sources than TV or radio.[621] More specifically, analyses devoted to the use of information sources by German consumers regarding CEP revealed that interpersonal communication (with relatives and friends, as well as shopping assistants) ranks first.[622] Likewise, Lunsford/Burnett

618 Cf. Gilly, M. C./Zeithaml, V. A. (1985), p. 356.

619 Cf. Federsel-Lieb, C. (1992), pp. 205; Phillips, L. W./Sternthal, B. (1977), p. 450.

620 Cf. Rogers, E. M. (1983), pp. 198.

621 Cf. GfK (1993), p. 65.

622 Cf. Medialine (2001), p. 25.

(1992) argue that the given decrease in the variety of social contacts causes older persons to become aware of innovations through communication with adult children or grandchildren, or through exposure to novelties purchased by them.[623] Reasons for high acceptance of recommendations from trusted, competent individuals are that they have pre-screened, evaluated and synthesized the desired information in an unbiased fashion.[624] Therewith, the advisors reduce the cognitive efforts which would normally be necessary in the adoption decision process for complex innovations from the CEP range. Similarly a study devoted to the subject of intergenerational influence compared the effect of family assistance, non-family sources, and mass media sources on the adoption of technological innovations by older persons.[625] The model could confirm that with age family assistance increases and that support from family members has a positive effect on adoption. Brown/Reingen (1987) support the assumption of the relevance of family members and close friends.[626] In their field study the authors differentiated between information seeking from primary and secondary groups (strong and weak ties). When both sorts of referrals were available, strong-tie relationships were preferred over weak-tie relationships and the former were perceived as more influential on the decision making.[627] It may also be worth considering that the assistance of close family members surpasses giving simply recommendations and helps to solve problems in the purchase (decision) process of elderly persons. With focus on complex electronic devices, it may be the children or grandchildren who take charge of screening product solutions, deciding for an alternative as well as purchasing and installing the innovative item in place of their parents or grandparents.[628]

It can be summarized that social assistance is expected to play a considerable role in CEP adoption of (elderly) consumers. Nonexistent consumer electronic knowledge can be compensated by children, grandchildren, other relatives or friends. Especially the availability of technology-interested, knowledgeable contact persons is decisive. It is more likely that they continuously inform themselves and thus dispose of deeper knowledge about innovative CEP items and therewith are more qualified in giving product choice recommendations. It is assumed that:

---

623 Cf. Lunsford, D. A./Burnett, M. (1992), p. 59.

624 Cf. Menzel, H. (1981), p. 157.

625 Cf. Mathur, A. (1999).

626 Cf. Brown, J. J./Reingen, P. H. (1987).

627 Cf. Ibid., pp. 357.

628 Related to this is research which is done in the area of "gift giving." Cf. Eichler, H.-V. (1991).

H_{19}: The older an individual the lower the received social assistance in purchasing innovative CEP.

H_{31}: The higher the received social assistance in purchasing innovative CEP the more innovative CEP are possessed (soc_assist $\xrightarrow{+}$ CEPO_w).

RQ 6: If social assistance is a significant factor of CEPO_w: With which of the social contacts do the different groups stay in touch most? Who is especially helpful in assisting an innovative CEP purchase?

4.2.4 Assessing the Relevance of Situational Variables

In the discussion of situational variables, diverse product categories have been under investigation.[629] The comparison of different products revealed that the effect of situational influences varies among others depending on product category characteristics. Belk (1975) found that especially consumer non-durables (e.g. beverages, meat products, snack foods) are susceptible to situational influences.[630] In contrast to that, consumer durable decisions are assumed to be less influenced. This type of product is purchased less frequently and more extensive decision processes have preceded the purchase.[631]

Nevertheless, for the purchase of innovative technological devices, it can also be expected that some situational effects may influence the (non-) purchase of a single CEP. For example in contrast to non-durable products which constitute necessities in everyday life, durable goods are mainly not seen as essential and, thus, are more sensitive to temporary financial restrictions.[632] The same should apply for innovative *electronic* devices. Furthermore, a situation of product malfunction has a positive consequence on innovation adoption. Malfunction of the current CEP, of course, stimulates activities to replace the old product and, thus, enhances the purchase of a substitute. When consumers replace the broken-down device, it is very likely that they upgrade the technology and buy an innovative consumer electronic solution.[633]

No age related differences have been found in the revised literature, which would indicate that older consumers are more or less impacted by situational factors than younger individuals. Furthermore, with the CEPO focus, the owner-

629 See the studies cited in Belk, R. W. (1975), p. 162.

630 Cf. Belk, R. W. (1975), p. 160.

631 Cf. Trommsdorff, V. (2004), pp. 320; Meffert, H. (2000), pp. 102.

632 Cf. Wilkie, W. L. (1986), pp. 275.

633 Cf. Gerhards, M./Klingler, W. (2006), p. 78.

ship not of one but several products is taken into consideration. Therefore it can be assumed that in sum the impact of situational factors on the ownership of CEP is only marginal and no concrete hypotheses are put forward. Still, a descriptive analysis of the situational variables might allow deriving recommendations to improve CEP offers (this aspect is covered in RQ 7).

5 Empirical Analysis

The literature review displays an inconsistent picture of the general innovativeness and innovation behavior of the elderly compared to younger consumers. Regarding innovative CEP, a first assessment of secondary data revealed that older consumers possess less CEP – especially less *innovative* CEP. Therefore the aim of this study is to thoroughly analyze whether the cues of age related differences in CEPO can be confirmed and, if yes, what reasons can be found to explain these differences.

First of all, the hypotheses derived from gerontological and innovation related studies are summarized and the research strategy used to test these hypotheses is outlined (Ch. 5.1). The focus is on the operationalization of theoretical constructs (s. Ch. 5.2) as well as on the questionnaire development and data collection (Ch. 5.3) prior to discussing the theoretical background and assumptions of the data analysis methods which are needed to answer the research questions (Ch. 5.4). In Chapter 5.5 the collected data is analyzed. Furthermore the survey participants are described and the ownership of CEP is analyzed. The scale development is subject of Chapter 5.6. In the last section of Chapter 5 the concrete hypotheses are tested and research questions are answered by the use of the collected data (Chapter 5.7).

5.1 Summary of Hypotheses and Research Strategy

As will be shown later in more detail, the respondents are grouped into three age groups: The "young" group (G1) includes individuals aged 18-34 years, followed by the "middle-aged" group (G2) which includes subjects between 35 and 59 years of age. The "old age" group (G3) consists of individuals who are 60 years and older. In the next two sub-chapters the hypotheses that have been proposed throughout the previous sections will be summarized and the research strategy will be elucidated.

5.1.1 Summary of Hypotheses

The review of gerontological, innovation and consumer behavior literature resulted in research hypotheses associated with age and possible adoption determinants of CEP innovations. This chapter provides an overview of this study's research objectives and hypotheses.

Firstly, it is intended to examine whether the introductory assumption of a negative relationship between age and innovations exists. As discussed earlier, a product category focus is indispensable to narrow the research study scope and to obtain results which are transferable to other products of the same category. Therefore, one objective is to explain age differences in ownership of CEP (CEPO), more specifically differences in the scores of the weighted index of CEP ownership (CEPO_w).[634] Accordingly the first hypotheses deal with the relationship between age and several definitions of CEPO:

H_1	The ownership of CEP (CEPO) is negatively related to age
H_2	The ownership of *innovative* CEP (CEPO_i) is negatively related to age
H_3	The *weighted* ownership of CEP (CEPO_w) is negatively related to age

Then, in order to delve deeper into the subject and to get an idea about the consumers' general assessment of innovative CEP, it will be examined whether significant differences between the three age groups exist regarding the assessment of several characteristics of innovative CEP.

Are there significant age group differences in the assessment of...

RQ 1_1	...the technical superiority of innovative CEP?
RQ 1_2	...the functional superiority of innovative CEP?
RQ 1_3	...the compatibility of older CEP with innovative CEP?
RQ 1_4	...the importance that older CEP are compatible with innovative CEP?
RQ 1_5	...the size of buttons and labels of innovative CEP?
RQ 1_6	...the conformance of innovative CEP with the needs?

The third block of hypotheses is devoted to the issue in how far age has an effect on the determinants which have been found in the literature to influence the adoption of innovations.[635]Together with the findings from CEP and age related studies the following hypotheses were put forward:

H_4	The older an individual the lower the ownership of CEP (CEPO)

634 The decision in favor of this weighted index and its exact composition will be explained in detail later (s. Ch. 5.2.4.1, pp. 170).

635 The effect of age on the three CEPO indices is also measured. This time, however, a dependency relationship will be tested.

H_5	The older an individual the lower the ownership of *innovative* CEP
H_6	The older an individual the lower the *weighted* ownership of CEP
H_7	Cognitive age is a better predictor of the innovative CEP adoption determinants than the chronological age
H_8	The older an individual the lower the perceived relative advantage of innovative CEP
H_9	The older an individual the higher the perceived product complexity of innovative CEP
H_{10}	The older an individual the higher the perceived in-store assortment complexity of CEP
H_{11}	The older an individual the higher the perceived performance risk of innovative CEP
H_{12}	The older an individual the higher the perceived financial risk of innovative CEP
H_{13}	The older an individual the better the perceived overall financial situation
H_{14}	The older an individual the lower the interest in CEP
H_{15}	The older an individual the lower the perceived knowledge compatibility of older and innovative CEP
H_{16}	The older an individual the higher the perceived psychological risk of innovative CEP
H_{17}	The older an individual the lower the self-efficacy in dealing with innovative CEP
H_{18}	The older an individual the lower the perceived social pressure of possessing innovative CEP
H_{19}	The older an individual the lower the received social assistance in purchasing innovative CEP

The next major block of hypotheses deals with the influence of the adoption determinants on the weighted CEPO index (CEPO_w). It is investigated in how far the theoretically derived set of variables is suitable to explaining differences in CEPO_w and in how far the single determinants contribute to changes in

CEPO_w scores. This will be done separately for the three age groups in order to unveil eventual age related differences of these groups.

Regarding the three age groups ("young," "middle-aged," and "old") it is expected that each of the following adoption determinants is **a)** a significant factor in determining CEPO, CEPO_i, and CEPO_w, and **b)** has the tendency of the relationship as proposed subsequently (that means either a positive (+) or negative effect (-) on CEPO_w).

H_{20}	The higher the perceived relative advantage of innovative CEP the more innovative CEP are possessed (rel $\xrightarrow{+}$ CEPO_w)
H_{21}	The higher the perceived complexity of innovative CEP the fewer/less innovative CEP are possessed (cx_prod $\xrightarrow{-}$ CEPO_w)
H_{22}	The higher the perceived in-store assortment complexity of CEP the fewer/less innovative CEP are possessed (cx_is $\xrightarrow{-}$ CEPO_w)
H_{23}	The higher the perceived performance risk of innovative CEP the fewer/less innovative CEP are possessed (r_perf $\xrightarrow{-}$ CEPO_w)
H_{24}	The higher the perceived financial risk of innovative CEP the fewer/less innovative CEP are possessed (r_fin $\xrightarrow{-}$ CEPO_w)
H_{25}	The better the perceived overall financial situation the more innovative CEP are possessed (fin_situ $\xrightarrow{+}$ CEPO_w)
H_{26}	The higher the interest in CEP the more innovative CEP are possessed (int $\xrightarrow{+}$ CEPO_w)
H_{27}	The higher the knowledge compatibility of older CEP with innovative CEP the more innovative CEP are possessed (compa $\xrightarrow{+}$ CEPO_w)
H_{28}	The higher the perceived psychological risk of innovative CEP the fewer/less innovative CEP are possessed (r_psy $\xrightarrow{-}$ CEPO_w)
H_{29}	The higher the perceived self-efficacy in dealing with innovative CEP the more innovative CEP are possessed (s_eff $\xrightarrow{+}$ CEPO_w)
H_{30}	The higher the perceived social pressure of possessing innovative CEP the more innovative CEP are possessed (soc_pres $\xrightarrow{+}$ CEPO_w)
H_{31}	The higher the received social assistance in purchasing innovative CEP the more innovative CEP are possessed (soc_assist $\xrightarrow{+}$ CEPO_w)

Next to these concrete hypotheses several other subjects are of interest for which it was not possible to formulate concrete hypotheses. Theses topics will be analyzed exploratively.

Exploratory analysis in model testing:

RQ 2	How well do the adoption determinants predict CEPO_w for each age group?
RQ 3	Within each age group, what is the order of importance for the adoption determinants in influencing the weighted index CEPO_w?
RQ 4	If not all of the theoretically derived twelve determinants have a significant influence on CEPO_w, which alternative models can be proposed for the different age groups?

Further Research questions:

RQ 5	Which are the most important information sources used by younger, middle-aged, and old consumers regarding the purchase of innovative CEP?
RQ 6	If social assistance is a significant factor of CEPO_w: With which of the social contacts do the different groups stay connected most? Who is especially helpful in assisting in an innovative CEP purchase?
RQ 7	What (relevant) insight can the situational variables give into the adoption behavior regarding innovative CEP?
RQ 8	Are any significant gender differences observable?

5.1.2 Research Strategy

In order to affirm or reject the hypotheses and to examine the exploratory research questions, this study conducts an empirical survey. In a quantitative cross-sectional analysis participants of the different age groups are inquired about CEP innovations as well as about their opinions and personal attitudes. However, before performing such a cross-sectional analysis, it is required to develop a research strategy and select the proper statistical methods that are capable of testing the respective hypotheses and answering the research questions. Prior to being able to answer the concrete hypotheses, it is necessary to translate them into single survey questions. Hypothesis H_9 which deals with the "perceived product complexity" regarding innovative CEP can be taken as an example. As product complexity is not visible and not directly measurable, it needs to be translated into several survey questions – also called items – which can be answered by the respondents (s. Ch. 5.2 "operationalization"). This is done by searching for existing operationalization approaches in other research studies. The theory-driven conceptualization of the research issues that have been performed, supplies input for this operationalization.

With the completion of the operationalization, the final questionnaire items are obtained, so that the questionnaire structure can also be finalized and the target population, as well as the data collection strategy, can be determined (s. Ch. 5.3). In this chapter, the particularities of surveying older individuals are also discussed. Experiences are included in the final questionnaire design and determine the final data collection procedure.

Existing studies are considered during the operationalization process. It is very likely that the identified scales need to be adapted to the actual research context. In order to filter out improper items that do not adequately represent what is intended to be measured, researchers use the statistical method of factor analysis. This method will be explained in Chapter 5.4.1 and applied using the collected data in Chapter 5.6. The results of the factor analysis are the final scales representing the theoretical constructs.

While the hypotheses regarding the relationship between age and the different CEPO indices (CEPO, CEPO_i, and CEPO_w) can be answered by simple correlational analyses in a first step, it is more complicated to test for significant differences when the three age groups formulate their opinion about some general aspects of innovative CEP. When it is intended to compare more than two groups, an analysis of variance (ANOVA) is necessary which is theoretically explained in Chapter 5.4.2 and applied to the actual research context in Chapter 5.7.2.

On the one hand the central hypotheses deal with the influence age has on the adoption determinants and, on the other hand, with the appropriateness of these determinants to explain differences in the CEPO of the three age groups. Both types of hypotheses can be answered with the method of regression analysis which estimates the relationship of one or more independent variables (IVs) with one dependent variable (DV). Next to assessing the influence of age on the determinants with single regressions (one IV and one DV), the other hypotheses are tested "en bloc" in presence of all other determinants (multiple IVs and one DV). The latter method allows the identification of the most important adoption determinants. Furthermore, as it can be assumed that not all adoption determinants exert a significant influence on CEPO_w (e.g. when adoption determinants correlate with each other), alternative models will be proposed by using the automated selection processes of the statistical software.[636]

5.2 Operationalization

Before being able to empirically test the postulated hypotheses, it is necessary to "translate" the hypotheses into measurable questionnaire items. The theoretical background of this procedure will be discussed briefly below (s. Ch. 5.2.1). The research questions differ in complexity. While variables aiming to primarily *describe* the respondents can be measured via the single item approach (Ch. 5.2.2), hypotheses including more complex variables or theoretical constructs[637] can only be tested when multi-item measures (Ch. 5.2.3) are used or (weighted) indices (Ch. 5.2.4) are constructed. As the latter two need a deeper theoretical foundation, a literature review of existing operationalization approaches is conducted prior to choosing the final items that will be used for constructing the scales.[638]

5.2.1 Theoretical Background of the Operationalization Process

Standard sociodemographic variables like educational level, employment status, or gender are easy to measure – they can be captured by asking single questions (single-item measure). The determinants of CEPO, however, are of a more complex nature. Different to the demographic variables, the concepts of relative advantage or perceived risk e.g. are not directly observable and thus cannot be measured directly. They cannot be captured adequately by asking one single question. These complex concepts – also termed constructs or latent variables – are abstract phenomena which are only accepted to explain consumer behavior when they can be empirically anchored.[639] Hence, a "bridging process" is neces-

636 The results obtained by the automated selection processes are merely propositions of alternative models. They cannot be confirmed or rejected with the same data set that will be collected in this study. These propositions rather deliver ideas for further research.

637 Constructs are concepts that have been defined by researchers, but which are not directly observable or measurable.

638 The final survey was conducted in German. The original questionnaire can be found in Appendix 4 (s. pp. 287). To ensure that English and German items were equivalent, the English items found in the literature have been translated into German and retranslated (s. also Ch. 5.3.2, pp. 179).

639 Cf. Kroeber-Riel, W./Weinberg, P. (2003), p. 29; Homburg, C./Giering, A. (1996), p. 6.

sary to "translate" the abstract, unobservable variable (construct) into more concrete, observable variables (indicators/items).[640] This process of assigning observable variables – of which it is assumed that they are in a causal relationship with the respective theoretical construct – is called operationalization.[641]

Due to their complexity, constructs are operationalized by means of several items. The multiple item approach permits to capture the different facets of the theoretical construct. This procedure combines several questions to ask the same thing. The measurement, then, also benefits from the fact that systematic and random errors which every indicator (item) possesses balance each other out.[642] As a matter of fact, even a very good single indicator cannot fully seize any construct. This is because a perfect measurement would imply a correlation of 1 between indicator and construct which is extremely unlikely.[643]

The operationalization of constructs is usually performed by developing a scale or an index, whereby scales can be interpreted as a special case of indices.[644] Different than for the development of an index, criteria exist for the development of a scale, which allow assessing whether an item does or does not belong to the scale.[645] Furthermore, indices are predominantly used for measuring theoretical concepts which address several dimensions of one latent variable.[646] Instead, scales measure along one dimension only.[647] Both, however, consist of multiple items which – especially in the case of indices – can be weighted if necessary.

For many theoretical constructs, scales or indices that have been validated through repeated use over time exist. Their unmodified use allows comparing results across different studies. In the domain of innovation analyses with special focus on comparing younger and older consumers' ownership of CEP, established scales or indices do not exist yet. In this instance, Malhotra (2004) recommends a scale development procedure that starts with an analysis of the underlying theory of the constructs to be measured, helping to specify these con-

640 Cf. Byrne, B. M. (2001), pp. 4; Brewer, M. B. (2000), pp. 6.

641 Cf. Christophersen, T./Grape, C. (2006), p. 116; Meffert, H. (2000), p. 147.

642 Cf. Trommsdorff, V. (2004), p. 42.

643 Cf. Christophersen, T./Grape, C. (2006), p. 117; Aaker, D. A./Bagozzi, R. P. (1979), p. 147.

644 Cf. Schnell, R./Hill, P. B./Esser, E. (2008), p. 127.

645 Cf. Ibid., p. 181; Holm, K. (1970), p. 356.

646 Cf. Schnell, R./Hill, P. B./Esser, E. (2008), pp. 166.

647 Cf. Holm, K. (1970), p. 356.

structs and to build the final scale and interpret the results.[648] Hence, an initial item pool needs to be generated based on theory, secondary data, or qualitative research. Next, this item pool must be purified in a qualitative and quantitative manner[649] – to obtain a reduced, final set of items.[650] Christophersen/Grape (2006) point to one critical phenomenon regarding the correct operationalization of constructs:[651] The authors stumbled across numerous research studies that boasted high correlations between indicators and construct[652] which were, however, achieved by merely rewording the very same indicator. Opposed to that, indicators should represent differing consequences of the latent construct, which is considered in the construction of multi-item scales.

5.2.2 Single-Item Measures

The items described below serve to describe the survey sample. These are standard variables which are operationalized via single-item measures. Some additional items allow to get a first impression and round off the picture regarding the respondents' opinion about several CEP characteristics. Next to the concrete verbalizations of the used questionnaire items, the answering options and the SPSS notations are also indicated.

5.2.2.1 Sociodemographic Variables

The sociodemographic variables provide valuable information about the characteristics of the respondent group. As the variables are standard items to describe the population under investigation, only few annotations need to be made here. Sociodemographic variables are directly measurable so that they will be integrated as single questions (single-item approach) in the questionnaire. The operationalization differs from the common operationalization of sociodemographic variables in most other surveys only in one aspect: The financial situation will be measured by means of the *perceived* financial situation.[653] There are two major reasons for this step. First, measuring gross income would be inappropriate in predicting actual consumption behaviors as e.g. different household

648 Cf. Malhotra, N. K. (2004), pp. 263.

649 That means that the scales should be tested for their reliability and validity.

650 Cf. Malhotra, N. K. (2004), pp. 263.

651 For the following see Christophersen, T./Grape, C. (2006), p. 117.

652 The high level of correlation also leads to high values of other key criteria like the Cronbach's alpha (which is explained in more detail later).

653 This is analogous to Topritzhofer (1969) who differs between objective and subjective liquidity. The author points out the importance of the latter concept. Cf. Topritzhofer, E. (1969), p. 65.

sizes or income taxes distort the picture. Second, disposable income would bypass these problems, but exact amounts are rarely known. Furthermore, disposable income does not reveal anything about the actual spending behavior of individuals. Despite their wealth, many older consumers spend less compared to younger individuals. A reason for this might be that it is not the objectively available financial assets which determine their consumption levels, but more specifically the subjective financial situation. Therefore, the *perceived* financial situation is preferred as a measure. As a positive side-effect, this also allows to bypass the critical question of directly asking about the interviewee's actual income or the disposable household income, and receive more reliable answers. Many respondents are reluctant to answer this question, or give wrong responses as they feel insecure about sharing sensitive information.

Age is one of the central variables which should indirectly help to explain innovation behavior regarding CEP. Just like other sociodemographic variables, chronological age is easy to measure – differences only exist with regard to the level of its specificity. On the one hand, interviewees may be asked to choose the age bracket they belong to. On the other hand, the respondents may be asked more specifically for their actual age (in years) or their date of birth (expressed in year, month and day).[654] As age brackets are defined arbitrarily and valuable information is lost with their construction, the more specific measure is favored here (s. Table 13). Measuring the respondents' actual age in years (not day and month) seems sufficient for this survey (open-ended / fill-in question).

SPSS notation	Item
age	How old are you?
	Fill-in
age_cog	How old do you feel?
	Fill-in

Table 13: Questionnaire items to measure age

Finally, the operationalization of the level of education is adapted to the German educational system which is differentiated into six categories. The final set of sociodemographic questionnaire items is shown in Table 14. For the German wording of the sociodemographic variables please refer to the final questionnaire in Appendix 4.

SPSS notation	Item
edu	What is your highest educational degree/graduation (level)?
	No degree / elementary school or secondary school/secondary school leaving certificate/general qualification for university

654 Cf. Settersten, Jr. R. A./Mayer, K. U. (1997), pp. 238.

	entrance/university or college/other (fill-in)
job	What is your employment status?
	Apprenticeship/student/employed or public official/freelancer or self-employed person/professional retraining/retiree or pensioner/housewife or homemaker/unemployed /other (fill-in)
gender	Gender
	Female/male
mart_stat	What is your current martial status?
	Single/married/divorced/widowed/separated
HH	Besides you, how many persons permanently live in your household?
	Fill-in
fin_situ	How would you describe your overall financial situation?
	Very negative – very positive (seven point Likert scale)

Table 14: Further sociodemographic questionnaire items

5.2.2.2 Further Descriptive Variables

Apart from the sociodemographics, several other variables are considered in this study, which further help to describe some preferences and attitudes of the survey participants regarding innovative CEP (s. Table 15). Some originate from the technological compatibility aspect of innovations. Furthermore, an analysis of the age specific information source selection helps to understand where consumers search to gather relevant information prior to purchasing an innovative CEP.

SPSS notation	Item
tech_sup	I think that innovative CEP compared to their preceding generation of devices are technically superior
funct_sup	I think that innovative CEP compared to their preceding generation of devices possess more features
compa_exist	I think that innovative CEP are compatible with older devices
compa_imp	I think that it is important that innovative CEP are compatible with older devices
compa_small	I think that innovative CEP have buttons and labels which are too small
compa_need	I think that innovative CEP meet my needs

Table 15: Questionnaire items to measure further preferences and attitudes of the respondents

The respondents are able to choose among the information sources displayed in Table 16. Further, respondents have the possibility to indicate that they would not buy the product by themselves, but leave the purchase to other persons.

SPSS notation	Item
inf	Which information sources would you mainly use prior to the purchase of an innovative CEP? - conversations with family/relatives/friends[a] - leaflets containing product information[b] - own experience[a] - in-store ads[b] - in-store sales personnel[a] - information on the product packaging[b] - internet[b] - TV-ads[b] - newspapers and periodicals[b] - testing journals[b] - consumer advice[a] - catalogues[b]
p_fam	I would let family members (e.g. children, grand-children, etc) or friends purchase the device in my place
a = personal information source; b = non-personal information source/mass media	

Table 16: Questionnaire items to measure information source selection of the respondents

It was intended to gain further insight into situational influence factors of innovative CEP (non-)purchase and into the willingness to spend money for innovative CEP. Therefore, on the one hand, the respondents were given the possibility to express why they did (not) buy an innovative CEP in recent time (last twelve months). This was done by integrating a list into a selection of the most probable situations for a CEP (non-)purchase in the questionnaire. As with CEP the focus is on a product *category* standing for a larger number of devices, the statements regarding the reasons for (non-)purchase were combined with a five point Likert scale ranging from “did not happen” to “happened often” (s. Table 17).

SPSS notation	Item
	How often did you purchase an innovative CEP in the last 12 months?
situ_1	…because your old device broke?
situ_2	…because service was discontinued for your old device
situ_3	…because you wanted to own a more modern device
	How often did you decide against purchasing an innovative CEP in the last 12 months?
situ_n_1	…because of your current financial situation
situ_n_2	…because the desired product was not available
situ_n_3	…because of lack of time to deal with the potential purchase
situ_n_4	…because you wanted to wait for a cheaper price for the device

Table 17: Questionnaire items to measure situational influences

On the other hand a survey question on the willingness to spend money for innovative CEP was included in the questionnaire. However, positioned directly after questions on the consumers' interest in CEP, it can be assumed that a sequence effect occurred, expressed through an unlikely high correlation between spending willingness and interest. Therefore, this item has been excluded from further analysis.

5.2.3 Multi-Item Measures: Scales

As noted in the introduction, a preliminary literature review helps to create an initial item pool which preferably draws on items of established scales. As most of the determinants have been borrowed from Rogers' fundamental publications about innovation diffusion, it would be consistent to align them with the operationalizations made there. However, Rogers does not provide any suggestions about the measurement of the designated variables because, according to him, operationalization in the domain of innovation adoption and diffusion should not be generalized, but depend on the specific research issue.[655] Scales should therefore be newly created for each investigation. To guarantee at least some comparability of results between different research projects, a common set of scales and measures would be desirable. After an extensive review of the relevant literature, it needs to be stated that this project is difficult to realize. On the one hand, as a meta-analysis by Tornatzky/Klein (1982) showed, in a large number of articles researchers infer characteristics of the independent variable and do not directly measure the innovation attributes as perceived by the decision maker.[656] On the other hand, many research studies do not reveal their item sets and employed scales.[657]

Interestingly, despite the large number of individual innovation adoption and innovation diffusion research studies, no generally applicable set of measures of the determinants has evolved over time. Furthermore, studies which do not infer innovation characteristics, and consider large sets of adoption determinants in their studies, mainly focus on adoption in organizations. These studies can be

655 Cf. Rogers, E. M. (2003), p. 225.

656 Cf. Tornatzky, L. G./Klein, K. J. (1982), pp. 31.

657 To name just one example, in Ostlund's (1973) study about predictors of innovative behavior the researcher performs a factor analysis to eliminate redundant variables. Unfortunately, the questionnaire items are kept under wraps and do not contribute to the further scale development process. Cf. Ostlund, L. E. (1973).

assigned to the "technology acceptance models" (TAM) and items can only be partially used for the current research topic.[658]

5.2.3.1 Relative Advantage

Tornatzky/Klein (1982) acknowledge that the relative advantage of a novelty is hard to measure.[659] The authors hold the "too broad and amorphous" conceptualization responsible for this difficulty and likewise criticize the many studies which infer the relative advantage from the innovations themselves instead of directly asking potential adopters about their perception. The latter issue is resolved by letting the potential and actual adopters rate the extent of their individually *perceived* relative advantage of innovative CEP.

The difficulty that the relative advantage determinant is too broad in its conceptualization is countered with finding a good level of specificity. The approach of Ettlie/Vellenga (1979) who employed one single question by asking to rate the innovation's relative advantage on a 1-9 point scale with the starting point "the innovation does not offer any relative advantage over previous ideas" and with "the innovation is significantly better than the idea it supersedes" at the other end of the scale seems too unspecific.[660] Litfin (2000), in turn, determines relative advantage on the basis of a vast number of specified product functions.[661] Such an approach is not suitable because of the product category focus in the current research study. Adapting Litfin's approach by surveying particular functions of each innovation would make the questionnaire too lengthy. More sophisticated measures can also be found in the domain of information technology acceptance or TAM research. Though, their major drawback is that adoption in organizations is analyzed and, therefore, much of the relative advantage is related to job performance and work productivity. A complete adoption of scales which have been applied in those studies is, again, not possible.

The relative advantage is therefore inquired by asking the respondents directly in how far it is important for them that newly purchased CEP are technologically superior to older devices. Apart from the technical superiority it will be questioned how important it is for the respondent that newly purchased CEP have more functions – which is a characteristic of the majority of innovative CEP. As

658 See for example Karahanna, E./Agarwal, R./Angst, C. M. (2006); Venkatesh, V. et al. (2003); Karahanna, E./Straub, D. W./Chervany, N. L. (1999); Moore, G. C./Benbasat, I. (1991); Davis, F. D. (1989).

659 Cf. Tornatzky, L. G./Klein, K. J. (1982), pp. 34.

660 Cf. Ettlie, J. E./Vellenga, D. B. (1979), p. 435.

661 Cf. Litfin, T. (2000), pp. 142.

proposed by Venkatraman (1991) the respondent's evaluation of the value-for-money ratio is added.[662] The final item set used in the questionnaire to measure relative advantage of CEP innovations is merged in Table 18.

SPSS notation	Item
rel_tech	It is important for me that newly purchased CEP are technologically superior
rel_funct	It is important for me that newly purchased CEP possess more functions
rel_coper	I think that innovative CEP have a good value-for-money ratio compared to their preceding generation of devices

Table 18: Operationalization of relative advantage (rel)

5.2.3.2 Product Complexity

As the discussion before has revealed, it makes sense to view the complexity determinant more holistically than merely as perceived product complexity. Therefore, two components of complexity have been identified that may play a major role in determining CEPO. Apart from product complexity, the chosen conceptualization includes the perception of the in-store complexity of CEP.

Different approaches exist to measure product complexity. While there are studies which infer complexity by measuring the extent to which the novelty requires the potential adopter to learn new things,[663] other researchers directly ask how complex consumers perceive the product or particular features of the product.[664] It is also possible to use more sophisticated operationalizations from the domains of ease-of-use measurement, information overload, consumer confusion, or cognitive complexity research. One approach with which the latter research area captures product complexity is by means of production rules. These are the steps needed to perform product related tasks in the man-machine interaction and thus seem to be an interesting approach at first glance.[665] However, Ziefle/Bay (2005) discovered that production rules are not reliable predictors for complexity measurements as difficulties experienced during product usage were higher than production rule analysis purported and also because age related performance differences were neglected.[666] Thus, the other approaches seem more

662 Cf. Venkatraman, M. P. (1991), p. 60.

663 See for example Mukherjee/Hoyer (2001) who infer complexity from learning costs by measuring how much time/effort/energy a consumer needs to invest to effectively use the features of an innovative product. Cf. Mukherjee, A./Hoyer, W. D. (2001), p. 466.

664 Cf. Venkatraman, M. P. (1991), p. 60.

665 For an overview see Bovair, S./Kieras, D. E./Polson, P. G. (1990), pp. 4.

666 Cf. Ziefle, M./Bay, S. (2005), p. 382.

promising in the CEP innovation context and should therefore be integrated in the final questionnaire.

TAM predominantly include the determinant "perceived ease of use" to measure, as the name already portends, how easy an innovation is to use.[667] High levels of ease of use suggest that low cognitive efforts are required, and therewith the adoption of novelties is facilitated. As it connotes the opposite of complexity, the scale of perceived ease of use items needs to be inversed to measure complexity. While the advantage is their extensive validation, one drawback of using scales of TAM is, again, that they refer to technology adoption in the workplace. Based upon the operationalization proposed by the TAM-researchers, numerous other studies explored the ease of use in non-organizational contexts and refined the initial item set.[668] Further suggestions come from Ziefle/Bay (2005) who operationalized ease of use of a technological innovation with an age related research background as well as from Mukherjee/Hoyer (2001).[669] These studies help to find adequate measures for the product complexity of innovative CEP covering both aspects, the difficulty to *understand* and *use* this innovation (see the selected items in Table 19). Taken from studies of consumer confusion (primarily discussed in the next sub-chapter of in-store complexity) as third item to measure product complexity the survey question in how far consumers perceive the number of functions of innovative CEP to be confusing is added.

SPSS notation	Item
cx_und_r	I think that innovative CEP are easy to understand
cx_use_r	I think that innovative CEP are easy to use
cx_conf	I think that the number of features of innovative CEP is confusing

Table 19: Operationalization of product complexity (cx_prod)

5.2.3.3 In-Store Complexity

Studies in the area of information overload and consumer confusion are helpful in finding adequate items to capture the second aspect of purchase complexity perception – the perception of complexity perceived in stores. As the concept of

667 E.g. Karahanna, E./Agarwal, R./Angst, C. M. (2006); Venkatesh, V. et al. (2003); Karahanna, E./Straub, D. W./Chervany, N. L. (1999); Moore, G. C./Benbasat, I. (1991); Davis, F. D. (1989).

668 See for example Porter, C. E./Donthu, N. (2006); Brown, S. A./Venkatesh, V. (2005); Venkatesh, V./Davis, F. D. (1996).

669 Cf. Ziefle, M./Bay, S. (2005), p. 377; Mukherjee, A./Hoyer, W. D. (2001), p. 469.

consumer confusion and related theoretical constructs is broad,[670] a lot of differing approaches exist for its operationalization. For example it can be referred to measurements of Sproles/Kendall (1986) and Schweizer/Kotouc/Wagner (2006) the latter of which pooled several scales used in different preceding research studies to measure their appropriateness.[671] The adapted questions which are included in the questionnaire are listed in Table 20. They cover all three sources of confusion as identified by Mitchell/Papavassiliou (1999): perceived similarity of competing products, perceived overchoice of alternative product solutions, and ambiguous or inadequate information.[672] For the appropriate wording of these items in German, it can be referred to Schweizer (2004) and Walsh/Henning-Thurau (2002) who also translated some of the items used here in their empirical research about consumer confusion into German.[673]

SPSS notation	Item
cx_sim	I think that it is difficult to choose between similar CEP variants
cx_choice	I think that the choice of CEP in stores is too large
cx_isinf	I think that product information of CEP in stores is difficult to understand (e.g. leaflets, packaging, store displays)

Table 20: Operationalization of in-store complexity (cx_is)

5.2.3.4 Performance Risk

Even though objective risks do exist, specific measurement tools to accurately measure them are unavailable.[674] Thus, analogous to proceeding so far, it is focused on the subjective, perceived risk which is regarded as a multifaceted construct adapted to the CEP context. Opposed to unidimensional measures of risk, that merely quantify the "overall perceived" risk or pick a single aspect of risk, multidimensional approaches take into consideration different risk types. Although it has been reproached that these risk types are not independent from each other,[675] this study follows the justification of Shimp/Bearden (1982) who

670 For an overview see Mitchell, V.-W./Walsh, G./Mo Yamin (2005), pp. 143; Eppler, M. J./Mengis, J. (2004); Schweizer, P. (2004), pp. 22; Wiedmann, K.-P./Walsh, G./Klee, A. (2001), p. 85.

671 Cf. Schweizer, M./Kotouc, A. J./Wagner, T. (2006), pp. 187; Walsh, G./Henning-Thurau, T. (2002), pp. 99; Sproles, G. B./Kendall, E. L. (1986), p. 273.

672 Cf. Mitchell, V.-W./Papavassiliou, V. (1999), p. 320.

673 Cf. Schweizer, P. (2004), p. 338; Walsh, G./Henning-Thurau, T. (2002), pp. 105.

674 Cf. Mitchell, V.-W. (1999), pp. 164.

675 Cf. Dowling, G. R. (1986), p. 200; and see, for example, the research results from Stone, R. N./Gronhaug, K. (1993), pp. 45; Brooker, G. (1984), p. 440; Kaplan, L. B./Szybillo, G. J./Jacoby, J. (1974), p. 289; Jacoby, J./Kaplan, L. B. (1972).

choose to treat the dimension separately in recognition of existing conceptual and practical distinctions.[676] Furthermore, the multidimensional approach enriches the perceived risk discussion as it expands the, until then, pursued *global* risk measurement. Therefore, this study considers the risk dimensions of performance and financial risk in the product context as well as psychological and social risks/pressures in the personal variables context.

As seen during the conceptualization, early studies in risk behavior research established two component approaches of risk measurement, integrating either the components "uncertainty" and "adverse consequences," or "probability" and "importance of loss." After the selection of components and dimensions of risk, risk indices were created either by adding or multiplying the various elements. Intuitively, it seems reasonable to opt for a multiplicative combination of the components for two reasons. First, "the absence of either variable would eliminate risk," and secondly, "the influence of a nonsalient adverse consequence on overall perceived risk is reduced."[677] Unlike this proposition, Mitchell (1999) reports of better results when additive models are applied and advises to use this approach also because of its long-standing tradition which allows comparing the different research results with each other.[678] More general, he affirms component models to be superior to measurement approaches which, for example, have been put forward by Pras/Summers (1976; 1978) as they are simple to use and easy to understand for respondents, compared to the large amount of data that needs to be surveyed in the competing approach.[679]

Researchers have predominantly criticized two component approaches for their way of risk element combination. They argue that this expectancy-value orientation stems from other disciplines like mathematics and economics instead of psychology and that it can consequentially be doubted whether consumers perform these kinds of risk calculations in the way it has been proposed for long.[680] Moreover, Horton (1976) and Bettman (1973) discovered positive correlations between these two components of perceived risk.[681] Another argument why omitting importance can be legitimated is because of an inflation of expectations that is likely to evolve in these situations.[682] Therefore, the following operationalization of "performance risk" is proposed (s. Table 21).

---

676 Cf. Shimp, T. A./Bearden, W. O. (1982), pp. 38.

677 Dowling, G. R. (1986), p. 199.

678 Cf. Mitchell, V.-W. (1999), pp. 178 and pp. 183.

679 Cf. Ibid., p. 184 and p. 187.

680 Cf. Ibid., p. 168; Stone, R. N./Gronhaug, K. (1993), p. 41.

681 Cf. Horton, R. L. (1976), p. 702; Bettman, J. R. (1973), p. 189.

682 Cf. Herrmann, A. K. (2003), pp. 136.

SPSS notation	Item
perf_expec	If I bought an innovative CEP, I would worry the device not rendering the benefit I had expected
perf_troub	If I bought an innovative CEP, I would worry the device not functioning reliably due to teething problems

Table 21: Operationalization of performance risk (r_perf)

5.2.3.5 Financial Risk

In line with the discussion above, it is refrained from the component approach also regarding the financial risk. Instead the financial risk will be measured according to Stone/Gronhaug (1993).[683] They defined risk as subjective expectations of loss and assumed that the more certain this loss, the greater the risk perceived by the individual. Their approach has been successfully applied, among others by Hirunyawipada/Paswan (2006), Bauer/Sauer/Becker (2003), and Dholakia (1997).[684] Thus, the general idea of the operationalization approach is followed here as well. The risk aspects regarding price reduction and fear that the CEP are not worth their money stem from Stone/Gronhaug (1993) and Venkatraman (1991).[685] These two items are completed by a third item in order to measure the fear of obsolescence. The final questionnaire items are as follows (s. Table 22).

SPSS notation	Item
fin_price	If I bought an innovative CEP, I would worry about the prices falling soon after purchase
fin_worth	If I bought an innovative CEP, I would worry the device not being worth its money
fin_obs	If I bought an innovative CEP, I would worry the device will be obsolete soon

Table 22: Operationalization of financial risk (r_fin)

5.2.3.6 Interest

It is difficult to find validated multi-item scales which could be easily applied to measure an individual's interest in the product class of CEP. On the one hand, this is because studies do not report the items which were chosen for the ques-

683 Cf. Stone, R. N./Gronhaug, K. (1993), p. 42.

684 Cf. Hirunyawipada, T./Paswan, A. K. (2006); Bauer, H. H./Sauer, N. E./Becker, S. (2003); Dholakia, U. M. (1997).

685 Cf. Stone, R. N./Gronhaug, K. (1993), pp. 49; Venkatraman, M. P. (1991), p. 60.

tionnaires.[686] On the other hand, it is not possible to transfer all scale items from related constructs of interest (e.g. involvement) as they usually integrate items belonging to competing constructs of this dissertation's conceptualization approach. Related to interest, involvement is a broader concept which also includes aspects such as risk, product usage, product knowledge, and product experience.[687] Hence, a substantial overlap with determinants used here to predict CEPO exists. Still, taken separately, some items seem to be apt to satisfactorily capture interest in the CEP category. Selected items shown in Table 23 were taken from the mentioned consumer involvement and product category involvement studies.

SPSS notation	Item
int_ce	I am generally interested in CEP devices
in_inno	I find it exciting to inform myself about novelties in the area of CEP

Table 23: Operationalization of interest (int)

5.2.3.7 Psychological Risk

The operationalization of the psychological risk orientates towards its conceptualization made earlier (s. Ch. 3.5.2.2, pp. 103 and Ch. 4.2.3.2, pp. 139). Principally the perceived discomfort before, during and after the CEP purchase in the form of lost time has been discussed there. As a consequence, the psychological risk will be measured by taking into account four aspects: the fear of loss of time needed for information gathering, the discomfort of going into details of such a purchase decision, the fear to choose the wrong technology standard, and the fear that after the purchase a lot of time resources need to be invested to properly operate that innovative CEP. The SPSS notation of the items and the final items representing psychological risk are given below in Table 24.

SPSS notation	Item
psy_info	If I bought an innovative CEP, I would worry that I need to spend a lot of time gathering the most important information before the purchase
psy_detail	It would be unpleasant for me to have to deal with the details of an innovative CEP purchase
psy_wrong	If I bought an innovative CEP, I would worry making a wrong decision regarding the different technology standards

686 For example Holak (1988) measures product category interest across four product categories, among which are several consumer electronic innovations, by use of 49 statements which he does not report. Cf. Holak, S. L. (1988).

687 See for example Jain, K./Srinivasan, N. (1990); Mittal, B./Lee, M.-S. (1988); Kapferer, J.-N./Laurent, G. (1985); Bloch, P. H. (1981).

psy_opera	If I bought an innovative CEP, I would worry about needing a lot of time in order to learn how to operate the device

Table 24: Operationalization of the psychological risk (r_psy)

5.2.3.8 Self-Efficacy

In the discussion of self-efficacy it was found that the level of a person's self-efficacy has an influence on the perceived risk during the decision process of an innovative CEP. This means that self-efficacy may automatically be evidenced in the perceived risk determinant to a certain amount. Still, this variable will be integrated in the questionnaire. The item which captures the self-efficacy of an individual to operate an innovative CEP even when no one else is around to help was adapted from Brown/Venkatesh (2005).[688] The second item picks up another aspect of that construct by considering also the situation during the purchase decision process (s. Table 25).

SPSS notation	Item
s_eff_oper	I would feel comfortable to operate an innovative CEP alone
s_eff_buy	I am very confident in purchasing an innovative CEP, even without support during the purchase decision process

Table 25: Operationalization of self-efficacy (s_eff)

5.2.3.9 Social Pressure

Due to the fact that CEP are visible to others, the social pressure determinant has also been chosen as one of the adoption variables that may influence innovative CEP adoption. It has been decided to use two wordings – a "positive" and a "negative" one. While the latter asks about the worry of not being perceived as innovative by friends when no innovative CEP are possessed, the other item is aimed at finding out in how far the possession of innovative CEP is rewarded by other individuals (s. Table 26).[689] Both aspects, however, are expected to be sources of social pressure.

SPSS notation	Item
pres_unprog	If I did *not* buy any innovative CEP, I would worry some of my friends thinking I am not up-to-date
pres_prest	I think that the possession of innovative CEP increases the esteem among friends or other persons.

Table 26: Operationalization of social pressure (soc_pres)

688 Cf. Brown, S. A./Venkatesh, V. (2005), p. 425.

689 Cf. Stone, R. N./Gronhaug, K. (1993), p. 49.

5.2.4 Multi-Item Measures: Weighted Indices

So far, the chosen variables have been operationalized via single- and multi-item scales. Next to this way of operationalization, the different aspects of a concept may be measured with an index and then weighted depending on the importance of the different aspects. Such weighted indices combine a set of variables into a composite measure and,thus, are more precise than one-item or unweighted multi-item measures.[690] The following three determinants – "CEPO," "social integration," and "knowledge compatibility" – are operationalized via weighted indices. After having captured the aspects of a concept with the help of several items and weighting them, the single weighted *items* are summed up to build the specific *index*.[691] In general, such an index may take the following form:

$$\text{Index} = w_1 * \text{indicator}_1 + w_2 * \text{indicator}_2 + w_3 * \text{indicator}_3 + _{\dots} + w_i * \text{indicator}_i$$

where $w_1,\dots w_i$ are the specific weights of the indicators $1,\dots i$
and i = is the specific indicator i

Weights of the indices may be derived either from theoretical considerations or empirically by expert ratings.[692]

5.2.4.1 Weighted CEPO Index (CEPO_w)

Earlier it has been realized that "innovations" per se are too broad as a research object to be analyzed in this context and that a product category focus on CEP is a reasonable step (s. Ch. 4.1, pp. 117). Putting the ownership of CEP at the center of interest in order to determine the innovation behavior of a person has several advantages over alternative approaches which are outlined briefly below.

McDonald/Corkindale/Sharp (2003) critically remark that studies exclusively focus on time of innovation adoption and therewith neglect the oftentimes large group of non-adopters.[693] In the same vein, Gatignon/Robertson (1989) address this problem in their study about organizational innovation behavior by remarking that "some level of information is lost by grouping all non-adopters as a single category. Organizations may still be in the process of evaluating whether or not they should adopt. Therefore it would be erroneous to classify them as having made a decision to reject the innovation."[694] Analogous to the case of or-

690 Cf. Zikmund, W. G. (2003), pp. 299.
691 Cf. Schnell, R./Hill, P. B./Esser, E. (2008), p. 173.
692 Cf. Schnell, R./Hill, P. B./Esser, E. (2008), p 173.
693 Cf. McDonald, H./Corkindale, D./Sharp, B. (2003), pp. 86.
694 Gatignon, H./Robertson, T. S. (1989) p. 42.

ganizations, it is important to be more specific in studies which are dealing with individual adoption processes. By simply classifying individuals into two groups – namely adopters of the novel product on the one hand and non-adopters of that innovation on the other hand – valuable information is lost. Additionally this approach demands the respondents to exactly recall the time of innovation purchase.[695] Hence, the dichotomous categorization of respondents exhibits major drawbacks.

More accurate measures could determine the precise position in the adoption process.[696] Non-adopters would then be further split into people who are not aware; people who are aware, but not interested; people who are aware and interested; people who are in the stage (of the adoption process) of evaluating the innovation; and people who decided in favor of an innovation, but have not purchased it yet. Those potential adopters who are at a later stage (closer to adoption) could be characterized as more innovative than individuals at earlier stages. However, as no standardized operationalization for the results of the innovation adoption process stages "acceptance," "adoption," "rejection," "resistance," and "postponement" exists, their measurement is complicated.[697]

In addition to these approaches, a few studies use purchase intention in order to measure innovativeness.[698] But this technique equally poses problems. Ownership in a category reduces the willingness to purchase an innovation in the same category (inventory effect). Also a new product which disposes of similar functionalities as the existing product will have a negative impact on its purchase intention.[699] An innovativeness definition drawing solely on purchase intention will, hence, be an inadequate measure.

Neither the above mentioned approaches, nor the focus on a single product (low level of abstraction), nor broadening the view to several product categories[700] seems rewarding. Instead, the most appropriate solution appears to be choosing the middle level of generalization of the extremes and measure innovative behavior by "determining how many of a prespecified list of new products a particular individual has purchased at the time of the survey."[701]

---

695 Cf. Goldsmith, R. E./Hofacker, C. F. (1991), p. 210.

696 See for example Gilly, M. C./Zeithaml, V. A. (1985), p. 355.

697 Cf. Nabih, M. I./Bloem, S. G./Poiesz, T. B. (1997), pp. 191.

698 Cf. Midgley, D. F./Dowling, G. R. (1978), p. 231.

699 Cf. Hardie, B. G./Robertson, T. S./Ross, Jr. W. T. (1996), p. 359.

700 This method is also referred to as cross-sectional method. Cf. Robertson, T. S./Myers, J. H. (1969), p. 165.

701 Midgley, D. F./Dowling, G. R. (1978), pp. 230.

Despite Goldsmith's/Hofacker's (1991) outline that most of the criticisms related to the time-of-adoption method would also apply for methods using prespecified lists of innovations, this approach offers some advantages with regard to the other proposals.[702] While the effect of situational influences may be considerable if surveys analyze adoption of a single product, these effects are controlled when analyzing a large battery of products within a particular product category.[703] That also means that if a consumer has not yet adopted an innovation although the purchase is planned for the near future, he or she will not be falsely categorized as "non-innovative." The error is compensated to some degree by the individual's actual possession of other innovations. The rationale behind these assumptions is given by Robertson (1971) who states that "it may be expected that at a given point in time, individuals who are typically among the first to adopt will also have adopted more items."[704]

To be more precise in the ownership measurement, it is not sufficient to merely take the total number of *innovative* CEP into consideration. First, non-owners of innovative CEP would be neglected due to exclusion. Second, it is not easy to determine a clear cut-off between innovative and non-innovative CEP (a classification attempt has been made in Table 27), and, third, even if innovative CEP were easy to identify, they vary in their *level* of newness. Putting all innovations on the same level would lead to an erroneous picture of the consumers' innovation behavior. Thus, a straightforward solution is to weight all CEP specified in the questionnaire by their novelty level. By integrating all CEP (innovative and non-innovative) and weighting them, a more sophisticated CEPO index (CEPO_w) will be created which facilitates the differentiation of persons showing a higher or lower level of innovative behavior regarding the CEP category.

Innovative CEP	Non-innovative CEP
Flatscreen TV (plasma-/LCD-TV)	Flatscreen (computer)
Home cinema system (Dolby Digital, DTS, 5.1,..)	Computer
Navigation system	Laptop/notebook
DVD-recorder (stand-alone device)	Beamer
Hard disk recorder (digital videorecorder)	DVD-player (stand-alone device – no computer drive)
Blu-ray disc player (stand-alone device)	VCR (video cassette recorder)
HD-DVD player (stand-alone device)	Multi-function video device (DVD-player + video cassette recorder)

702 Cf. Goldsmith, R. E./Hofacker, C. F. (1991), p. 210.

703 Cf. Midgley, D. F./Dowling, G. R. (1978), p. 231.

704 Robertson, T. S. (1971), p. 90.

Digital camera (with at least 8 megapixels)	Mobile phone with camera
Digital photo frame	Mobile phone without camera
Camcorder with hard-drive or memory chip	Digital camera (with less than 8 megapixels)
iPhone	Other camcorder
iPod	MP3 player (stand-alone device)
Mobile organizer	
Mobile videoplayer	
Wii or Sony Playstation 3	
Nintendo DS or Sony Playstation Portable	

Table 27: Classification attempt of innovative and non-innovative CEP

As proposed by Schnell/Hill/Esser (2008) the product-specific innovativeness scores that are used as weighting factors will be empirically gathered.[705] They are calculated from an additional survey with the respondents rating all products listed in the questionnaire on a five point scale ranging from 1 ("not at all innovative") to 5 ("very innovative").[706] This procedure has a major advantage compared to applying the date of invention or introduction into the market of the specific innovation in relation to other products: Though more subjective, the chosen measurement procedure includes more than just a time dimension. Implicitly, next to the mere temporal newness aspect (date of invention or market introduction), the level of technological newness (disruptive vs. continuous innovation), the level of diffusion, price, and complexity are taken into account implicitly. Therewith the operationalization is oriented towards the innovation view IV of Kotzbauer's categorization. Instead of the individual novelty perception of each respondent the degree of market novelty will be determined. This allows comparisons between groups of individuals as it is intended in this study.

As shown in Table 28, Blu-ray disc players are rated most innovative and are very close to the theoretical maximum score of 5 with 4.27. HD-DVD players (3.73), iPhones (3.6), digital video-recorders (3.6), digital cameras with more than 8 megapixels (3.13), and digital photo frames (3.13) can also be accounted to the group of highly innovative products. As expected, devices such as mobile phones without camera (1.67), regular camcorders (1.67), followed by computers (1.40) and VCRs (1.13) can be found at the very bottom of the table.

---

705 Cf. Schnell, R./Hill, P. B./Esser, E. (2008), p. 173.

706 For the specification of weights 15 persons of a doctoral seminar have been inquired via a paper and pencil questionnaire to rate the innovativeness of the CEP as objectively as possible.

These devices have low innovativeness scores and, hence, add relatively less to the CEPO_w index.

Rank	Innovativeness score (inno_score)	CEP
1.	4.27	Blu-ray disc player (stand-alone device)
2.	3.73	HD-DVD player (stand-alone device)
3.	3.60	iPhone
	3.60	Hard disk recorder (digital videorecorder)
5.	3.13	Digital camera (with at least 8 megapixels)
	3.13	Digital photo frame
7.	3.07	Nintendo DS or Sony Playstation Portable
8.	3.00	Wii or Sony Playstation 3
	3.00	Flatscreen TV (plasma-/LCD-TV)
10.	2.93	Navigation system
	2.93	Camcorder with hard-drive or memory chip
12.	2.73	Home cinema system (Dolby Digital, DTS, 5.1,..)
13.	2.67	Mobile organizer
	2.67	iPod
15.	2.60	Mobile videoplayer
	2.60	DVD-recorder (stand-alone device)
17.	2.53	Flatscreen (computer)
18.	2.40	Beamer
19.	2.33	Mobile phone with camera
20.	2.27	MP3 player (stand-alone device)
	2.27	Digital camera (with at less than 8 megapixels)
22.	2.13	Laptop/notebook
23.	2.00	DVD-player (stand-alone device – no computer drive)
24.	1.93	Multi-function video device (DVD-player + VCR)
25.	1.67	Mobile phone without camera
	1.67	Other camcorder
27.	1.40	Computer
28.	1.13	VCR (video cassette recorder)

Table 28: Innovativeness scores of the CEP listed in the questionnaire

The CEPO_w score is now calculated by summing up the number of owned technical devices of the list multiplied with the corresponding innovativeness score (row two of Table 28). Personal weighted CEPO score of person i:

$$\text{CEPO\_w}_i = \sum_{p=1}^{p=28} \text{CEPO}_p * \text{inno\_score}_p$$

with $\text{CEPO}_p = 1$ (if person i possesses p) or 0 (if person i does not possess p)
inno_score_p = the specific weight/innovativeness score of product p
p = the specific product (ranges from 1 to 28)

In consequence, the CEPO_w index ranges from 0 points (does not own any device) to the theoretical maximum of 73.39 points (possesses all listed devices).

5.2.4.2 Social Assistance Index

A person's social integration and the WOM-communication resulting from these interactions with people from the social surrounding is assumed to promote the awareness of novelties and reception of relevant information. These interactions increase the likelihood that potential adopters obtain support throughout the specific adoption decision process. The operationalization of social networks or social integration often focuses on simply counting the number of contact persons from primary or secondary groups. While this method is eligible to demonstrate in how far social networks decrease in size when people age (e.g. due to retirement), it disregards the frequency of contacts with persons from different membership groups and particularly the quality of the lost and of the remaining contacts (as well as the relevance for CEP innovation adoption).

Elderly people often compensate the loss of weak-tie contacts with a stronger focus on few but close relationships with persons from their social surrounding. Thus, the extent of WOM- or, more generally, interpersonal communication can remain practically unchanged even though there are fewer persons older individuals have contact with. It therefore seems more meaningful to directly measure the frequency of interpersonal communication (quantitative aspect) and, additionally, integrate some sort of measure which captures the qualitative aspects of the specific communication channels.[707] A simple approach to integrate qualitative aspects is to directly ask the respondents how helpful the specific reference groups is in getting advice regarding the purchase of an innovative CEP. Thus, the social assistance index will consist of two components, (a) social integration and (b) helpfulness of different person groups.

The categorization of contacts will be based on the differentiation made before which divided social membership groups into four rough categories (s. Table 29). Secondary formal groups are omitted because they only encompass weak-tie relationships with individuals who do not see each-other very often and

---

707 In a study about the adoption of technological innovations by elderly consumers, Mathur (1999) operationalized social assistance from family members by asking the respondents to name one younger family member other than the spouse and rate the amount of present assistance on a four-point scale ("no assistance at all" to "a lot of assistance"). Cf. Mathur, A. (1999), p. 27. This operationalization neglects the multidimensional facets of social assistance and, for this reason, is not pursuit here.

hence are less important for marketers.[708] Finally, the frequency of communication contacts and potential assistance received through WOM activities are measured for nine sub-groups of contacts. The contact frequency is divided up into six categories: nearly every day (5), 2-3 times per week (4), several times per month (3), several times per year (2), less often (1), no contact/does not apply (0). The second part of the index is operationalized by assessing how helpful an individual perceives these person groups regarding the purchase of a new CEP (to cover the qualitative aspect of social assistance). The helpfulness possesses five realizations on a Likert type scale with the extremes "not at all helpful" (1) and "very helpful" (5). Also the middle realization is described as "partly" and it is given the possibility to indicate that a certain person group "does not apply." The qualitative aspect serves as weight to create the final index.

SPSS notation	Item
si	How often do you usually have personal or telephone contact with the following persons?
	Parents / siblings / children / grand-children / other relatives / friends / neighbors / former colleagues or fellow students / friends from clubs or political parties or leisure time
hp	How helpful would the following persons be if you bought an innovative CEP?
	Parents / siblings / children / grand-children / other relatives / friends / neighbors / former colleagues or fellow students / friends from clubs or political parties or leisure time

Table 29: Questionnaire items for social assistance index (soc_assist) calculation

Accordingly, the overall social assistance index of an individual is the sum of the extent of social interaction and degree of expected support by all contact person groups. In other words, the extent of "social assistance" (soc_assist) of an individual will be calculated by multiplying the contact frequency (si, [1-5;0]) and the contact quality (hp, [1-5;0]) of the corresponding reference group, dividing the result by the maximum score of five, and summing up the scores of all nine reference groups. Personal social assistance score of person k:

$$\text{soc\_assist}_k = \sum_{i=1}^{i=9} \frac{si_i * hp_i}{5}$$

with si_i = the extent of social integration with person group i (contact frequency)
hp_i = the degree of help of person group i (contact quality)

708 Cf. Assael, H. (1992), pp. 405.

i = the specific person group (ranges from 1 to 9)

The social assistance index ranges from a theoretical minimum 0 (would not receive any helpful assistance) to a theoretical maximum of 45 (would receive full assistance by all person groups). Higher numbers, thus, represent better social assistance regarding the purchase decision of new CEP. It is important to notice that even high levels of contact frequency may result in low total scores of relevant social assistance if the respective reference group's help potential is low as regards to CEP. Furthermore, low scores may be achieved even if numerous reference groups have high levels of support potential. This is the case when the contact frequency is very low.

5.2.4.3 Knowledge Compatibility Index

It was decided to operationalize knowledge compatibility with the help of a weighted index. This is legitimated with the fact that surveying alone the extent to which the respondents believe that knowledge about older CEP is compatible with CEP innovations (meaning that it can easily be transferred) is not sufficient. Apart from the mere "perceived compatibility" aspect, the amount of knowledge about older CEP will be taken into consideration as well, resulting in an index with the following form (s. also Table 30). Personal knowledge compatibility score of person i:

$$compa_i = \frac{compa\_tran_i * compa\_know_i}{5}$$

SPSS notation	Item
compa_tran	I find that knowledge about *older* CEP is transferable to innovative devices
compa_know	I find that my knowledge about *older* CEP is good

Table 30: Questionnaire items for knowledge compatibility index (compa) calculation

5.3 Questionnaire Development and Data Collection Strategy

As the formerly abstract concepts are operationalized, the final questionnaire structure and design will be developed (s. Ch. 5.3.2). Next, the targeted respondents are defined (Ch. 5.3.3) and the data collection procedure will be elucidated (s. Ch. 5.3.4). Prior to defining the final layouts and the study sample, some findings of particularities in surveying older consumers are highlighted which should further help to accurately design the questionnaire (Ch. 5.3.1).

5.3.1 Particularities in Surveying Older Respondents

Although older consumers are more and more often the target group in surveys, hardly any particularities are reported that may exist compared to surveying younger individuals. Still, it is very likely that younger and older age groups differ in their willingness to accept interview requests, or in understanding the interview questions. Phillips/Sternthal (1977) for example refer to several studies which found that the older persons' non-response rates to survey questionnaires are fairly high – they are more likely to refuse participation than younger individuals.[709] Moreover, the authors speak of difficulties in understanding and following instructions.[710] Further investigation revealed that, if available, elderly respondents make use of the "no opinion" answering option more frequently than younger respondents. Gergen/Back (1966) trace this phenomenon back to a decreased attention towards the survey questions and recommend to avoid finer gradations and intermediate response categories.[711]

For these reasons it seems even more important to strictly follow the general credo to keep the questionnaire short, simple and clear.[712] The introduction should create interest for the study's topic and point out that, even though it is focused on CEP innovations, no specialist knowledge is necessary to complete the questionnaire. To avoid false responses due to misapprehension regarding CEP innovations a trade-off needs to be done regarding the integration of a "no opinion" alternative versus the use of a forced rating scales where such an option is not given. Although Malhotra (2004) reports from the risk that undecided respondents mark, if available, the middle scale position which then would distort the measure of central tendency and variance, a "no opinion" answering option will not be provided here.[713] This measure can be justified with the finding that older respondents over-utilize the "no opinion" category either to avoid complex decision making, or due to the elderlies' cautiousness in answering questions about something they do not *fully* comprehend.[714]

The necessity to label all points of the used scale rather than merely labeling the endpoints of the scale in age related surveys is pointed out by

709 Cf. Phillips, L. W./Sternthal, B. (1977), pp. 452. However, there are also studies with high rates of questionnaire completion (92%) by elderly individuals which has been mainly owed to their personal interest in the research issue. Cf. Zelan, J. (1985).

710 Cf. Phillips, L. W./Sternthal, B. (1977), p. 453.

711 Cf. Gergen, K. J./Back, K. W. (1966), pp. 388.

712 Cf. Kühn, K./Porst, R. (1999), p. 23.

713 Cf. Malhotra, N. K. (2004), p. 263.

714 Cf. Jayanti, R. K./McManamon, M. K./Whipple, T. W. (2004), pp. 265.

Jayanti/McManamon/Whipple (2004) and explained with the tendency of older aged individuals to show an extremity response bias, meaning that they prefer extreme answers when only endpoints are labeled verbally.[715] To guarantee a certain level of simplicity not all points are labeled but three – both extreme ones and the middle score. The mode of self-administered questionnaires was preferred. They dispose of the advantage that there is less time pressure to answer the questions compared to face-to-face or telephone interviews. Furthermore, older persons might be unwilling to let interviewers enter their apartment, or feel straitened when an unexpected phone call reaches them during which their personal opinions are polled. These constraints can be hold responsible for the stated reluctance of the elderly to participate in face-to-face and telephone interviews.[716]

5.3.2 Questionnaire Structure

The final questionnaire consists of the items that were discussed in the previous chapters. As the items were predominantly taken from English research papers, they were translated into German and handed out to several younger and older individuals to check wording and clarity of the questions. The same was done with the German translations – they were translated back into English and cross-checked with the initial English questionnaire items. After minor changes in the wording of several items, the final layout and structure was developed by dividing the questionnaire into several sections. The questionnaire starts with a short description of the research background and with emphasizing that no special knowledge is necessary to respond to the survey questions. Next to affirming that the anonymity of the respondents is preserved, it is mentioned that only one questionnaire per household is allowed to avoid duplication in the collection of statistical information. The background information ends with the note that for each completed questionnaire a certain amount (2 €) will be donated to a charitable association[717] that can be selected on the last page of the questionnaire.[718] Before the research variables can be evaluated by the respondents, some directions are given on how to fill out the questionnaire. Respondents should spontaneously answer the questions and express what is closest to their opinion and

715 Cf. Ibid., pp. 267.

716 See the cited literature in Kühn, K./Porst, R. (1999), p. 13 and p. 25.

717 Three options were offered: "SOS Kinderdörfer," "Ärzte ohne Grenzen," and "Deutscher Tierschutzbund e.V.". The money was donated after finalization of the study.

718 This kind of incentive also seems to increase the response rate. Zikmund (2003) mentions that response rates are higher when the offered monetary incentives are donated to a charity instead of directly to the respondents. Cf. Zikmund, W. G. (2003), p. 217.

that there are neither "correct" nor "wrong" answers. Three CEP are listed to give an example for innovative CEP in order to avoid any misunderstandings of what is meant by an "innovative" CEP. The constructs of interest were put in the following order: The questionnaire starts with the items belonging to the relative advantage constructs, followed by product- and in-store complexity. Then in a larger battery of items, the financial risk, performance risk, social pressure, psychological risk, self-efficacy and knowledge compatibility as well as the general CEP assessment variables are inquired. After the interest construct and the situational variables, the items belonging to the social assistance index and the list of CEP are integrated, followed by a list with information sources which are eventually used prior to the purchase of a new CEP. The last page of the questionnaire deals with the sociodemographic variables and includes the questions regarding the perceived financial situation. The survey closes with the announced possibility to select among three organizations to address the donation.

All adoption determinants which have been operationalized by multiple items are measured with Likert scales. Likert scales are the most widespread scaling technique used in empirical social research.[719] All items are measured on a five point (pt.) scale, except for financial situation which is measured on a seven pt. scale. Though seven pt. scales are more precise in measuring a concept as e.g. "perceived risk," the preliminary wording test has shown that elderly encounter more problems when they are asked to answer questions with a higher number of answer categories. The statistical disadvantage of smaller variance in the answers is, partly, balanced by utilizing several questions per determinant – that means using multi-item measures. In those cases, five categories have been reported to be sufficient.[720] The financial situation measure is more differentiated by giving the respondent seven answer categories, as the financial situation is operationalized by a single item only. While the five pt. scale items start with 1 = "strongly disagree"[721] and end with 5 = "strongly agree,"[722] the middle point 3 is labeled verbally with "partly."[723] The financial situation is the only exception as *all* scale points ranging from 1 to 7 are labeled verbally starting with 1 (very negative) and finishing with 7 (very positive). The social assistance, innovativeness and information source selection are all surveyed by a list where the respondent should tick the appropriate answer (s. Appendix 4). Simple fill-in

719 Cf. Schnell, R./Hill, P. B./Esser, E. (2008), pp. 187.

720 Cf. Malhotra, N. K. (2004), p. 262.

721 In German: "stimme gar nicht zu."

722 In German: "stimme voll zu."

723 In German: "teils/teils."

questions have been used to ask for the respondents' chronological and cognitive age, and in order to determine the household size.

The length of the questionnaire has been reduced to a minimum. This step accommodates not only elderly respondents who may be overstrained otherwise. Shorter questionnaires will be more attractive for younger people or employed persons, who have time restrictions and would be reluctant to participate at a lengthily survey. Many facets of the adoption theory are analyzed in this study yet only two or three items can be used for measuring each determinant in order to keep the questionnaire to an acceptable length.

5.3.3 Study Sample

So far, no further differentiation of individuals has been made except for talking about younger, middle-aged, and older consumers. All individuals who are older than 18 years of age and keep their own household are allowed to participate at the survey. The younger age group as well as the middle aged consumers will contain students. As they represent a special group of individuals, the student group needs to be discussed a bit more in detail here.

In many respects students differ from non-students at comparable age. For example they dispose over significantly less financial resources and have heterogeneous working situations (part-time working vs. non-working). Thus, they can supposedly express their innovativeness only partially through the purchase of innovative CEP which is, however, crucial to this analysis. As a researcher one would now be tempted to find statistical proof for that type of presumption by testing for significant mean differences of the student and non-student subsample. If no significant mean differences existed, students would behave comparable to non-students of the same age and thus could be kept in the dataset. The mean difference analysis must integrate the age variable, though, in order to avoid age effects related to the specific item. The obvious method to choose would be the analysis of covariance (ANCOVA) which is generally able to eliminate confounds.[724] Yet, one important assumption of ANCOVA cannot be met – the independence of the covariate (age) and the treatment effect (student vs. non-student).[725] Even when taking a sub-sample of this study with respondents younger than 28, the data shows a strong correlation between age and treatment effect (the younger the respondent, the more likely he or she will be a student). Students are more likely to be younger and, thus, a remarkable amount

724 Cf. Fields, A. P. (2009), p. 397; Herrmann, A./Landwehr, J. R. (2008), pp. 603.

725 Cf. Fields, A. P. (2009), pp. 397.

of variance would spuriously be deleted by adding age as a covariate in ANCOVA. It is unfortunate that there is no analytic method available that can accomplish a "control" for those group differences on the covariate.[726] In consequence, it is not possible to prove that the student group shows the same characteristics as non-students. As there is the danger that they may distort the picture of subsequent analyses, students that answered the questionnaire are excluded from further analyses.

5.3.4 Data Collection Procedure

Though it has been tried to reduce the length of the questionnaire to a minimum, the questionnaire consists of seven pages including the introducing page. Due to the still relatively lengthy questionnaire and the explicit consideration of older respondents, refusal was feared to be high. Therefore, neither telephone interviews nor on-the-street interviews seemed appropriate. Although internet surveys represent an increasingly widespread method of primary data collection (e.g. easy to set up, fast, economic, and no data entry necessary – data is directly available for statistical software), the idea of *exclusively* using the internet method needed to be dropped. The sample would contain a bias in disfavor of older individuals. While internet use among younger people is widespread (e.g. 94.8% of the 20-29 year old persons), only 26.4% of the individuals who are 60 years of age and older have access to the internet and use it at least from time to time.[727] For these reasons, next to an online survey[728] a self-administered questionnaire was developed. In order to increase the response quote and enhance participation, instead a of regular, impersonal mail survey, the drop off method was used and easily available persons where inquired. These samples are also referred to as "convenience samples."[729] Personal contact with the potential respondents has been regarded as helpful in order to point out the research goals and the importance of participating in the survey.[730] In the following cities the paper and pencil survey has been distributed: Bonn, Mannheim, Breisach (Rhine), and Arnsheim (in order of number of respondents). Due to organizational reasons, the inquiry was split into two waves, with the first one taking place from April 18th until May 25th 2008 and the second wave only several

726 Cf. Miller, G. A./Chapman, J. P. (2001), p. 40.

727 The mentioned percentages refer to internet usage in 2008 in Germany. Cf. ARD/ZDF-Onlinestudien (1998–2010) cited in van Eimeren, B./Frees, B. (2010).

728 Unipark EFS Survey Reporting (Globalpark) Version 1.5.

729 Cf. Zikmund, W. G. (2003), pp. 380.

730 Further advantages of the selected survey method have been described earlier (see pp. 178) Ch. 5.3.1.

months later from October, 16th until November 29th in 2008 (details about the number of participants are presented later). To reach larger groups of participants, apart from dropping off questionnaires at random households in the mentioned cities, sport courses, choirs etc. were visited and distributed questionnaires were collected three to seven days later when the course or event took place again.

5.4 Statistical Data Analysis Methods

This chapter provides an overview of the statistical research methods that are applied in Chapter 5.6 and Chapter 5.7 and necessary to answer the concrete research hypotheses as well as the exploratory research questions. The first subchapter deals with the Exploratory Factor Analysis, a statistical method used to select adequate items from the operationalization process in order to obtain the final scales (Ch. 5.4.1). Next, the Analysis of Variance is explained which allows to make statements about eventual differences in the general CEPO assessment of the three age groups (Ch. 5.4.2). Finally, the idea of the Regression Analysis is described (Ch. 5.4.3). With this latter method not only hypotheses about the impact of age on the adoption determinants but also the impact of the adoption determinants on CEPO can be evaluated.

5.4.1 Exploratory Factor Analysis

5.4.1.1 Theoretical Background

The final scales are developed by use of the Factor Analysis (FA) method – more specifically by the Exploratory Factor Analysis (EFA). FA "is a generic name given to a class of techniques whose purpose often consists of data reduction and summarization. Used in this way, the objective is to represent a set of observed variables (or persons or occasions) in terms of a smaller number of hypothetical, underlying, and unknown dimensions called factors."[731] That means that FAs find the fewest number of factors represented by items and assign items to factors.[732] Contrary to Confirmatory Factor Analysis (CFA), no concrete knowledge about the underlying factor structure in EFA exists. Instead, EFA is a method to *detect* structures in the dataset and thus to find out which factor the individual items belong to.[733] Although some assumptions have been made about the affiliation of items to specific theoretical constructs during the opera-

731 Smith, S. M./Albaum, G. S. (2005), p. 708.

732 Cf. Gorsuch, R. L. (2003), p. 143.

733 Cf. Backhaus, K., et al. (2008), pp. 380.

tionalization process, it is still unclear, whether the affiliation can also be confirmed by empirical data as no pre-existing scales could be employed. Furthermore, it is not clear whether all nine multi-item constructs derived from theory correspond to distinct factors in reality – in other words whether the respondents distinguish between them. So, another premise for EFA is given, as the final number of factors is not known a priori.

There are two main methods of extracting factors from a pool of items: Principal Component Analysis (PCA) and Principal Axis Factoring (PAF). In their article about the different choices in EFA, Costello/Osborne (2005) recommend to use the PAF method instead of PCA.[734] While component analysis is a mere data reduction method, axis factoring distinguishes between shared, unique and error variance. The latter method only considers the shared variance of the variables in its solution and avoids inflated values of variance that can arise under certain circumstances.[735] Another reason why PAF is preferred is its superiority in situations when data is not multivariate normally distributed which is the case in this study and will be shown later in Chapter 5.6.1.2 (pp. 227).[736] The rotation which helps to make the output interpretable will be conducted by applying the oblique rotation method "Direct Oblimin." While its counterpart of orthogonal rotation methods is widely accepted – as its results are very easy to interpret – they are obtained under the unrealistic assumption that factors are uncorrelated. In contrast to that, oblique methods better portray reality of human behavior by allowing factors to correlate.[737] The degree to which factors are allowed to correlate can be determined by the delta value. While larger positive delta values mean that the factors are also allowed to correlate to a larger extent, large negative delta values limit the correlation to a minimum.[738]

After the extraction of factors, the communality of each item needs to be calculated. Communalities specify the amount of variance of each item that can be explained by all factors.[739] Communality values range from a minimum of 0 (the extracted factors do not explain any variance of the specific item) to a maximum of 1 (the extracted factors explain the entire variance of the specific item).[740] Consequently, variables with low communalities need to be eliminated and should not be considered in the further scale development process.

---

734 Cf. Costello, A. B./Osborne, J. W. (2005), p. 2.
735 Cf. Gorsuch, R. L. (1997), pp. 542.
736 Cf. Costello, A. B./Osborne, J. W. (2005), p. 2.
737 Cf. Ibid., p. 3.
738 Cf. Pett, M. A./Lackey, N. R./Sullivan, J. J. (2003), pp. 155.
739 Cf. Martens, J. (2003), p. 225.
740 Cf. Pett, M. A./Lackey, N. R./Sullivan, J. J. (2003), p. 88.

On the other hand, a measure called "eigenvalue" that "represents the amount of variance in all of the items that can be explained by a given principal component or factor" also exists.[741] In order to determine the number of factors to be extracted, the most widely used criterion is if eigenvalues are greater than one (also called "Kaiser criterion").[742] However, this has been criticized not to be the optimum procedure. Bühner (2011) for example mentions that extraction does not need to strictly follow the Kaiser criterion and that extracting more factors should be preferred over extracting too few factors.[743] In the same vein, Field (2009) argues against orientating towards the eigenvalue greater than one rule. He suggests referring to the Kaiser criterion only in a first step for an initial analysis by allowing other numbers of extracted factors to be appropriate as well.[744]

5.4.1.2 Assumptions

Prior to performing the EFA, several assumptions need to be met:[745] (a) appropriateness of the selected method "EFA," (b) adequate number of items, (c) multivariate normality, (d) unbiased intercorrelations, (e) substantial intercorrelations between items of the item pool, (f) sufficiently large sample.

(a) Appropriateness of the Selected Method "EFA"
Bühner (2011) points to the fact that oftentimes not FA but component analysis or cluster analysis are the correct methods of choice – namely when the mere description of variables or the reduction of dimensions is the research goal.[746]However, when scales are developed the correlations of the items should be explained through underlying factors and, thus, EFA is the appropriate method.

(b) Adequate Number of Items
Furthermore, Bühner (2011) recommends employing at least four items per factor in an EFA, especially when broader concepts are measured.[747] Even before

---

741 Pett, M. A./Lackey, N. R./Sullivan, J. J. (2003), p. 91.

742 Cf. Gorsuch, R. L. (2003), p. 157.

743 Cf. Bühner, M. (2011), p. 349.

744 Cf. Field, A. P. (2009), p. 652; see also Backhaus (2008) for different rules of extraction. Cf. Backhaus, K. et al. (2008), p. 369.

745 Cf. Bühner, M. (2011), pp. 342; Field, A. P. (2009), pp. 645. Whether or not a satisfactory level of item reliability can be achieved, will be examined later (s. Ch. 5.6.3, pp. 233).

746 Cf. Bühner, M. (2011), p. 343.

747 Cf. Bühner, M. (2011), p. 344.

executing the EFA-routine later, during which further items may need to be omitted, this requirement cannot be met in the present study. Most theoretical constructs that will be analyzed are represented by two or three items only. However, this drawback is a result of the large number of determinants that are considered as potential influene factors on CEPO – with this step the questionnaire length was kept at a manageable level for the respondents. In addition, broader concepts were split into conceptually narrower entities. For example "complexity" was subdivided into "product complexity" and "in-store complexity" which may relativize the recommendation of minimum items per factor.

(c) Multivariate Normality

Multivariate normality represents another requirement in order to perform EFAs. Multivariate normality assumes that "all variables and all linear combinations of those variables are normally distributed."[748] Therefore, if the single variables violate the assumption of normality, then multivariate normality is not given either. To check the normality of the single variables the Kolmogorov-Smirnov-Test and the Shapiro-Wilk-Test can be employed, which are both available in SPSS.[749] As will be shown later, the variables used here are not multivariate-normally distributed. In this case it is possible to transform the data before continuing the analysis. However, it is refrained from transforming the data as this would severely complicate the interpretation.[750] Instead, in these situations the PAF method to extract factors should be used as recommended by Costello/Osborne (2005).[751] This test is robust against deviations from multivariate normality.

(d) Unbiased Intercorrelations

Prior to examining the assumption of substantial correlations in more detail, it is important to note that the correlations are unbiased since outliers may have a considerable impact on correlation strength.[752] They can influence correlations positively or negatively. The logic behind the exclusion of outliers is the assumption that these outliers do not belong to the underlying population.[753] The items destined for the EFA possess few realizations as they all range from a minimum of 1 to a maximum of 5 in (steps of) whole numbers. Therefore, an univariate outlier analysis is inappropriate. More relevant than extreme values of

748 Fidell, L. S./Tabachnick, B. G. (2003), p. 128.

749 For further details about normality tests see the literature indicated in Bortz, J./Schuster, C. (2010), p. 145.

750 Cf. Tabachnick, B. G./Fidell, L. S. (2010), p. 86.

751 Cf. Costello, A. B./Osborne, J. W. (2005), p. 2.

752 Cf. Bühner, M. (2011), p. 343.

753 Cf. Tabachnick, B. G./Fidell, L. S. (2010), p. 73.

single items are, for this reason, multivariate outliers. These are cases that have an unusual combination of values for a number of variables. Thus, the Mahalanobis distance is drawn on, a statistic that calculates in how far a single case differs from the average of all other cases.[754] Fidell/Tabachnick (2003) suggest a probability estimate of $p<0.001$ for the Mahalanobis distance to unveil outlier cases.[755] As a precaution the so detected cases will be deleted from the dataset before performing the EFA. Later, when the final scales have been determined and regression analysis is performed, another multivariate outlier analysis will be conducted including the calculation of the respective leverages to determine the final outliers (s. Ch. 5.7.3.1, pp. 241).

(e) Substantial Intercorrelations

A necessary prerequisite for performing an EFA is that the selected items show a certain level of correlation. If low correlations between the indicators were stated, it would be very unlikely that those items are represented by the same factor.[756] Several approaches exist which allow to inspect on two levels whether the selected items are suited to build the theoretically derived scales. First, in conformance with Backhaus et al. (2008), the entire item pool is tested for its general suitability to perform an EFA with the help of three tests: (aa) the Bartlett's test of sphericity, (bb) the anti image covariance matrix, and (cc) the Kaiser-Meyer-Olkin measure of sampling adequacy.[757] Second, the single correlations are observed in more detail. Each item should correlate with items of the same construct (factor) within a corridor that will be defined later. Items that do not fall into this corridor or correlate with other than the expected items are candidates for exclusion.

(1) General Suitability of the Item Pool

All three statistical tests that investigate the general suitability of the item pool can be performed with the SPSS software package.

(aa) Bartlett's Test of Sphericity

With Bartlett's test of sphericity one can find out whether the sample originates from a population in which the items are uncorrelated.[758] Small Bartlett scores

754 Cf. Schendera, C. F. (2007), p. 185. For the equation see Bortz, J./Schuster, C. (2010), p. 457.

755 Cf. Fidell, L. S./Tabachnick, B. G. (2003), p. 130.

756 Cf. Backhaus, K. et al. (2008), pp. 330.

757 Cf. Backhaus, K. et al. (2008), pp. 335.

758 Cf. Dziuban, C. D./Shirkey, E. C. (1974), pp. 358.

are an indicator that the null hypothesis of uncorrelated items should be accepted.[759]

(bb) Anti-Image Covariance Matrix

Guttmann (1953) shaped the concept of the "anti-image covariance matrix."[760] Items are eligible for EFA if the anti-image is small. This implies that non-diagonal elements should be close to zero.[761] Dziuban/Shirkey (1974) introduced a critical value of 25%, bearing on the ratio of non-diagonal elements and the total amount of elements of the anti-image covariance matrix, that are allowed to be larger than $|0.09|$.[762] If it more than 25% of the non-diagonal elements are larger than $|0.09|$, the item pool in question will not qualify for EFA.

(cc) Kaiser-Meyer-Olkin Measure of Sampling Adequacy

Additionally, using the "anti-image correlation matrix" the Kaiser-Meyer-Olkin (KMO)-test investigates whether the data truly reflect correlations between items and factors. The test statistic, also called "measure of sampling adequacy" (MSA), allows judging the item pool as a whole, as well as the single items.[763] Kaiser/Rice (1974) defined cut-off values and established a classification which allows judging the quality of the item pool and the single items, respectively (s. Table 31).[764]

MSA score	Classification
≥ 0.9	marvelous
≥ 0.8	meritorious
≥ 0.7	middling
≥ 0.6	mediocre
≥ 0.5	miserable
< 0.5	unacceptable

Table 31: Classification of MSA scores
(Source: **Kaiser, H. F./Rice, J.** (1974), p. 112.)

(2) Examination of Single Items

With the KMO measure an examination of both, the dataset as a whole and the single items can be performed. Now, the single items and their correlations are examined in closer detail. As mentioned already, substantial correlations be-

759 Cf. Pett, M. A./Lackey, N. R./Sullivan, J. J. (2003), pp. 73.

760 Cf. Guttman, L. (1953).

761 Cf. Backhaus, K. et al. (2008), pp. 335; Pett, M. A./Lackey, N. R./Sullivan, J. J. (2003), pp. 80.

762 Cf. Dziuban, C. D./Shirkey, E. C. (1974), p. 359.

763 Cf. Backhaus, K. et al. (2008), pp. 336.

764 Cf. Kaiser, H. F./Rice, J. (1974), p. 112.

tween the selected items are a necessary prerequisite of performing an EFA[765].[766] But, on the other hand, the correlations between the items must not be too high in order to avoid multicollinearity.[767] In practical terms this means that the correlation between items that belong to the same factor should lie between 0.3 and 0.8.[768] Items outside this range will consequently not be considered in the following scale development. Furthermore, by displaying the correlation matrix, it can be checked whether the items highly correlate with the "correct" other items belonging to the same factor.

(f) Sufficiently Large Sample
Many rules have been proposed by researchers what the minimum sample size should be in order to conduct an EFA.[769] According to Guadagnoli and Velicer (1988) studies including factors with few loadings should be based on at least 300 cases.[770] Similarly, MacCallum et al. (1999) suggest interpreting studies with small communalities and many factors if at least 500 individuals are in the dataset.[771]

5.4.1.3 Scale Reliability

The operationalization of theoretical constructs based on an in-depth literature review. Numerous approaches of other studies have been screened and existing scale items have been adapted to the current research topic. Adaptations were necessary as in the research area dealing with age-specific differences of adoption determinants and innovative CEP no pre-existing scales were available. As a consequence, the selected items which are believed to form the respective scales are tested regarding their reliability and validity and, if necessary, purified by eliminating inadequate items.

Reliability refers to the extent to which measures are unaffected by random error. While systematic sources of error do not have an unfavorable impact on reliability, random errors produce inconsistency and, hence, cause lower reliabil-

765 This is evident as those variables that should be represented by a factor/latent variable stand for the same underlying concept and thus should strongly correlate.

766 Cf. Backhaus, K. et al. (2008), p. 330.

767 In the same way as in regression analysis, multicollinearity causes problems in seperating the effects of the single variables in factor analysis and may be responsible for unstable coefficients. Cf. Anderson, D. R./Sweeney, D. J./Williams, T. A. (2009), pp. 644; Allen, M. P. (1997), pp. 179.

768 Cf. Field, A. P. (2009), p. 647.

769 Cf. Field, A. P. (2009), p. 645.

770 Cf. Guadagnoli, E./Velicer, W. F. (1988), p. 274.

771 Cf. MacCallum, R. C. et al. (1999), p. 96.

ities. Validity assessments, in contrast, analyze whether the selected scales really measure what is intended to be measured.[772] Therewith, validity includes the aforementioned concept of reliability. Next to random errors, validity also considers systematic errors. A perfectly valid measurement, hence, requires the absence of all kinds of measurement errors.[773] Different forms of validity exist: construct validity, content validity, and criterion validity.[774] However, here, the focus lies on assessing the scales' reliabilities.[775]

Three approaches for assessing the reliability of a scale exist: the test-retest, alternative-forms, and internal consistency method.[776] The latter has received much attention because the test-retest and alternative-form methods are difficult to implement and complicate research designs.[777] Therefore, internal consistency is the most widespread method to examine reliability and will, thus, be used in this study, too.[778]

The simplest form of internal consistency measurement is the split-half reliability. With this method, scale items are split into two halves and correlation between those two item groups are computed.[779] High correlation scores stand for high internal consistency. Cronbach's alpha[780] offers a solution in order to be independent of how scales are spilt in two halves. Cronbach's alpha "is the average of all possible split-half coefficients resulting from different ways of splitting the scale items."[781] Values of Cronbach's alpha range from 0 to 1 whereas their level depends on the number of employed scale items. Scales with more items automatically tend to produce higher Cronbach's alpha scores. Furthermore, established scales that have already been validated before should have higher scores. Nunnally (1994) proposes 0.7 as minimum level for Cronbach's alpha for established scales.[782] New scales and scales with few items, however,

772 Cf. Zikmund, W. G. (2002), pp. 301.

773 Cf. Malhotra, N. K. (2004), p. 269.

774 For a detailed overview see for example Hartig, J./Frey, A./Jude, N. (2007), pp. 140.

775 Due to the study's exploratory character, it is difficult to determine the validity of the developed scales with the present dataset. In particular assessing construct validity, which consists of convergent and discriminant validity (Cf. Schmidt, M./Hollensen, S. (2006), p. 130), are beyond the scope of this work and would, thus, be an interesting starting point for future research.

776 Cf. Malhotra, N. K. (2004), p. 267.

777 Cf. Hildebrandt, L. (1998), p. 88.

778 Cf. Gerbing, D. W./Anderson, J. C. (1988), p. 190.

779 Cf. Schermelleh-Engel, K./Werner, C. (2007), pp. 122.

780 In the literature, also the name "coefficient alpha" is used.

781 Malhotra, N. K. (2004), p. 268.

782 Cf. Nunnally, J. C./Bernstein, I. H. (1994), pp. 264.

dispose of a satisfactorily large Cronbach's alpha when values of about 0.5 - 0.6 are achieved.[783]

Two further test statistics available in the standard SPSS output help to purify scales. If the "corrected item-total correlation"-value[784] should fall under the threshold of 0.5 the item is a candidate for exclusion.[785] In combination with the "Cronbach's alpha if item deleted"-statistic all items will be inspected. If item deletion should largely improve Cronbach's alpha of the scale, that item will be deleted.

5.4.2 Analysis of Variance

The first research question (RQ 1) deals with the consumers' general assessment of innovative CEP. It is interesting to find out whether differences between the "young," "middle-aged," and the "old" age group turn up. As the research question includes investigating differences between several (age-) groups, applying the analysis of variance (ANOVA) method will be necessary.

5.4.2.1 Theoretical Background

The research question requires performing a one-factor univariate ANOVA for each aspect of innovative CEP that will be assessed. The univariate ANOVA is a special case of the univariate general linear model (GLM) which uncovers the effects of a categorical variable on *one* dependent interval-scaled variable.[786] In the prevailing study the categorical independent variable, also called factor, is age group membership with three factor levels (also called treatments):[787] "young," "middle-aged," and "old." As there are more than two factor levels in the independent variable, a single t-test cannot be used to test for statistical significance.[788] Here, the dependent interval-scaled variables are the single aspects of innovative CEP (e.g. the perceived technical superiority, or the importance of compatibility with older CEP).

---

783 Cf. Lienert, G. A./Raatz, U. (1998), p. 14 and p. 269.

784 The "corrected item-total correlation" can be defined as the correlation of this item with the sum of all items which are assigned to the same factor. Cf. Pett, M. A./Lackey, N. R./Sullivan, J. J. (2003), p. 194.

785 Cf. Zaichkowsky, J. L. (1985), p. 343.

786 Cf. Berekoven, L./Eckert, W./Ellenrieder, P. (2009), pp. 204; Herrmann, A./Landwehr, J. R. (2008), p. 585; Homburg, C./Krohmer, H. (2003), pp. 288.

787 Cf. Malhotra, N. K. (2004), p. 469.

788 Cf. Field, A. P. (2009), pp. 348.

The ANOVA puts into relation the amount of systematic variance (between group variance) and random error (within-group variance). The first kind of variance represents the mean deviation of the different age groups regarding the dependent variable – it corresponds to the variance that can be explained by the independent variable. The latter sort of variance represents the variation that cannot be explained by age group membership. Instead this variance is owed to individual differences of the group members.[789]

Comparing the systematic variance with the error variance allows obtaining information about whether or not group means significantly differ from each other. The key statistic in ANOVA, therefore, is an F-test which measures whether the relation of systematic variance compared to the error variance occurred randomly or not.[790] Thus, large F-values stand for a high proportion of explained variance so that it is likely that significant mean differences exist. As the null hypothesis implies that the independent variable (age groups) does not have an influence on the dependent variable (the different aspects of CEP assessment), this hypothesis needs to be rejected if the empirical F-value is larger than the theoretical F-value.[791] Analogous, if the empirical p-value of the test statistic is smaller than the defined error probability, the null hypothesis needs to be rejected and mean differences do exist.

By use of the F-test it can be determined whether any group mean differs significantly from the other group means regarding the same dependent variable. However, it does not help in specifying exactly which one of the means differ. Therefore, post-hoc tests have been established. With their help it is possible to analyze differences in the means of all existing group combinations. For that reason, theses tests are also called multiple comparison tests. A large variety of tests and procedures exists.[792] Their choice depends in some part on whether or not the assumptions of ANOVA are met.

5.4.2.2 Assumptions

Prior to performing the ANOVA with the actual data set, the conditions that need to be fulfilled should be listed:[793]

- Independence of errors

789 Cf. Herrmann, A./Landwehr, J. R. (2008), p. 583.

790 Cf. Zikmund, W. G. (2003), pp. 530.

791 Cf. Backhaus, K. et al. (2008), pp. 159.

792 For an overview of multiple comparison tests see Kirk, R. E. (1995).

793 Cf. Bortz, J./Schuster, C. (2010), pp. 212; Böhler, H. (2004), p. 196.

- Homogeneity of error variances
- Normally distributed errors

The first assumption requires that "errors of measurement are independent of one another – that the size of the error for one case is unrelated to the size of the error for cases near in time, space or whatever."[794] Scariano/Davenport (1987) report serious consequences when this condition is not met.[795] Error terms may increase and misleading results may be produced.[796] However, when respondents are randomly assigned to factor levels or different samples are investigated, then the requirement of independent errors is not violated.[797]

Next, the error variances should be homogenous – meaning the dependent variable should display the same variance in each factor level of the independent variable. This is necessary because the F-ratio denominator includes the average of group variances as well as the different group sizes. Large differences in the group variances would make the F-ratio a poor summary measure resulting in an increased chance of wrongly rejecting the null hypothesis that group means are equal.[798] The homogeneity of variances can either be examined with the Levene or Bartlett-Test. From these two, the first is preferred as the Bartlett-Test is sensitive in cases where the normality assumption is violated, while the Levene-test is not.[799] The Levene-test tests the hypothesis that the variances in the groups are equal. If Levene's test is significant at $p \leq 0.05$, the variances significantly differ and the homogeneity requirement is not fulfilled.[800] Testing for homogeneous variances is only needful if the samples in the factor levels are unequal in size and small. In situations where sample sizes (n) are equal (ratio of smallest to largest n of the factor levels is smaller than 4:1) and relatively large, ANOVA is robust in respect of a violation of this assumption.[801] Although it has just been reported that F-tests are robust if the homogeneity requirement is not met, deviations from homogeneity still are important to consider. This is because homogeneity influences which post-hoc test is selected (in order to determine which group means differ). If the test confirms homogeneity, the Scheffé post-hoc test will be employed. By contrast, it will be referred to the Tamhane-T2 Test in

794 Fidell, L. S./Tabachnick, B. G. (2003), p. 125.
795 Cf. Scariano, S. M./Davenport, J. M. (1987), pp. 123.
796 Cf. Fidell, L. S./Tabachnick, B. G. (2003), p. 125.
797 Cf. Bortz, J./Schuster, C. (2010), pp. 212.
798 Cf. Garson, G. D. (2011b).
799 Cf. Bortz, J./Schuster, C. (2010), p. 129; Bortz, J. (2005), pp. 285.
800 Cf. Janssen, J./Laatz, W. (2007), pp. 245.
801 Cf. Field, A. P. (2009), p. 360; Fidell, L. S./Tabachnick, B. G. (2003), p. 124.

those cases where the Levene-test detects heterogeneity, as this post-hoc test is tailored for data where no equal variances are on hand.[802]

Finally, as third assumption, the errors should be normally distributed within the groups (factor levels). As the originally measured values are related to the errors by adding the constant term of the group mean, the normality assumption also holds true for the measured values for each separate factor level.[803] When group sizes are equal, however, ANOVA is also robust in terms of a violation of this assumption.[804] Similar to Bender/Ziegler/Lange (2007) who recommend a sample size of ten times the number of factor levels minus one, Bortz/Schuster (2010) state that samples with less than ten cases per factor level are seen as critical.[805]

5.4.3 Regression Analysis

Hypotheses H_8 - H_{19} deal with the impact that age has on the adoption determinants. Together with the hypotheses H_{20} - H_{31} dealing with the impact of the adoption determinants on CEPO_w they have in common that they make statements about dependency relationships. A statistical technique that investigates this sort of dependency relationships is the regression analysis. While the first block of hypotheses needs to be assessed by a series of bivariate dependency analyses with single regressions (age → single adoption determinants), the latter hypotheses ask for a multiple regression analysis (set of adoption determinants → CEPO_w). As the regression analysis is used to test the study's central hypotheses, this method is explained in more detail compared to the other research methods.

5.4.3.1 Theoretical Background

The purpose of the *simple* (bivariate) linear regression is to estimate the linear relationship of a metric scaled independent variable (also called exogenous, predictor, or explanatory variable) with a metric scaled dependent variable (also called endogenous, criterion, or response variable).[806] In the *multivariate* case (multiple regression analysis), not one but several independent variables – which

802 Cf. Janssen, J./Laatz, W. (2007), pp. 367; Nachtigall, C./Wirtz, M. (2006), pp. 192.

803 Cf. Bortz, J./Schuster, C. (2010), pp. 213.

804 Cf. Field, A. P. (2009), pp. 359 and the cited literature.

805 Cf. Bortz, J./Schuster, C. (2010), p. 214; Bender, R./Ziegler, A./Lange, S. (2007), p. e60.

806 Cf. Skiera, B./Albers, S. (2008), p. 469, Malhotra, N. K. (2004), p. 503.

have an influence on the dependent variable – are examined simultaneously.[807] Although metric scaled data for the dependent as well as the independent variables are often reported to be a necessary premise, qualitative or nominal scaled variables can also be integrated into dummy-variables via transformation.[808] Due to its versatility and flexible use, the multiple regression analysis is a widespread data analytic system which has found its way into many research areas.[809] One major benefit of multiple regression analysis is that it offers ways of preventing spurious relationships from distorting the relationships of the other variables.[810]

The equation of the *simple* (bivariate) regression line is:

$$Y = \alpha + \beta \cdot X + \varepsilon$$

where:
Y is the dependent variable
X is the independent variable
α is the intercept of the regression line
β is the slope coefficient of the regression line
ε is the error term

As it is the case in the example diagram (Figure 27) usually most data points do not lie on the regression line. In fact, unless there is a perfect correlation between the variables, no regression line exists that captures all observations.[811] The deviations from the regression line are, thus, errors in predicting Y from X. The errors of the single observations (e_i) – also called residuals – can be expressed as the difference between the actual, observed value of the dependent variable (y_i) and value predicted by the regression line ($\hat{y}_i$).[812] The objective is to propose a regression line that reduces the total error in predicting the y_i from the x_i. This is done by using the least-square method. Technically, the least-square method minimizes the sum of squared deviations[813] of the observed y-values from the predicted $\hat{y}$-values:[814]

807 Cf. Homburg, C. et al. (2008), p. 166.
808 For an example see Backhaus, K. et al. (2008), pp. 12.
809 Cf. Cohen, J. et al. (2003), p. 1.
810 Cf. Ibid., pp. 158.
811 Cf. Zikmund, W. G. (2003), p. 557.
812 The errors which cannot be explained by the regression line are also called "residuals."
813 Through squaring, larger deviations are weighted stronger than smaller deviations that may have occurred by random measurement error.
814 Cf. Bortz, J./Schuster, C. (2010), p. 186.

$$\sum_{i=1}^{n} e_i^2 = \sum_{i=1}^{n} (y_i - \hat{y}_i)^2 = \sum_{i=1}^{n} (y_i - \alpha - \beta \cdot x_i)^2 \rightarrow \min.$$

where:
e_i is the unexplained error/residual of observation i
y_i is the observed value of the dependent variable Y for x_i
$\hat{y}_i$ is the estimated value of the dependent variable Y for x_i
i is the number of the particular observation (with i = 1, 2, 3,…, n)
n is the number of observations

The application of the least-square method results in the final prediction equation:

$$\hat{Y} = a + b \cdot X$$

where:
$\hat{Y}$ is the estimation of the dependent variable Y, and
a and b are estimators for α and β
In simple terms: $Y = \hat{Y} + \varepsilon$

The *multiple* regression analysis is an extension of the just mentioned simple regression. Instead of solely observing the effect of one independent variable, in the multiple cases, the effect of several independents on the dependent variable is analyzed. The resulting b_J are named partial regression coefficients. As usually, the independent variables correlate with each other, each partial regression coefficient is calculated by holding constant the effect of the other independent variables.[815]

The equation of multiple regressions is as follows:816
$$Y = \alpha + \beta_1 \cdot X_1 + \beta_2 \cdot X_2 + \ldots + \beta_j \cdot X_j \ldots + \ldots \beta_J \cdot X_J + \varepsilon$$

where:
J is the number of independent variables

Here also, the least-square approach is the method to find the best regression line:

$$\sum_{i=1}^{n} e_i^2 = \sum_{i=1}^{n} [y_i - (\alpha + \beta_1 \cdot x_{1i} + \beta_2 \cdot x_{2i} + \beta_j \cdot x_{ji} + \ldots + \beta_J \cdot x_{Ji}]^2 \rightarrow \min.$$

815 Cf. Malhotra, N. K. (2004), pp. 513.
816 See for example ibid., p. 512; Aaker, D. A./Kumar, V./Day, G. S. (2001), p. 514.

Analogously, the use of the least-square method results in the final prediction equation of the form:

$$\hat{Y} = a + b_1 \cdot X_1 + b_2 \cdot X_2 + \ldots + b_j \cdot X_j \ldots + \ldots b_J \cdot X_J$$

where:
b_j is the partial regression coefficient of the independent variable j

The just introduced equations allow determining a regression line that fits best to the existing data. Yet, none of the mentioned equations allowed drawing any conclusions about the appropriateness of the proposed model so far – neither regarding the overall performance of the regression function, nor the individual predictors. Therefore, first, the global assessment of the regression equation is examined with aid of the:[817]

- Coefficient of multiple determination (R^2)
- F-test

In the second step, the regression coefficients are assessed. On the one hand it is tested whether the single variables of the regression model are significant in explaining the dependent variable Y, and, on the other hand, the strength of their influence is determined with the calculation of the:

- t-value
- Coefficient Beta

Coefficient of multiple determination

The coefficient of multiple determination (R^2) measures the "goodness of fit" of the regression function compared to the empirically gathered data. Though the best regression line can be calculated as described above, it can still be a poor fit to the data. The result of the regression analysis needs to be put into relation to something else. Therefore, the regression line is assessed against the most basic model available, which simply is the mean of the dependent variable ("$\bar{y}$").[818]

As mentioned above, the least-square method is used to minimize the sum of squared deviations. In this process, the total sum of squares (SS_t) – which stands for the total variation in the dataset – can be partitioned into the regression sum of squares (SS_r) and the unexplained error sum of squares (SS_e).[819] In other words the total deviation from the group mean (y_i - "$\bar{y}$") can be partitioned into a deviation which is explained by the regression ($\hat{y}_i$ - "$\bar{y}$") and in a deviation

817 Cf. Backhaus, K. et al. (2008), pp. 67.

818 Cf. Field, A. P. (2009), pp. 201.

819 Cf. Malhotra, N. K. (2004), pp. 515.

which is not explained by the regression (y_i - $\hat{y}_i$).[820] As illustrated in Figure 27 the deviations from the mean are of interest.

$SS_t = SS_r + SS_e$

where

$$SS_t = \sum_{i=1}^{n} (y_i - \text{“}\bar{y}\text{”})^2$$

$$SS_r = \sum_{i=1}^{n} (\hat{y}_i - \text{“}\bar{y}\text{”})^2$$

$$SS_e = \sum_{i=1}^{n} (y_i - \hat{y}_i)^2$$

On the basis of this decomposition of variation, the coefficient of multiple determination can easily be measured. It is obtained by putting into relation the explained variance and the total variance:

$R^2 = SS_r / SS_t = 1 - SS_e / SS_t$

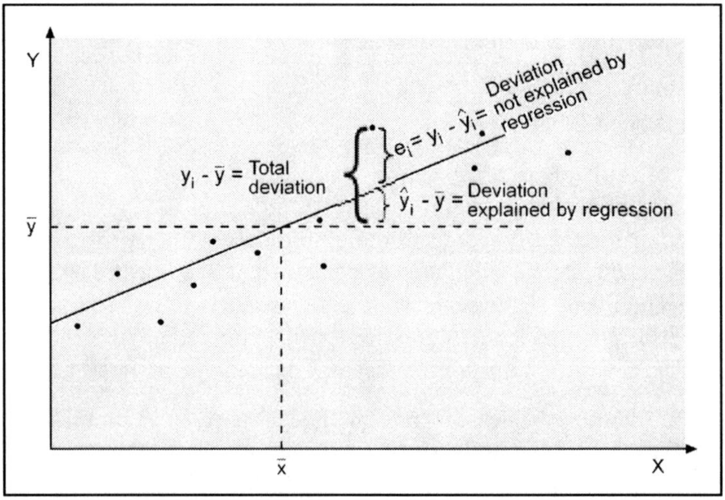

Figure 27: Scatter diagram of explained and unexplained variation (Source: **Brosius, F.** (2006), p. 549; **Zikmund, W. G.** (2003), p. 562.)

This coefficient is standardized and can range from a minimum of 0 (the regression function does not explain any variation from the mean) to a maximum of 1

820 Cf. Backhaus, K. et al. (2008), pp. 67.

(the regression function does explain the complete variation from the mean). The latter case implies that the estimated values of the independent variables correspond to the observed values.[821] By extracting the root of R^2, the multiple correlation coefficient (R) can be obtained. R can be viewed as a simple correlation coefficient which indicates the strength of association between Y and $\hat{Y}$.[822] With the computation of the multiple correlation coefficient, the adequacy of the independent variables selected for the model can be assessed. In behavioral sciences, R^2- and R-values are usually considerably lower than in textbook examples. In the case of *simple* regressions (coefficient of determination, r^2) and *simple* correlations (correlation coefficient, r) Cohen (1988) talks about small effect sizes when $r<0.30$. According to the author, medium effect sizes range from 0.30 to 0.50 and large effects start at $r\geq 0.30$.[823] An even more elaborated assessment is given by Diaz-Bone (2006) with the following classification:[824]

$0.00 \leq r < 0.05$ no correlation
$0.05 \leq r < 0.20$ low correlation
$0.20 \leq r < 0.50$ moderate correlation
$0.50 \leq r < 0.70$ strong correlation
$0.70 \leq r \leq 1.00$ very strong correlation

With the above mentioned relationship between simple regression and multiple regression, R and R^2 can now also be assessed. Among others, the interpretation of the coefficient of determination remains the same regardless of the number of variables in the regression equation.[825] In consequence, the listed r-values simply need to be squared in order to determine the regression models' goodness of fit and its corresponding R^2-values. While, for example, a medium R^2 value ranges from 0.04 to 0.25, a large R^2 value varies between 0.25 and 0.49. This is in line with other researchers who discussed the size of desirable R^2-values in behavioral studies.[826]

F-Test

The multiple correlation coefficient expresses how well the regression function fits the observed data. The nature of the coefficient is, thus, merely descriptive so far and refers to the specific sample. To generalize the obtained results for the

821 Cf. Skiera, B./Albers, S. (2008), pp. 472.
822 Cf. Malhotra, N. K. (2004), pp. 515.
823 Cf. Cohen, J. (1988), pp 78 and pp. 412.
824 Cf. Diaz-Bone, R. (2006), p. 91.
825 Cf. Black, K. (2010), p. 540; Allen, M. P. (1997), p. 92.
826 Cf. Pagano, R. R. (2009), p. 130; Fromm, S. (2008), p. 357; Cohen, J. et al. (2003), p. 92.

total population, an F-test is conducted. The F-test tests the null hypothesis that the coefficients of multiple determination in the total population (R^2_{pop}) are zero.[827]

H_0: $R^2_{pop} = 0$ ($\beta_1 = \beta_2 = \beta_3 = \ldots = \beta_J = 0$)
H_1: $R^2_{pop} \neq 0$ (not all β are equal to zero)

If the F-test is significant (i.e. the SPSS output for the F-test is $p<0.05$), the null hypothesis needs to be rejected. That means that one or more partial regression coefficients in the total population have values other than zero. In order to determine which of the β are nonzero further tests are necessary. For this reason, t-tests are performed for each β.

t-values
In the same way as in the F-test, the t-test null hypothesis presumes that the partial regression coefficients in the total population (t_{pop}) are zero.[828]

H_0: $\beta_j = 0$
H_1: $\beta_j \neq 0$

Analogous, the null hypothesis needs to be rejected if the results of the t-tests become significant (i.e. the SPSS output for the t-test is $p<0.05$). That implies that the regression coefficient at hand is different from zero and has an effect on Y in the total population X_j.

Coefficient beta
The coefficient b is referred to as the nonstandardized regression coefficient. This implies that the b coefficients cannot be compared to each other when the independent variables use different units of measurement.[829] To guarantee a meaningful comparability, the estimated regression coefficients need to be standardized. This is accomplished by transforming the initial variables into new variables which have a mean of 0 and a variance of 1.[830] Those standardized regression coefficients are called coefficient beta or beta weights ($\hat{b}$)[831] and allow to assess the relative impact of the different independent variables within an

827 Cf. Backhaus, K. et al. (2008), pp. 71.
828 Cf. Skiera, B./Albers, S. (2008), pp. 474.
829 Cf. Cohen, J. et al. (2003), p. 154.
830 Cf. Malhotra, N. K. (2004), p. 507; Allen, M. P. (1997), pp. 46.
831 To simplify, the standardized coefficient $\hat{b}$ is called "Beta" in the following.

equation.[832] With the help of the beta coefficients, the order of importance of the predictors can be determined.[833]

$\hat{b}_j = b_j \cdot \sigma_{Xj}/\sigma_Y$

where:
$\hat{b}_j$ is the standardized coefficient of b
σ_{Xj} is the standard deviation of the independent variable X_j
σ_Y is the standard deviation of the dependent variable Y

By use of the calculated Beta values, comparisons between all regression coefficients can easily be made. High absolute values indicate a stronger influence and vice versa. With the calculation of the Beta coefficients even the relative importance can be determined.[834]

Regression methods
The final estimation of the model's regression coefficients can be performed in multiple ways – depending on the sequence at which the independent variables are entered into the analysis. Using different methods which determine the sequence of integration, a variety of regression models can be constructed from the same set of variables. The decision criterion of integration or disintegration of an independent variable is usually its effect on the F-value discussed above or its strength of correlation with the dependent variable. The statistical software SPSS, offers the following methods to enter the independent variables:[835]

- Enter (Regression): A procedure for variable selection in which all variables in a block are entered in a single step.
- Stepwise: At each step, the independent variable not in the equation that has the smallest probability of F is entered if that probability is sufficiently small. Variables already in the regression equation are removed if their probability of F becomes sufficiently large. The method terminates when no more variables are eligible for inclusion or removal.
- Remove: A procedure for variable selection in which all variables in a block are removed in a single step.

832 Cf. Newton, R. R./Rudestam, K. E. (1999), pp. 267.
833 Unless the independent variables are highly correlated. Cf. Siegel, A. F. (2012), pp. 369; Keit, T. Z./Reynolds, C. R. (2003), p. 95.
834 Taking a regression example with two independent variables with the respective Betas of 2 (variable 1) and 2.4 (variable 2) means that variable 2 explains 1.2 times the amount of the variation in the dependent variable compared to the first variable. Cf. Brosius, F. (2006), pp. 569.
835 In the following see SPSS Inc. (2007), pp. 323 (quoted verbatim).

- Backward Elimination: A variable selection procedure in which all variables are entered into the equation and then sequentially removed. The variable with the smallest partial correlation with the dependent variable is considered first for removal. If it meets the criterion for elimination, it is removed. After the first variable is removed, the variable remaining in the equation with the smallest partial correlation is considered next. The procedure stops when there are no variables in the equation that satisfy the removal criteria.
- Forward Selection: A stepwise variable selection procedure in which variables are sequentially entered into the model. The first variable considered for entry into the equation is the one with the largest positive or negative correlation with the dependent variable. This variable is entered into the equation only if it satisfies the criterion for entry. If the first variable is entered, the independent variable not in the equation that has the largest partial correlation is considered next. The procedure stops when there are no variables that meet the entry criterion.

Though reported to be one of the popular methods,[836] the stepwise variable selection in particular, but also the other automated selection processes in general, have largely been criticized when used alone. The main critics focus on the fact that by use of those techniques the software instead of the researcher decides which variables are relevant for the final model. Especially in studies with a large number of potential independent variables and in research areas where little theory exists to guide the selection process, automated selection processes offered by statistical software "are a sore temptation"[837] – oftentimes used to avoid preliminary theoretical work.[838] Instead, the variable selection should be based on their specific importance and causal priority.[839] Yet, if contradictory results emerge regarding specific independent variables, researchers should not hesitate excluding the variables in questions, even though the explained variance of the model might be reduced thereby.[840]

5.4.3.2 Assumptions

So far, the focus has been on the method and possible results of the regression analysis. However, prior to actually performing the regression analysis, the un-

836 Cf. Bühl, A. (2008), p. 368.
837 Cohen, J. et al. (2003), p. 161.
838 Cf. Urban, D./Mayerl, J. (2006), p. 112.
839 Cf. Ibid., pp. 304; Cohen, J. et al. (2003), pp. 158.
840 Cf. Backhaus, K. et al. (2008), p. 101.

derlying assumptions of the multiple linear regression analysis need to be checked. This is important as violations of the assumptions may lead to biased estimations of the regression coefficients or a biased estimation of the standard errors of the regression coefficients.[841] In the first case, the estimated regression coefficients would not conform to the true values of the regression coefficients in the population. In the second case, if the estimation of the standard errors of the regression coefficients is biased, the estimated values of the regression coefficients would be correct, but hypothesis tests and confidence intervals, in turn, may not.

A number of requirements that the data must meet can be found in the literature:[842]

1) No outliers
2) Correct specification of the model
3) Mean of the residuals is zero
4) Mean independence
5) Homoscedasticity
6) No autocorrelation
7) No multicollinearity
8) Normally distributed residuals
9) Adequate sample size

1) Cases with unusual values for the variables may have a considerable influence on the results of regression analyses.[843] Different combinations of leverage and discrepancy effects may prevail in the data. While cases with high leverage are cases which are far from the mean (univariate situation) or centroid (multivariate situation), cases with high discrepancy are those with a large distance between the observed value and the value predicted by the linear regression model.[844] Influential cases, which are the main interest here, combine high leverage with high discrepancy. Cases that display these two characteristics most likely affect the regression coefficients.[845] The detection of unusual values is performed carefully in several steps.

841 Cf. Cohen, J. et al. (2003), p. 117.

842 Cf. Field, A. P. (2009), pp. 214; Backhaus, K. et al. (2008), pp. 80; Skiera, B./Albers, S. (2008), pp. 478; Janssen, J./Laatz, W. (2007), pp. 450; Cohen, J. et al. (2003), pp. 117.

843 Cf. Janssen, J./Laatz, W. (2007), p. 454.

844 Cf. Schendera, C. F. (2007), pp. 183.

845 Cf. Fox, J. (2008), pp. 21.

First, univariate outliers are diagnosed. Unlike the items used for factor analysis which only ranged from a minimum of 1 to a maximum of 5 in whole number steps, the final scales have more values a they consist of several items. So now, a univariate outlier analysis also makes sense. Among others, univariate outliers can be detected via boxplots, a graphical method to display data. The box, also called interquartile range, represents the 50% of the data which lie between the first and third quartile. The vertical lines – the "whiskers" – delimit an area which covers all data up to the 1.5-fold of the interquartile range. All cases with values outside this area are considered as outliers.[846]

Prior to finally excluding such cases with extreme values, it is checked whether they are also influential – meaning that the extreme values are able to alter the parameters of the model.[847] Following Fox (1991), the centered leverage values (CLV) will be plotted against the absolute studentized deleted residuals (sdr_abs) in order to find multivariate outliers which dispose of high leverage and discrepancy values.[848] Cases with a $CLV>2*(k + 1)/n$ (where n is the number of cases and k is the number of independent variables) and $sdr\_abs>2$ are seen as outliers.[849]

Supplementary, Cook's Distance (D) is a third measure taken into consideration. It evaluates the changes of the estimated regression parameters when deleting a given observation.[850] Cases with a $D>4/[n-k-1]$[851] dispose of high leverage and discrepancy – absolute values as for instance $D>1$ have been found too conservative and are not recommended.[852]

2) The correct specification of the independent variables is one of the central steps in regression analysis, but it is at the same time a difficult task.[853] The model should contain all relevant influence factors of Y. Usually it is not possible to include all relevant variables either because it is technically impossible, too complex, or not all relevant factors are known. The situation where relevant variables have been left out is called "underfitting" while "overfitting" refers to a model which consists of too many variables. Both situations may have an un-

846 Cf. Anderson, D. R./Sweeney, D. J./Williams, T. A. (2009), p. 106.
847 Cf. Field, A. P. (2009), pp. 217.
848 Cf. Fox, J. (1991), pp. 37.
849 Cf. Ibid., pp. 33.
850 Cf. Schendera, C. F. (2007), pp. 186; Fox, J. (1991), pp. 29.
851 "n" is the number of cases and "k" is the number of independent variables.
852 Cf. Fox, J. (1991), p. 34.
853 Cf. Cohen, J. et al. (2003), p. 119.

favorable effect on regression results.[854] As a consequence, variables should not be selected at random but through theoretical deduction. This helps to integrate important and exclude extraneous variables.[855]

Furthermore, the regression coefficients' signs should be checked.[856] A comparison between the theory induced relationship tendency of the regression coefficients and the calculated ones is easily feasible and can be a hint for misspecifications. If signs do not show he same tendency as proposed by the hypotheses (s. Table 32), researchers may be alerted. The concerned variables should be excluded from the model and reasons for the divergence should be discussed.

Independent variables	Hypothesized direction of signs
rel	+
cx_prod	-
cx_is	-
r_perf	-
r_fin	-
r_psy	-
soc_pres	+
int	+
s_eff	+
compa	+
fin_situ	+
soc_assist	+

Table 32: Hypothesized tendency of the relationship between the specific independent variables and the dependent variable (CEPO_w)

The regression equations mentioned earlier are all based upon *linear* relationships between independent and dependent variables. However, in many situations it is justified to assume nonlinear relationships. They can be accounted for by applying nonlinear transformations.[857] Problems occur when linearity is assumed while the variables are nonlinear. Such an incorrect model specification can produce misleading results.[858] Linearity can be detected by visualizing the relationship between each independent variable and the dependent variable in separate scatterplots and comparing the conformity of the linear relationship line with the Loess curve which follows the trend of the data.[859] Alternatively, the

854 Cf. Backhaus, K. et al. (2008), pp. 83.
855 Cf. Urban, D./Mayerl, J. (2006), pp. 17.
856 Cf. Skiera, B./Albers, S. (2008), p. 476.
857 See for example Cohen, J. et al. (2003), pp. 193.
858 Cf. Fox, J. (1991), pp. 53.
859 Cf. Cohen, J. et al. (2003), p. 125.

plots which will be utilized in the heteroscedasticity analysis (see point 5) can be used. These plots include the residuals as well as the predicted values. When linearity exists in the data, then the mean of the residuals should be zero regardless of the independent variable value (see also requirement 3). The Loess curve should, therefore, not deviate largely from the "zero-mean" line.[860]

3) Regression models assume that the mean of the residuals is zero – more specifically that the "expected value or mean of the errors of prediction in the population regression function is equal to zero."[861] This assumption is violated when the residuals contain systematic measurement errors.[862] Violations of this assumption cannot be tested because the expected value in question refers to residuals of the population model which is unknown.[863] However, if the regression model is specified correctly and contains all systematic influence factors of Y, the residuals (ε) contain only random errors. In this case deviations offset each other and assumption 3 can be accepted.[864] Furthermore, if the mean of the residuals is non-zero, this will only have an effect on the estimation of the relatively unimportant regression intercept, but not on the regression coefficients or standard errors.[865]

4) The mean independence requirement demands the independent variables to be unrelated to the residuals. The reasons for a correlation can be: omitted independent variables in the model, reverse causation (meaning Y has an effect on X), or measurement errors of the independent variables.[866] The problem of this assumption is that no statistical test exists to prove or refute mean independence. However, when data stem from randomized experiments, residuals should be unrelated to the independent variables.[867]

5) Homoscedasticity stands for the constant variance of residuals. This assumption assures that the residuals are dispersed randomly throughout the range of the estimated dependent variable.[868] The lack of homoscedasticity, called heteroscedasticity, may be caused by (a) a prevailing interaction effect between a

---

860 Cf. Field, A. P. (2009), pp. 247; Urban, D./Mayerl, J. (2006), pp. 205; Cohen, J. et al. (2003), pp. 125.

861 Allen, M. P. (1997), p. 182.

862 Cf. Backhaus, K. et al. (2008), p. 83.

863 Cf. Urban, D./Mayerl, J. (2006), p. 201.

864 Cf. Backhaus, K. et al. (2008), p. 83.

865 Cf. Allen, M. P. (1997), pp. 182.

866 Cf. Allison, P. D. (1999), pp. 124.

867 Cf. Ibid.

868 Cf. Field, A. P. (2009), p. 220; Skiera, B./Albers, S. (2008), p. 490.

measured independent variable and an unmeasured independent variable which is not considered in the model; or (b) the fact that some independent variables are skewed while others are not.[869] In the least-square procedure heteroscedasticity favors the prediction of cases with high variances in the optimization process so that inefficient results are obtained.[870] The consequence of heteroscedasticity is that the calculated standard errors will be erroneous which means that significance tests and confidence intervals will also be erroneous.[871]

The presence or absence of heteroscedasticity can, among others, be determined via a graphical test.[872] It is performed by plotting the standardized predicted values (named *ZPRED in SPSS) on the x-axis and the standardized residuals (named *ZRESID in SPSS) on the y-axis.[873] If the plot takes the form as depicted on upper graphic of Figure 28, there is homoscedasticity, while on the other hand, in the lower graphic this assumption needs to be rejected, and the data is heteroscedastic.[874] While the graphical analysis has been reported to be the faster and more reliable way to detect heteroscedasticity,[875] data not always take an ideal-typical shape as presented in the examples which can impede an easy interpretation. Far more serious is the fact that for multiple regressions simple graphical tests are no longer possible.[876] In these situations the Glejser-Test can for example be used which involves a regression of the non-standardized predicted values (independent variable) and the absolute non-standardized residuals (dependent variable). As the null hypothesis postulates constant variances of the residuals, a non-significant coefficient stands for homoscedasticity ($p>0,05$).[877]

---

869 Cf. Garson, G. D. (2011a).
870 Cf. Pindyck, R. S./Rubinfeld, D. L. (1998), p. 128.
871 Cf. Cohen, J. et al. (2003), p. 120.
872 For a variety of further tests check Cohen, J. et al. (2003), p. 133.
873 Cf. Field, A. P. (2009), pp. 247; Backhaus, K. et al. (2008), pp. 99.
874 For an intepretation of the residual plots see e.g. Cohen, J. et al. (2003), pp. 130.
875 Cf. Urban, D./Mayerl, J. (2006), p. 248.
876 Cf. von Auer, L. (2005) p. 364.
877 Cf. Glejser, H. (1969), pp. 316.

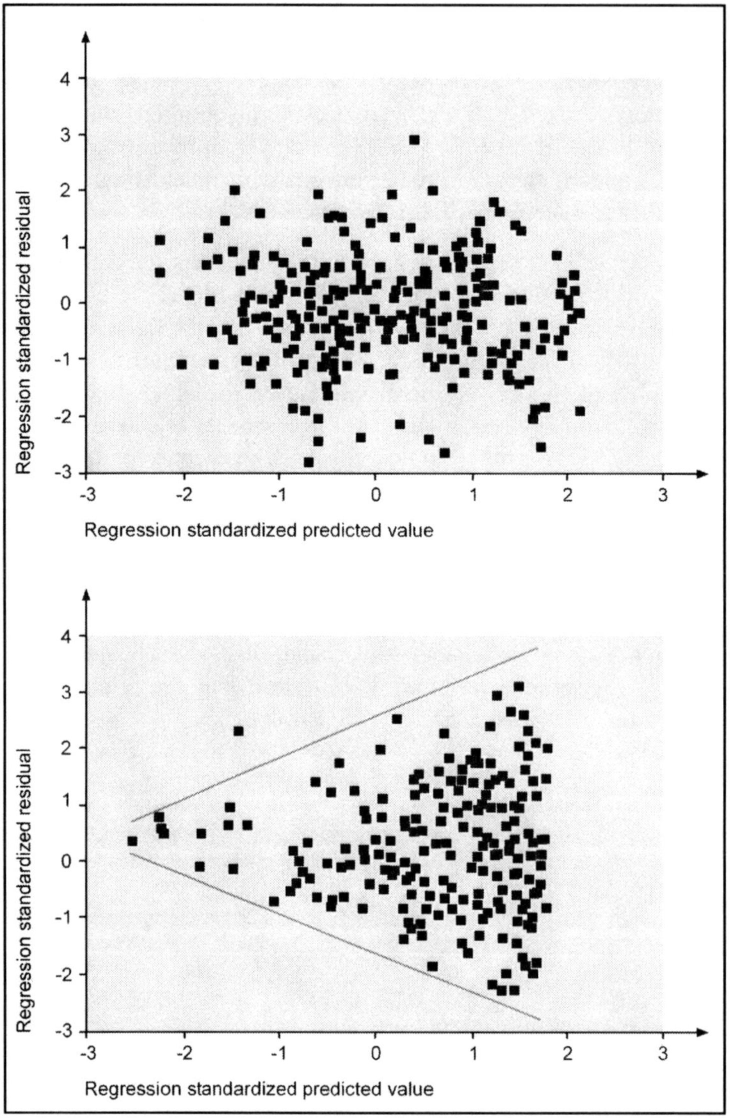

Figure 28: Graphical example of homoscedasticity and heteroscedasticity
(Source: E.g. according to **Field, A. P.** (2009), p. 248; **Backhaus, K. et al.**(2008), p. 85; **Bortz, J.** (2005), p. 217.)

If heteroscedasticity is detected, transformations may be helpful to bypass this problem.[878] It can, for example, be corrected by the weighted least-square (WLS) procedure. When estimating the parameters, this method gives less weight to those observations with large variances and thus contributes less to the fit of the regression line.[879] However, regression is relatively robust regarding the homoscedasticity assumption. Small to moderate violations of homoscedasticity only have a minor impact on estimating the regression equation.[880] Therefore, Fox (1997) recommends that only in cases of strong heteroscedasticity, this procedure should be applied.[881]

6) Autocorrelation prevails when the residuals of the observations are not independent of one another. In consequence, this is merely a problem in designs with repeated measures of the same group of persons such as in longitudinal studies.[882] For this type of studies it is reasonable to perform a Durbin-Watson test which is designed to detect this autocorrelated data.[883]

7) In another step of testing the assumptions of regression analyses, the dataset must be checked for multicollinearity. Multicollinearity refers to an excessive correlation of the model's independent variables.[884] In case of non-perfect multicollinearity, the fit of the regression model (R^2) will remain unaffected. The problem is rather that the influence on the dependent variable cannot clearly be assigned to either one of the independent variables.[885] Screening the correlation matrix of all independent variables can help detecting multicollinearity. However, multicollinearity can also prevail even if correlation coefficients between independent variables are low, as only the bi-variate correlations are mapped. Instead, the linear relationship causing the bias may exist for a combination of several independents.[886] Therefore, the linear dependency of the independent variables should be tested. The corresponding indicator is the variance inflation

---

878 For an overview of different transformation approaches see Janssen, J./Laatz, W. (2007), p. 453.

879 Cf. Urban, D./Mayerl, J. (2006), pp. 250. Further details about WLS can be found in Cohen, J. et al. (2003), p. 147.

880 Cf. Fox, J. (2005), p. 516.

881 Cf. Fox, J. (1997), p. 306.

882 Cf. Cohen, J. et al. (2003), p. 120.

883 Cf. Brosius, F. (2006), pp. 574.

884 Cf. Backhaus, K. et al. (2008), pp. 87.

885 Cf. Brosius, F. (2006), p. 577.

886 Cf. Brosius, F. (2006), pp. 578.

factor (VIF),[887] whereas high VIF-values (>5) indicate the presence of multicollinearity.[888]

8) The normality assumption of the residuals (ε) is irrelevant for the calculations of the least-squares. Estimates of the regression coefficients remain unbiased even when this assumption is violated.[889] Nevertheless, normality is important when t-tests and F-tests are performed. Due to the central limit theorem, no normality test of the residuals is necessary in those situations where datasets with n>40 exist.[890]

9) One additional aspect that has already been mentioned several times throughout this assumption chapter needs to be pointed out. A minimum number of observations are needed in order to be able to perform the regression calculations and to be allowed to generalize from the regression model results into reality. At least three or better five times as many observations as the number of regression coefficients are necessary to ascertain significant relationships.[891] Field (2009) suggests ten to 15 times as many observations per predictor.[892]

5.5 Description of the Survey Sample

In the subsequent Chapters 5.5.1 - 5.5.4, the collected data is analyzed before starting to test the concrete hypotheses. The preliminary data analysis is performed in order to check for issues like the online dropout-rate, missing values, and possible answering biases (Ch. 5.5.1). This is followed by the sociodemographic description of the survey sample (Ch. 5.5.2). Next, the CEP ownership is generally analyzed (Ch. 5.5.3). Finally, Chapter 5.5.4 covers the descriptive statistics of the adoption determinants.

5.5.1 Preliminary Sample Analysis

5.5.1.1 Questionnaire Response

A total of 510 persons participated in the online questionnaire, while 332 (65.1%) of them fully completed the questionnaire and 178 stopped filling out

887 For a deeper analysis of the VIF see e.g. Hübler, O. (2005) pp. 283.

888 Cf. Urban, D./Mayerl, J. (2006), p. 232.

889 Cf. Cohen, J. et al. (2003), p. 120.

890 Cf. Backhaus, K. et al. (2008), pp. 90.

891 Cf. Skiera, B./Albers, S. (2008), p. 479.

892 Cf. Field, A. P. (2009), pp. 222.

the survey after a few questions. With 34.9%[893] the dropout-rate is on an acceptable level.[894] A relatively strict threshold has been set for the missing values analysis. While some studies allow that up to 10% of the questionnaire items may remain unanswered, here, all respondents having more than 3 missing values were excluded. In consequence, from the remaining 332 cases, nine more were removed, so that the final number of valid online respondents comes to 323.[895] In addition to the online survey, 359 persons participated in the written questionnaire. Reduced by 22 cases that had more than three missing values, the final number of "paper-and-pencil"-participants is 337. In total, 148 students participated in the survey. As explained before, students are not part of the subsequent analyses. Thus, the corresponding cases were deleted as well. Hence, prior to checking for inconsistent responses or eventual outliers, the survey resulted in a total of 512 valid cases with three or less missing values (s. Table 33). These cases were numbered consecutively in SPPS by assigning an ID.

The screening for missing values revealed that 62 of the 512 respondents had at least one, but no more than three missing values. In the following, two different approaches were combined to impute the missing data. As variables of reflective scales are highly correlated, it is reasonable to replace this first type of missing values by the remaining variables' mean value (which belong to the same scale). This means that for the proposed reflective scales the mean values of the non-missing *variables* are used instead of the mean values of the non-missing *cases*.[896] This procedure cannot be applied to descriptive variables or variables that belong to an index as these do not necessarily correlate with the other associated variables. This second type of missing values was replaced by the items' mean values of the corresponding age/gender group of the specific individual.[897] This procedure was preferred over other imputation variants.[898] The here chosen approach seems more accurate, as with considering the individual, consistency in answering questions related to the same scale and age-gender membership, respectively, allows to better approximate to the true values.

---

893 The drop-out rate of the first (34,1%) and second (35,8%) wave was almost identical.

894 Cf. Shih, T.-H./Fan, X. (2008), p. 257.

895 No bias of technology savvy older respondents in the form of too many elderly online-participants is expected. 14.9% of the respondents who were 60 years and older participated in the online survey compared to 26.4% who regularly use the internet at that age in Germany. Thus, elderly onliners are not overrepresented in this sample.

896 Cf. Graham, J. W./Cumsille, P. E./Elek-Fisk, E. (2003), p. 108.

897 See for example Backhaus, K. et al. (2008), p. 26; Göthlich, S. E. (2006), p. 139.

898 Cf. De Ayala R. J. (2009), pp. 149.

	Total (without students)	Online	Offline	Mean age
1. Wave	**258**	130	128	42.76
2. Wave	**254**	75	179	54.32
Total	**512**	205	307	48.49

Table 33: Questionnaire response

5.5.1.2 Plausibility Checks

In the following, the item responses are checked for plausibility. Careless responses by the survey participants may have a negative impact on questionnaire validity and lead to higher measurement errors. In the worst case, existing relationships will not be identified by the later applied statistical methods.[899] Therefore, three measures are taken which should detect eventually problematic cases:

Careless participants may, for example, fail to notice negatively coded items. The use of negatively coded items has been avoided as far as possible in this study as answering such questions requires higher cognitive efforts. Only one theoretically derived construct (product complexity) possesses two negatively coded items (cx_und_r / cx_use_r). After recoding, an individual should normally answer these two items similarly to the remaining third item of the construct (cx_conf). Therefore differences have been calculated between cx_und_r and cx_conf as well as between cx_use_r and cx_conf. Five respondents show a difference of "-3" in both comparisons, four respondents show a difference from "+3" in both comparisons (s. Table 34). It is very likely that these nine individuals (highlighted in bold) incorrectly filled out the negatively formulated questions. Values are corrected[900] whereas the item cx_ov (perceived overall complexity) serves as an additional control variable to check for consistency (s. Table 35).

Cases where difference between...			
cx_und_r and cx_conf is "-3"	**cx_use_r and cx_conf is "-3"**	**cx_und_r and cx_conf is "+ 3"**	**cx_use_r and cx_conf is "+3"**
176	164	200	**88**
186	**186**	**478**	399
277	346	**504**	**478**
299	**374**	**582**	**504**
306	**401**	**627**	**582**

899 Cf. Cote, J. A./Buckley, R. (1988), pp. 579.

900 The correction is also known as imputation (imp) as discussed earlier.

374	458		**627**
401	477		
427	**496**		
496	**531**		
531	569		
565			

Table 34: Analysis of negatively coded items

ID	Age	Gender	diff_und_conf	diff_use_conf	cx_und_r	imp	cx_use_r	imp	cx_conf	imp	cx_ov
186	28	female	-3.0	-3.0	2.0	4	2.0	4	5.0	-	4.0
374	55	female	-3.0	-3.0	2.0	4	2.0	4	5.0	-	3.0
401	36	male	-3.0	-3.0	2.0	4	2.0	4	5.0	-	5.0
496	59	male	-3.0	-3.0	2.0	4	2.0	4	5.0	-	3.0
531	67	female	-3.0	-3.0	1.0	5	1.0	5	4.0	-	4.0
478	55	male	3.0	3.0	5.0	-	5.0	-	2.0	4	3.0
504	60	female	3.0	3.0	4.0	-	4.0	-	1.0	5	5.0
582	63	male	3.0	3.0	4.0	2	4.0	2	1.0	-	3.0
627	71	male	3.0	3.0	4.0	-	4.0	-	1.0	5	5.0

Table 35: Value correction of negatively coded items

One of the advantages of online surveys is the possibility to track the time needed to fill out the questionnaire. This allows identifying online participants with extremely short and extremely long fill-out durations. In total, eleven respondents needed less than 400 seconds to complete the online survey which is equals less than 25 seconds per page on average (16 pages). The other extreme is displayed by seven participants who needed more than two minutes per page on average (1920 seconds to complete the whole questionnaire). Since one can only speculate on the reasons behind the required time to fill out the form (clicking through vs. paused during fill-out or questions too difficult), the cases listed in Table 36 are not deleted from the dataset. However, this analysis serves as an explanation to possible outlier cases which are analyzed later.

ID	Age	Gender	Duration in seconds
289	31	male	267
271	31	male	289
173	25	female	301
238	27	male	327
5	29	female	380
398	35	male	388
253	29	male	389

308	37	female	389
171	25	female	392
246	28	male	398
213	33	female	399
532	67	female	5907
311	38	female	3037
221	23	male	2957
215	34	female	2729
261	30	male	2443
437	43	male	2373

Table 36: Analysis of the time required to complete online questionnaire

Furthermore, respondents should be identified who went through the questionnaire without the necessary involvement and express this behavior by giving uniform responses – e.g. marking the "middle" option in order to just proceed through the entire questionnaire. Therefore, a case-wise diagnostic of the response variance was performed.[901] That means that for each individual the variance of his or her responses to the items of the proposed constructs rel, cx_prod, cx_is, r_perf, r_fin, r_psy, soc_pres, int, s_eff, and compa was calculated. The critical value was set at 0.7. All cases with a standard deviation (SD) equal to or blow the critical value are reported (s. Table 37). The so detected individuals are candidates for deletion. However, they should not automatically be excluded. Drawing on the time for completing the questionnaire (only the online respondents), with only 289 seconds case number 271 is deleted from the data set. As for all other cases so far no further critical values are evidenced (from the analysis of negatively coded items and missing values), they should not be deleted unless they become particularly striking in the outlier analysis for extreme cases (following in Ch. 5.6.1.1, pp. 227 and Ch. 5.7.3.1, pp. 241). As a consequence, the descriptive statistic of the survey sample is based on a data set of 511 participants.

ID	Age	Gender	Duration (online; in sec)	SD
114	23	male	-	0.68
173	25	female	301	0.64
211	33	female	-	0.68
245	28	male	473	0.52
271	**31**	**male**	**289**	**0.51**
303	34	male	542	0.64
314	39	female	-	0.68
442	44	male	-	0.62

901 Analogous to Christophersen, T. (2007), pp. 134.

485	57	male	-	0.70
554	75	female	-	0.47
620	69	male	-	0.66

Table 37: Analysis of construct variance (uniform responses)

5.5.2 Sociodemographic Description of the Survey Sample

Next, the sociodemographic characteristics of the survey sample are presented (s. Table 38 - Table 43). An overview is given over (a) the composition of age groups, (b) gender split, (c) education, (d) job situation, (e) martial status, and (f) household size.

a) Age Groups

In order to get a better understanding of the survey participants, the respondent sample is subdivided into different age groups. As the statistical methods applied in the following chapters require a minimum number of cases per subgroup, *three* age groups are built, subsequently referred to as the "younger," "middle-aged," and "older" group of individuals. In line with the conclusion made in Chapter 1.3 (pp. 8), a person of 60 years and older is considered an "older individual." As shown in Table 38, the respondents are nearly equally distributed among the three age groups.

Group number	Group description	Age range (in years)	N	%	Cumulative %
1	"younger" individuals: non-students younger than 35 years of age (employees, freelancers etc.)	18-34	150	29.4	29.4
2	"middle-aged" individuals: non-retired individuals 35 years and older, but younger than 60	35-59	197	38.6	67.9
3	"older" individuals: retired persons and individuals being 60 years of age or older[902]	60-85	164	32.1	100.0
Total		18-85	511	100.0	

Table 38: Age group definitions and sizes

902 Three individuals have been added to this group who are younger than 60 years, so-called early retirees.

b) Gender

Gender	Group 1		Group 2		Group 3	
	N	%	N	%	N	%
Female	56	37.3	90	45.7	71	43.3
Male	94	62.7	107	54.3	93	56.7
Total	150	100.0	197	100.0	164	100.0

Table 39: Gender split among the survey respondents

c) Education

Education	Group 1		Group 2		Group 3	
	N	%	N	%	N	%
No degree / graduation	0	0	0	0	0	0
Elementary school or secondary modern school	1	0.7	21	10.7	38	23.2
Secondary school leaving certificate	17	11.3	58	29.4	47	28.7
General qualification for university entrance	24	16.0	25	12.7	18	11.0
University or college	104	69.3	89	45.2	54	32.9
Other	4	2.7	4	2.0	7	4.3
Total	150	100,0	197	100,0	164	100,0

Table 40: Educational level of the survey respondents

d) Job

Job	Group 1		Group 2		Group 3	
	N	%	N	%	N	%
Apprenticeship	9	6.0	0	0	0	0
Employed or public official	117	78.0	138	70.1	15	9.1
Freelancer or self-employed person	15	10.0	35	17.8	4	2.4
Professional retraining	0	0	1	0.5	0	0
Retiree or pensioner	0	0	0	0	137	83.5
Housewife or househusband	1	0.7	9	4.6	8	4.9
Unemployed at the moment	1	0.7	6	3.0	0	0
Other	7	4.7	8	4.1	0	0
Total	150	100.0	197	100.0	164	100,0

Table 41: Job situation of the survey respondents

e) Martial Status

Martial status	Group 1		Group 2		Group 3	
	N	%	N	%	N	%
Single	124	82.7	50	25.4	9	5.5
Married	24	16.0	116	58.9	113	68.9
Divorced	2	1.3	25	12.7	14	8.5
Widowed	0	0	6	3.0	28	17.1
Total	150	100.0	197	100.0	164	100.0

Table 42: Martial status of the survey respondents

f) Household Size

Household size in persons	Group 1		Group 2		Group 3	
	N	%	N	%	N	%
1	56	37.3	39	19.8	41	25.0
2	57	38.0	70	35.5	87	53.0
3	27	18.0	47	23.9	29	17.7
4	8	5.3	29	14.7	3	1.8
5	2	1.3	9	4.6	2	1.2
6	0	0	1	0.5	0	0
7	0	0	1	0.5	2	1.2
8	0	0	1	0.5	0	0
Total	150	100.0	197	100.0	164	100.0

Table 43: Household size of the survey respondents

5.5.3 CEPO Analysis

5.5.3.1 Possession of Innovative and Non-Innovative CEP (Product View)

The CEP which are most spread among the respondents are (regular) desktop computers (67.1%), stand-alone DVD-Players (60.5%), and mobile phones with cameras (56.6%). Table 44 gives an overview of the top ten CEP possessions. Further, to get a first impression of the distribution of innovative CEP possessions, the CEP employed in the questionnaire have been dichotomously classified as either innovative or non-innovative devices (s. Table 45).[903]

CEP	N	%
Computer	343	67.1
DVD-Player (stand-alone device – no computer drive)	309	60.5
Mobile phone with camera	289	56.6
Laptop/notebook	283	55.4

903 For the classification see Table 27, p. 173.

VCR	272	53.2
Flatscreen (computer)	225	44.0
Digital camera (with less than 8 megapixels)	208	40.7
Navigation system	195	38.2
Mobile phone without camera	189	37.0
MP3 player (stand-alone device)	163	31.9

Table 44: Absolute and relative possessions of CEP (innovative and non-innovative CEP)

CEP_i	N	%
Navigation system	195	38.2
Digital camera (with at least 8 megapixels)	161	31.5
Flatscreen TV (plasma-/LCD-TV)	148	29.0
DVD-recorder (stand-alone device)	101	19.8
Hard disk recorder (digital videorecorder)	75	14.7
iPod	68	13.3
Home cinema system (Dolby Digital, DTS, 5.1,..)	54	10.6
Mobile organizer	47	9.2
Nintendo DS or Sony Playstation Portable	32	6.3
Wii or Sony Playstation 3	27	5.3

Table 45: Absolute and relative possessions of innovative CEP

It is interesting to see that the top five innovative CEP of groups 2 and 3 consist of the same products (s. Table 46). There are just differences in the order of CEP_i possession. While in group 3 flatscreens rank first (36%), navigation systems second (29.3%), and digital cameras with at least 8 megapixels third (26.2%), the order in group 2 is navigation systems (42.6%), digital cameras with at least 8 megapixels (34.5%), and flatscreens (27.9%). These three products are also found in the top five of the younger individuals (group 1), however instead of DVD-recorders and DVD-recorders, a large percentage of group members possess iPods (26.7%) and mobile organizers (16.7%).

Group 1	N	%
Navigation system	63	42.0
Digital camera (with at least 8 megapixels)	50	33.3
iPod	40	26.7
Flatscreen TV (plasma-/LCD-TV)	34	22.7

Mobile organizer	25	16.7
Group 2	**N**	**%**
Navigation system	84	42.6
Digital camera (with at least 8 megapixels)	68	34.5
Flatscreen TV (plasma-/LCD-TV)	55	27.9
DVD-recorder (stand-alone device)	50	25.4
Hard disk recorder (digital videorecorder)	35	17.8
Group 3	**N**	**%**
Flatscreen TV (plasma-/LCD-TV)	59	36.0
Navigation system	48	29.3
Digital camera (with at least 8 megapixels)	43	26.2
DVD-recorder (stand-alone device)	32	19.5
Hard disk recorder (digital videorecorder)	19	11.6

Table 46: Top five innovative CEP of each age group

Furthermore, it is interesting to take a look at some extremes. Table 47 illustrates the possession rates of the specific CEP of the oldest age group 3 in the first column, the difference between the percental ownership of group 1 and group 3 (group 1 minus group 3) in the second column, and the difference between the percental ownership of group 2 and group 3 (group 2 minus group 3) in the third column. The products are listed according to relative ownership of the oldest (G3) and youngest (G1) age group. The largest *negative* difference (G3 has a higher possession percentage compared to G1) can be found for mobile phones with camera

CEP	**G3**	**G1 - G3**	**G2 - G3**
	%	in percentage points (pp)	in percentage points (pp)
Mobile phone without camera	48.8	-28.8	-8.7
VCR	56.1	-17.4	5.8
Flatscreen TV (plasma-/LCD-TV)	36.0	-13.3	-8.1
Multi-function video device (DVD-Player + VCR)	13.4	-11.4	-3.8
DVD-recorder (stand-alone device)	19.5	-6.8	5.9
Other camcorder	9.8	-3.1	-0.1
HD-DVD player (stand-alone device)	4.3	-0.9	0.8
Blu-Ray disc player (stand-alone device)	1.2	-0.6	0.3

Mobile videoplayer	2.4	-0.4	-0.9
iPhone	3.0	-0.4	-0.5
Camcorder with hard-drive or memory chip	3.7	-0.3	4.0
Digital photo frame	3.0	1.0	-1.0
Beamer	3.0	1.6	4.1
Hard disk recorder (digital videorecorder)	11.6	2.4	6.2
Digital camera (with at least 8 megapixels)	26.2	7.1	8.3
Nintendo DS or Sony Play-station Portable	0.6	7.4	9.0
Computer	54.3	7.7	27.5
Flatscreen (computer)	35.4	8.0	16.4
Home cinema system (Dolby Digital, DTS, 5.1,...)	4.3	11.1	7.9
Wii or Sony Playstation 3	0.0	12.7	4.1
Navigation system	29.3	12.7	13.4
Mobile organizer	3.7	13.0	4.5
iPod	3.0	23.6	8.6
Digital camera (with less than 8 megapixels)	21.3	32.0	25.9
DVD-Player (stand-alone device – no computer drive)	40.9	36.5	23.1
Laptop/notebook	36.0	44.7	16.3
MP3 player (stand-alone device)	6.7	50.6	26.8
Mobile phone with camera	28.0	58.0	29.8

Table 47: Possession rates of the specific CEP – difference between the oldest and the younger age groups

(-28.8 pp[904]), VCRs (-17.4 pp), and flatscreen TVs (-13.3 pp), followed by multi-function video devices (-11.4 pp) and DVD-recorders (-6.8 pp). As expected, three out of five CEP are "old" technologies. However it is notable that with flatscreen TVs and DVD-recorders two innovative CEP are more common in group 3 households. Contrary to that, the youngest age group possesses largely more mobile phones with camera (58.0 pp), MP3-players (50.6 pp), laptops (44.7 pp), DVD-Players (36.5 pp), and digital cameras with less than 8 megapixels (32.0 pp). This is striking in so far as these products have all been classified as non-innovative. Still, very few older consumers own one of these devices implying that group 3 does not only lack possession of innovative CEP but also CEP that belong to the "standard" equipment of younger individuals.

904 Percentage points (pp).

5.5.3.2 Respondent Analysis Regarding CEPO (Personal View)

Leaving the view of the specific CEP, the following analyses deal with the total amount of CEP an individual owns. In this context Table 48 and

No. of CEP	N	%	Cumulative %
0	6	1.2	1.2
1	14	2.7	3.9
2	21	4.1	8.0
3	45	8.8	16.8
4	40	7.8	24.7
5	65	12.7	37.4
6	73	14.3	51.7
7	48	9.4	61.1
8	62	12.1	73.2
9	54	10.6	83.8
10	26	5.1	88.8
11	22	4.3	93.2
12	13	2.5	95.7
13	8	1.6	97.3
14	10	2.0	99.2
15	3	0.6	99.8
18	1	0.2	100.0
Total	511	100.0	

Table 48: Overview of individual CEP possessions

Figure 29 show that ownership of the CEP listed in the questionnaire ranges from 0 (possesses no CEP) to 18. Most people possess between five and six CEP (mode = 6, median = 6), the mean is 6.6. It is not surprising that all six individuals that do not possess any of the listed CEP belong to the oldest age group G3. In contrast to that the 14 persons who own 14 CEP or more can mainly be assigned to G2 (11 persons) whereas only two persons belong to G1 and one person to G3.

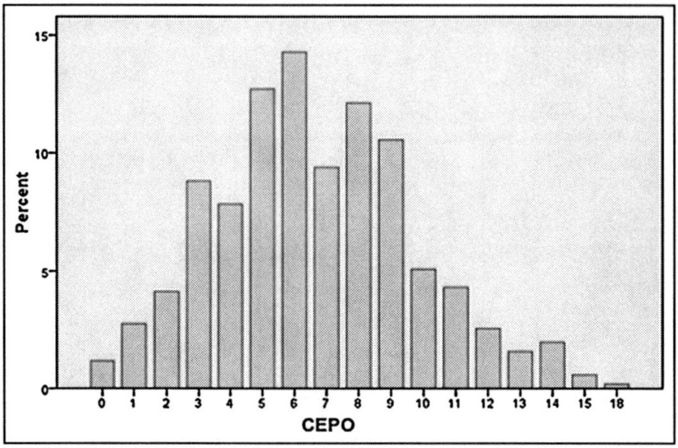

Figure 29: Graphical representation of individual CEP possessions

While the possessions of CEP[905] (s. Figure 29) and non-innovative CEP (s. Figure 30) resemble more or less a normal distribution, the distribution of innovative CEP possession is positively skewed (meaning there are high scores on the left of the bar chart) (s. Figure 31): 70% of the respondents own none, one or two innovative devices. In contrast to 1.2% of the respondents who do not own any CEP at all, a considerable 21.1% do not own a CEP classified innovative – whereas quite large differences exist between the three age groups. While the non-owner proportion of G1 (16.0%) and G2 (18.8%) are fairly close together, more than a fourth of G3 individuals (28.7%) do not possess any innovative CEP (s. Table 49). Similarly, the elderlies' CEPO_i mean (G3 = 1.52) deviates from G1 (2.20) and G2 (2.14) (s. Table 50).

905 Innovative and non-innovative CEP taken together.

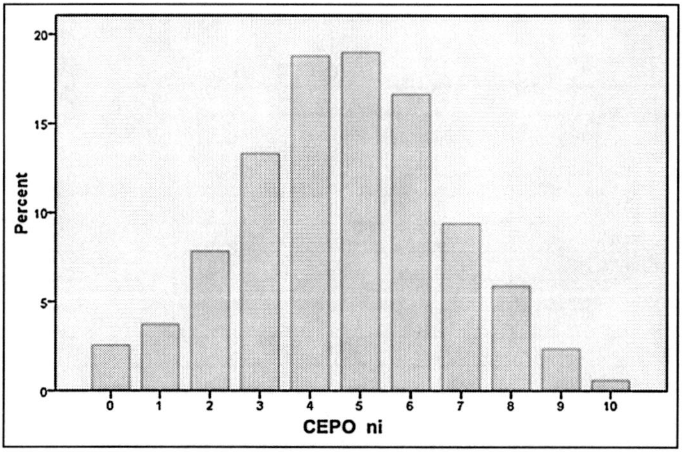

Figure 30: Possession of non-innovative CEP

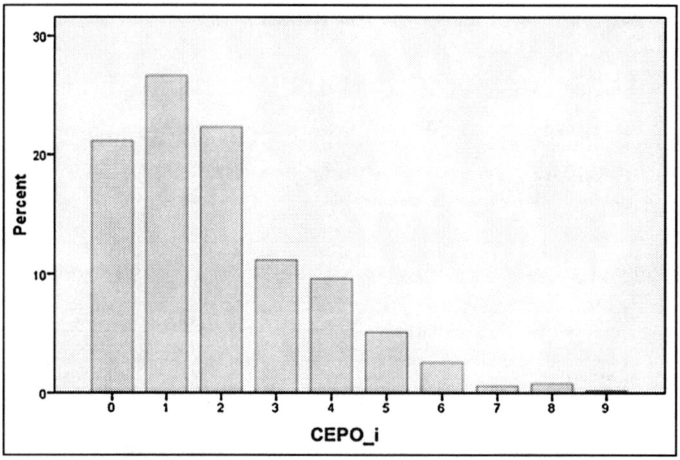

Figure 31: Possession of innovative CEP

CEPO_i	**Total**		**G1**		**G2**		**G3**	
N (CEPO_i)	N	%	N	%	N	%	N	%
0	108	21.1	24	16.0	37	18.8	47	28.7
1	136	26.6	45	30.0	47	23.9	44	26.8
2	114	22.3	29	19.3	47	23.9	38	23.2
3	57	11.2	16	10.7	22	11.2	19	11.6
4	49	9.6	19	12.7	21	10.7	9	5.5
5	26	5.1	7	4.7	13	6.6	6	3.7
6	13	2.5	6	4.0	6	3.0	1	0.6

7	3	0.6	1	0.7	2	1.0	0	0.0
8	4	0.8	2	1.3	2	1.0	0	0.0
9	1	0.2	1	0.7	0	0.0	0	0.0
Total	511	100.0	150	100.0	197	100.0	164	100.0

Table 49: Possession of innovative CEP (per age group)

CEPO_i	All	G1	G2	G3
N	511	150	197	164
Mean	1.96	2.20	2.14	1.52
Median	2.00	2.00	2.00	1.00
Mode	1	1	1;2	0
SD	1.73	1.89	1.80	1.40
Variance	3.00	3.57	3.22	1.96
Skewness	1.04	1.09	0.88	0.85
Kurtosis	0.96	1.04	0.39	0.22
Range	9	9	8	6
Minimum	0	0	0	0
Maximum	9	9	8	6

Table 50: Descriptive statistics of innovative CEP possession (per age group)

5.5.3.3 CEPO_w Analysis (Personal View)

While the analysis above could give a first insight into the ownership of CEP, it is even more interesting to see how the respondents can be characterized as regards to the weighted ownership index CEPO_w (s. Table 51 and Figure 32). This index respects the novelty degree of each CEP and avoids the arbitrary classification of CEP in either innovative or non-innovative devices.

CEPO_w	All	G1	G2	G3
Valid	511	150	197	164
Mean	14.92	17.29	16.33	11.06
Median	13.93	16.69	15.26	9.36
Mode	0.00	12.530	6.070	0.00
SD	7.82	7.02	7.69	7.27
Variance	61.22	49.33	59.09	52.80
Skewness	0.554	0.939	0.634	0.589
Std. Error of Skewness	0.108	0.198	0.173	0.190
Kurtosis	0.309	1.687	0.063	-0.405
Std. Error of Kurtosis	0.216	0.394	0.345	0.377
Range	47.32	45.19	37.39	32.33
Minimum	0.000	2.130	1.400	0.000
Maximum	47.32	47.32	38.79	32.33

Table 51: Descriptive statistics of the CEPO_w index

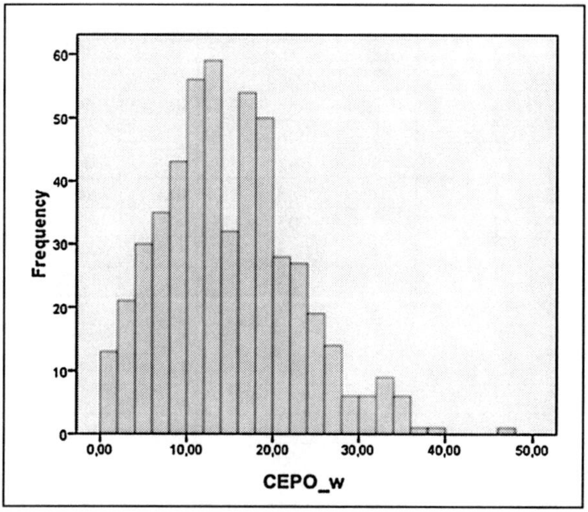

Figure 32: Histogram of the CEPO_w index scores (all respondents)

As assumed, the CEPO_w scores are lowest for the oldest age group. A look at the mean values of CEPO_w supports the assumption of the oldest age group scoring much lower than the other age groups with the G3 mean (11.06) lying largely below that of G1 (mean = 17.29) and G2 (mean = 16.33).

5.5.4 Descriptive Statistics of the Adoption Determinant Items

After the preliminary sample analysis and sociodemographic description of the respondents, the scale items and further variables which gave the respondents the possibility to assess the innovative CEP are at the center of attention. To get a first impression of the variables, Table 52 shows the most important descriptive statistics of those variables that are further used for the final scale development. Table 53 displays the statistics of single-item scales and indices.

	Mean	Med.	Mode	SD	Var.	Skew.	Kurtosis	Min.	Max.
rel_tech	4.02	4	5	1.103	1.217	-0.914	-0.040	1	5
rel_funct	3.32	3	3	1.172	1.373	-0.161	-0.760	1	5
rel_coper	3.19	3	3	0.981	0.962	0.125	-0.118	1	5
cx_und_r	3.35	3	3	0.978	0.957	0.039	-0.379	1	5
cx_use_r	3.33	3	3	0.943	0.890	0.132	-0.300	1	5
cx_conf	3.62	4	5	1.172	1.373	-0.392	-0.854	1	5
cx_choice	3.34	3	3	1.219	1.485	-0.284	-0.819	1	5

cx_sim	4.01	4	5	1.030	1.061	-0.917	0.151	1	5
cx_isinf	3.74	4	5	1.102	1.214	-0.548	-0.420	1	5
perf_expec	2.96	3	3	1.068	1.140	0.049	-0.681	1	5
perf_troub	2.99	3	3	1.055	1.114	-0.085	-0.461	1	5
fin_worth	3.11	3	3	1.048	1.099	-0.142	-0.558	1	5
fin_price	3.83	4	4	1.062	1.127	-0.836	0.195	1	5
fin_obs	3.68	4	4	1.133	1.284	-0.641	-0.365	1	5
psy_detail	3.09	3	3	1.295	1.678	-0.097	-1.019	1	5
psy_info	3.90	4	4	1.038	1.077	-0.784	-0.026	1	5
psy_wrong	3.42	3	3	1.163	1.354	-0.294	-0.706	1	5
psy_opera	3.57	4	5	1.263	1.594	-0.437	-0.952	1	5
pres_unprog	1.67	1	1	1.024	1.049	1.539	1.633	1	5
pres_prest	2.20	2	1	1.175	1.381	0.666	-0.481	1	5
s_eff_buy	2.99	3	3	1.311	1.718	-0.036	-1.115	1	5
s_eff_oper	3.88	4	5	1.157	1.338	-0.896	-0.054	1	5
int_ce	3.27	3	3	1.163	1.353	-0.261	-0.645	1	5
int_inno	2.86	3	3	1.281	1.640	0.078	-1.018	1	5
Med. = Median; Var. = Variance; Skew. = Skewness									

Table 52: Descriptive statistics of the variables which will be used for the scale development

	Mean	Med.	Mode	SD	Var.	Skew.	Kurtosis	Min.	Max.
fin_situ	4.91	5.00	5	1.105	1.221	-0.535	0.342	1	7
compa	2.29	2.40	2.4	1.089	1.185	0.521	-0.013	0.2	5
soc_assist	10.71	10.80	0	5.127	26.282	0.116	-0.083	0	28.4

Table 53: Descriptive statistics of the single-item scales and indices

5.6 Scale Development

After having explored the theoretical background of EFA in Chapter 5.4.1 (pp. 183), which is a method to help developing scales, the scales are now built by means of actual data. Before identifying the final item-structure of theoretical constructs and the number of extracted factors, the assumptions of performing an EFA are explored for the current dataset and inappropriate items are deleted (Ch. 5.6.1). The factor extraction and interpretation of the final structure are content of the next chapter (Ch. 5.6.2). In Chapter 5.6.3 the reliability of the newly developed scales is tested, followed by the descriptive statistics of the final scales (Ch. 5.6.4).

5.6.1 Testing for EFA Assumptions

Some of the mentioned assumptions associated with EFA have already been discussed (s. Ch. 5.4.1.2, pp. 185). The appropriateness of EFA as selected method is given as in the search for the final scales the underlying factors should be responsible for the correlation of the factors with the items. As presumed before, some scale items need to be dropped, still most of the factors are represented by two or three items which is a drawback that needs to be taken into account. The broad conceptualization of the study with its large number of potential adoption determinants on the one hand and the need to keep the questionnaire at a considerable length on the other made it necessary to go with fewer items per factor than the recommended number of at least four items per factor. With respect to the sufficiently large sample size for EFA, a minimum number of 300 to 500 respondents have been reported. After missing data check and outlier analysis the minimum number is reached so that this assumption is met. Whether or not the other assumptions of unbiased intercorrelations (Ch. 5.6.1.1), multivariate normality (5.6.1.2), and substantial intercorrelations (5.6.1.3) can be met will be tested with the help of the collected data.

5.6.1.1 Unbiased Intercorrelations

As outliers do not belong to the underlying population and may influence the results of EFA, the Mahalanobis distance is used for detecting multivariate outliers. The suggested cut-off value by Fidell/Tabachnick (2003) is used. Table 54 shows all cases with $p<0.001$ for the Mahalanobis distance.[906] As these five cases show an exceptional combination of values throughout several variables, they are removed from the data set. Consequently, the final scale development is based on the remaining 506 respondents.

Case ID	p-value	Gender	Age
544	0.0002	Female	71
286	0.0004	Male	31
514	0.0007	Female	63
635	0.0007	Male	73
364	0.0007	Female	52

Table 54: Cases with critical p-values of the Malhalanobis distance outlier analysis

5.6.1.2 Multivariate Normality

The existence of multivariate normality implies the existence of univariate normality. Therefore, the Kolmogorov-Smirnov-Test and the Shapiro-Wilk-Test, available in SPSS, are executed and all items are tested for univariate normality first (s. Table 55). Both tests show that no single item meets the assumption of univariate normality and, hence, equally fail regarding multivariate normality.

906 Cf. Fidell, L. S./Tabachnick, B. G. (2003), p. 130.

Still, the EFA can be continued. This is possible because PAF factor extraction is the preferred method when multivariate normality is absent in the data.

	Kolmogorov-Smirnov			Shapiro-Wilk		
	Statistic	df	Sig.	Statistic	df	Sig.
age	*0.101*	*506*	*0.000*	*0.952*	*506*	*0.000*
CEPO	*0.098*	*506*	*0.000*	*0.981*	*506*	*0.000*
CEPO_w	*0.054*	*506*	*0.001*	*0.979*	*506*	*0.000*
rel_tech	0.262	506	0.000	0.814	506	0.000
rel_funct	0.182	506	0.000	0.906	506	0.000
rel_coper	0.267	506	0.000	0.885	506	0.000
cx_und_r	0.246	506	0.000	0.892	506	0.000
cx_use_r	0.264	506	0.000	0.883	506	0.000
cx_conf	0.182	506	0.000	0.881	506	0.000
cx_choice	0.175	506	0.000	0.905	506	0.000
cx_sim	0.240	506	0.000	0.827	506	0.000
cx_isinf	0.192	506	0.000	0.873	506	0.000
perf_expec	0.172	506	0.000	0.913	506	0.000
perf_troub	0.207	506	0.000	0.912	506	0.000
fin_worth	0.179	506	0.000	0.912	506	0.000
fin_price	0.252	506	0.000	0.855	506	0.000
fin_obs	0.236	506	0.000	0.874	506	0.000
psy_detail	0.150	506	0.000	0.906	506	0.000
psy_info	0.244	506	0.000	0.853	506	0.000
psy_wrong	0.174	506	0.000	0.904	506	0.000
psy_opera	0.195	506	0.000	0.875	506	0.000
pres_unprog	0.358	506	0.000	0.695	506	0.000
pres_prest	0.216	506	0.000	0.854	506	0.000
s_eff_buy	0.164	506	0.000	0.906	506	0.000
s_eff_oper	0.238	506	0.000	0.832	506	0.000
int_ce	0.171	506	0.000	0.909	506	0.000
int_inno	0.148	506	0.000	0.907	506	0.000

Table 55: Results of the normality test

5.6.1.3 Substantial Intercorrelations

Several tests have been presented to examine whether the item pool as a whole and the individual items show sufficient levels of correlation. The Bartlett's test of sphericity verifies whether the sample originates from a population in which the items are uncorrelated. The present data affirms, however, that the items are significantly correlated in the population (sample without students: Chi-square of 4231.7; significance level of 0.000). Therefore, a first indication of an adequate item pool is obtained.

Furthermore, the item pool is generally suited to perform an EFA when the non-diagonal elements of the Anti-image covariance-matrix are close to zero. The matrix only contains 42 non-diagonal elements larger than $|0.09|$. In relation to the total number of 625 matrix elements this corresponds to a share of 6.72% which is lower than the cut-off value of 25% recommended not to be exceeded. From this perspective, the initial item set meets the requested condition.

With the help of the Kaiser-Meyer-Olkin measure of sampling adequacy the item pool as well as the single items can be judged for their appropriateness. Drawing on the classification of Kaiser/Rice (1974) the survey data achieves very good scores. The MSA of the entire dataset reaches a "meritorious" score of 0.828. The single items also show very good results. No item scores an MSA lower than 0.5 which would have indicated the need of exclusion. In fact, all scores lie above 0.6 with most items being "middling" or "meritorious" (s. Table 56).

MSA score	Classification	Items
≥ 0.9	marvelous	cx_conf, psy_detail, psy_wrong, psy_opera
≥ 0.8	meritorious	cx_und_r, cx_use_r, cx_choice, cx_isinf, perf_expec, perf_troub, fin_obs, psy_info, s_eff_buy
≥ 0.7	middling	rel_tech, rel_funct, cx_sim, fin_worth, fin_price, s_eff_oper, int_ce, int_inno
≥ 0.6	mediocre	rel_coper, pres_unprogr, pres_prest
≥ 0.5	miserable	-
< 0.5	unacceptable	-

Table 56: Item classification according to the Kaiser-Meyer-Olkin measure of sampling adequacy

Single variables should yield a minimum correlation of 0.3 with items that belong to the same latent variable. To avoid multicollinearity, however, the correlation between those items is restricted to a maximum of 0.8. A correlational analysis indicates whether the items are related to those items suggested by theory. As expected, high correlations between most of the items and the respective constructs can be confirmed (s. Appendix 5). Still, two items need be removed: rel_coper due to low correlations with the other "relative advantage" items (correlation of 0.214 with rel_tech and 0.183 with rel_funct respectively), and psy_opera due to strong correlations with a number of other than the expected "psychological risk" items (correlation of 0.478 with cx_und_r, 0.485 with cx_use_r, 0.520 with cx_conf, 0.307 with cx_sim, 0.338 with cx_isinf, and 0.312 with perf_expec).

5.6.2 Factor Extraction and Interpretation

The preliminary tests have shown that the item-pool generally qualifies to perform an EFA. After the assessment of the single items the factor extraction process is focused upon more closely. As discussed earlier (s. Ch. 5.4.1.1, pp. 183), principal axis factoring with the oblique rotation is preferred. The "Direct Oblimin" method allows factors to be correlated. The degree of correlation can be controlled via delta values. As correlations between factors are expected and should not be suppressed, the delta values are set to 0 which allows for a maximum correlation between factors. The initial extraction is performed according to the mentioned Kaiser criterion (eigenvalue greater than one). If, however, no clear factor structure arises, solutions with more or less factors are tested in order to obtain better results.

The final scale items and number of factors are determined in an iterative process. All items with communalities lower than 0.3 are deleted and the EFA is executed again.[907]. After the first iteration step (FX1), perf_troub, fin_price, and s_eff_oper are eliminated. In iteration step two (FX2), fin_obs and s_eff_buy exhibit communalities lower than 0.3 and are, thus, also deleted. The third EFA (FX3) already shows an interpretable structure. However, iteration continues as with a communality of 0.294 psy_info lies slightly under the cut-off value of 0.3 and, thus, is deleted. In the next calculation of the EFA (FX4), still no clear factor structure was obtained for the "psychological risk" construct (r_psy). This could be either due to a selection of unsuitable items to represent the construct or by the extraction of too many/too few factors. As discussed earlier, there is no irrefutable rule of how many factors should be extracted. So far, in the iteration process, the pre-selected SPSS option has been selected which automatically extracts as many factors as eigenvalues larger than one exist. This led to a solution with six extracted factors. In line with Bühner (2011), the subsequent EFA (FX5) allows the extraction of one additional factor, accumulating to seven factors.[908] After adding psy_info again (FX6), which now achieves a communality largely above the critical value, a clear factor structure becomes visible. Table 57 - Table 59 depict the results of the final EFA (FX6). The computed communalities, the total variance explained, and the pattern matrix with the relevant

---

907 Low communalities means that only a small percentage of a variable's variance can be explained by the extracted factors (s. Ch. 5.4.1.1, 183).

908 Cf. Bühner, M. (2011), p. 349.

factor-loadings of the seven factor solution (all factor-loadings greater 0.3 are represented) are displayed subsequently (s. Table 57, Table 58, and Table 59).

	Communalities	
	Initial	Extraction
rel_tech	0.295	0.547
rel_funct	0.331	0.508
cx_und_r	0.664	0.816
cx_use_r	0.643	0.745
cx_conf	0.462	0.508
cx_choice	0.321	0.364
cx_sim	0.424	0.809
cx_isinf	0.353	0.389
perf_expec	0.438	0.545
fin_worth	0.435	0.745
psy_detail	0.382	0.463
psy_wrong	0.379	0.458
psy_info	0.323	0.509
pres_unprog	0.278	0.554
pres_prest	0.274	0.407
int_ce	0.591	0.697
int_inno	0.624	0.836
Extraction Method: Principal Axis Factoring.		

Table 57: Communalities of the scale items

The extracted seven factors suitable to explain 74.6% of the remaining 17 items' variance.

Total Variance Explained							
Factor	Initial eigenvalues			Extraction Sums of Squared Loadings			Rotation Sums of Squared Loadings
No.	Total	% of Variance	Cumulative %	Total	% of Variance	Cumulative %	Total
1	4.581	26.947	26.947	4.193	24.663	24.663	2.999
2	2.270	13.355	40.301	1.848	10.872	35.536	1.285
3	1.559	9.170	49.472	1.222	7.186	42.721	1.871
4	1.200	7.060	56.532	0.888	5.226	47.947	2.393
5	1.129	6.640	63.172	0.760	4.469	52.416	2.447
6	1.061	6.242	69.415	0.586	3.448	55.864	1.646
7	0.875	5.145	74.559	0.404	2.377	58.240	2.593
8	0.654	3.845	78.405				
9	0.545	3.206	81.610				

10	0.535	3.145	84.756	
11	0.523	3.079	87.835	
12	0.471	2.768	90.603	
13	0.439	2.583	93.185	
14	0.383	2.251	95.436	
15	0.341	2.004	97.441	
16	0.229	1.348	98.788	
17	0.206	1.212	100.000	
Extraction Method: Principal Axis Factoring				

Table 58: Total Variance Explained of the EFA

The next step involves interpreting the factors. While in the final iteration rel, cx_prod, cx_is, r_psy, soc_pres, and int where factored as expected, neither a "financial risk" factor nor a factor standing for "self efficacy" could be confirmed. Moreover one of the "market risk" items (fin_worth) loads exclusively high on the same factor which belongs to "performance risk" (perf_expec). Though fin_worth focuses more on financial aspects ("If I bought an innovative CEP, I would fear that the device will not be worth its money") compared to perf_expec (…I would fear that the device will not provide the benefit I expected), both items refer to some sort of expected benefit.

	Pattern matrix[a]						
	Factor						
	1	2	3	4	5	6	7
rel_tech	0.751						
rel_funct	0.652						
cx_und_r		0.899					
cx_use_r		0.858					
cx_conf		0.508					
cx_choice			-0.518				
cx_sim			-0.929				
cx_isinf			-0.470				
perf_expec				-0.693			
fin_worth				-0.879			
psy_detail					0.445		0.325
psy_wrong					0.504		
psy_info					0.713		
pres_unprog						0.734	
pres_prest						0.603	
int_ce							-0.805
int_inno							-0.871
Extraction Method: Principal Axis Factoring. Rotation Method: Oblimin with							

Kaiser Normalization
a = Rotation converged in 9 iterations

Table 59: Pattern Matrix of the EFA

As the two items express the worry of obtaining a lower than expected benefit when purchasing an innovative CEP, the factor should be named "benefit risk" (r_ben). The hypotheses are adapted correspondingly:

H_{11a}: The older an individual the higher the perceived benefit risk of innovative CEP.

H_{23a}: The higher the perceived benefit risk of innovative CEP the fewer/ less innovative CEP are possessed (r_ben $\xrightarrow{-}$ CEPO_w).

5.6.3 Scale Reliability

The calculation of the Cronbach's alpha shows good results (s. Table 60) as all values are above the requested 0.5/0.6 with "rel" having the lowest Cronbach's alpha (0.673) and "int" having the largest Cronbach's alpha (0.855). The additional test statistics do not deliver any arguments to exclude further items. Though three items possess a "corrected item-total correlation"-value smaller 0.5 (psy_detail, pres_unprog, pres_prest), deleting them would not improve but reduce Cronbach's alpha. Further, a deletion of cx_conf would only slightly improve Cronbach's alpha. In addition, with 0.6 the "corrected item-total correlation" lies above the 0.5 threshold.

	Cronbach's alpha	Corrected item-total correlation	Cronbach's alpha if item deleted
rel	0.673		
rel_tech		0.508	n.a.
rel_funct		0.508	n.a.
cx_prod	0.831		
cx_und_r		0.757	0.704
cx_use_r		0.741	0.725
cx_conf		0.600	0.877
cx_is	0.696		
cx_choice		0.474	0.659
cx_sim		0.611	0.488
cx_isinf		0.463	0.662
r_ben	0.770		
perf_expec		0.627	n.a.
fin_worth		0.627	n.a.
r_psy	0.687		
psy_detail		0.492	0.618

psy_info		0.510	0.593
psy_wrong		0.517	0.575
soc_pres	0.625		
pres_unprog		0.458	n.a.
pres_prest		0.458	n.a.
int	0.855		
int_ce		0.750	n.a.
int_inno		0.750	n.a.
n.a.: No "Cronbach's alpha if item deleted"-value can be calculated for scales that consist only of two items			

Table 60: Scale Reliability of the created scales

5.6.4 Descriptive Statistics of the Final Scales and Indices

After having tested the items in the previous chapters, in a last step, the final scales are built by summing up the item scores. For each case, the item scores that belong to the same construct are summed up and divided by the number of items of the respective construct.[909] The scores of the multi-items scales will, thus, range from a minimum of 1 to a maximum of 5. Only one-item measures and indices will have a deviating range. Descriptive statistics of the final scales including all age groups are shown in Table 61. The respective statistics divided according to single age groups G1-G3 can be found in Appendix 6. Additionally, Table 62 shows a matrix with the correlation of the final scales and indices.

All age groups	Mean	Med.	Mode	SD	Var.	Skew.	Kur-tosis	Min.	Max.
rel	3.672	4.00	4.00	0.9846	0.969	-0.584	-0.113	1	5
cx_prod	3.435	3.33	3.00	0.8951	0.801	-0.057	-0.335	1	5
cx_is	3.696	3.67	4.33	0.8801	0.775	-0.555	-0.137	1	5
r_ben	3.044	3.00	3.00	0.9508	0.904	-0.047	-0.430	1	5
r_psy	3.461	3.67	3.67	0.9179	0.843	-0.218	-0.489	1	5
soc_pres	1.938	2.00	1.00	0.9359	0.876	0.940	0.274	1	5
int	3.059	3.00	3.00	1.1404	1.300	-0.063	-0.836	1	5
fin_situ	4.903	5.00	5.00	1.1047	1.220	-0.542	0.341	1	7
compa	2.295	2.40	2.40	1.0912	1.191	0.518	-0.018	0.2	5
soc_assist	10.734	10.80	0.00	5.1294	26.311	0.108	-0.072	0	28.4
Med. = Median; Var. = Variance; Skew. = Skewness									

Table 61: Descriptive statistics of the final scales and indices

909 For this reason the method of Likert-scales is also known under the name of "method of summated ratings." Cf. Schnell, R./Hill, P. B./Esser, E. (2008), p. 189.

	rel	cx_prod	cx_is	r_ben	r_psy	soc_pres	int	fin_situ	compa	soc_assist
rel	1.000	-0.291**	-0.123**	-0.028	-0.124**	0.231**	0.308**	0.050	0.182**	0.091*
cx_prod	-0.291**	1.000	0.439**	0.179**	0.441**	-0.049	-0.342**	0.095*	-0.264**	-0.074
cx_is	-0.123**	0.439**	1.000	0.196**	0.428**	0.071	-0.197**	0.099*	-0.042	-0.082
r_ben	-0.028	0.179**	0.196**	1.000	0.390**	0.139**	-0.148**	-0.060	-0.031	0.022
r_psy	-0.124**	0.441**	0.428**	0.390**	1.000	0.103*	-0.378**	-0.037	-0.185**	-0.049
soc_pres	0.231**	-0.049	0.071	0.139**	0.103*	1.000	0.234**	-0.104*	0.080	-0.001
int	0.308**	-0.342**	-0.197**	-0.148**	-0.378**	0.234**	1.000	0.009	0.323**	0.028
	rel	cx_prod	cx_is	r_ben	r_psy	soc_pres	int	fin_situ	compa	soc_assist
fin_situ	0.050	0.095*	0.099*	-0.060	-0.037	-0.104*	0.009	1.000	-0.024	0.029
compa	0.182**	-0.264**	-0.042	-0.031	-0.185**	0.080	0.323**	-0.024	1.000	-0.010
soc_assist	0.091*	-0.074	-0.082	0.022	-0.049	-0.001	0.028	0.029	-0.010	1.000

** = Correlation is significant at the 0.01 level (2-tailed)
*= Correlation is significant at the 0.05 level (2-tailed)

Table 62: Final scales and indices correlations

5.7 Hypotheses Testing and Empirical Analysis of the Research Questions

Secondary data suggested that older and younger age groups differ in their purchase behavior regarding (innovative) CEP. Therefore, in a first step, it is analyzed whether a negative relationship between age and CEP can be confirmed (Ch. 5.7.1). The examination of the other hypotheses and research questions continues with the respondents' assessment of some general aspects of innovative CEP (Ch. 5.7.2), followed by measuring the impact of age on adoption determinants (Ch. 5.7.3) and analyzing the effects of adoption determinants on the weighted index CEPO_w (Ch. 5.7.4). The end of Chapter 5.7provides answers to the "diverse research questions" (Ch. 5.7.5). With the help of the empirical data, the preferences of information source selection, the structure of social assistance, and the situational variables are analyzed. Further, a first investigation of potential gender differences is performed.

5.7.1 Relationship Between Age and CEP Ownership

This chapter is intended to give a first glimpse regarding the relationship of the consumers' age and the ownership of CEP. Secondary data already indicated that with higher age the ownership of innovations in general and innovative CEP in particular decreases. Thus, before continuing data analysis and testing the influence of possible adoption determinants, it is tested whether the collected data confirms a general negative relationship between age and ownership of CEP. Therefore, correlations are calculated between age and CEPO, CEPO_i, and the CEPO_w index (s. Table 63). As can be seen, age significantly correlates negatively with all three ownership measures. Thus, hypotheses H Own 1, H Own 2, and H Own 3 can be confirmed.

		Age
CEPO	Pearson Correlation	-**
	Sig. (2-tailed)	0.000
CEPO_w	Pearson Correlation	-**
	Sig. (2-tailed)	0.000
CEPO_i	Pearson Correlation	-**
	Sig. (2-tailed)	0.000
** = Correlation is significant at the 0.01 level (2-tailed)		

Table 63: Correlations between age and different CEPO definitions

5.7.2 Consumers' General Assessment of Innovative CEP

Introducing the topic of age related differences in CEPO, it is interesting to look at some general aspects of the assessment of innovative CEP, taking into consideration whether individuals are young, middle-aged or old. To answer the first research question (RQ 1_1 - RQ 1_6) several ANOVA need to be calculated.

5.7.2.1 Assumption Testing

Prior to calculating the ANOVA, the assumptions mentioned in Chapter 5.4.2.2 (pp. 192) need to be tested. In this study, the respondents are grouped on the basis of their age. The age groups that have been built ("young," "middle aged," and "old" individuals) represent three distinct sub-samples. No individual turns up in more than one factor level. Thus, the errors are independent from each other and the first assumption of independent errors is fulfilled.

The other two assumptions – the homogeneity of error variances and normally distributed errors – are also fulfilled. After the outlier analysis performed in

Chapter 5.6.1.1 (p. 227), the data set consists of a total of 506 respondents. With 149 individuals in the "young" age group, 196 individuals in the "middle-aged" group, and 161 individuals in the "old" age group, the sample size of the factor levels does not surpass the 4:1 ratio which would be critical (s. Table 64). Furthermore, each group consists of more than the required 40 cases. Thus, the three factor levels are fairly equal in size and large enough to proceed to the F-test.

Group	N	%	Cumulative %
1	149	29.4	29.4
2	196	38.7	68.2
3	161	31.8	100.0
Total	506	100.0	

Table 64: Group sizes of the respondent sample

Although the ANOVA would be robust regarding departures from normality of errors and homogeneity of variances in this study, the Levene test is still conducted.[910] This test is helpful in identifying which post-hoc test needs to be employed. The post-hoc test is necessary to determine between which factor levels (groups) significant differences exist for the CEPO determinants if the F-test indicates a general mean difference between any of the factor levels in a first step. The result of the Levene-test shows that no group variances are inhomogeneous, except for tech_sup (s. Table 65). In consequence, the Scheffé post-hoc test can be employed for funct_sup, compa_exist, compa_imp, compa_small, compa_need and the Tamhane T-2 procedure for the remaining tech_sup aspect which disposes over inhomogeneous variances.

Dependent variable	Levene test statistic	df1	df2	Sig.
tech_sup	4.708	2	503	0.009
funct_sup	0.985	2	503	0.374
compa_exist	1.354	2	503	0.259
compa_imp	1.370	2	503	0.255
compa_small	1.260	2	503	0.284
compa_need	0.455	2	503	0.635

Table 65: Levene's Test of Equality of Error Variances

910 The Levene's Test of Equality of Error Variances tests the null hypothesis that the error variance of the dependent variable is equal across groups.

5.7.2.2 Results

Next to further elucidating descriptive statistics, Table 66 contains the total and group means of the different CEP aspects which are of interest here. However, it cannot be said yet whether the specific age group differences are significant. The results of the ANOVA presented in Table 67 help in the interpretation of the data and show that the null hypothesis of equal means of the factor levels can be confirmed for funct_sup, compa_exist, and compa_imp, whereas the null hypothesis can be rejected for tech_sup, compa_small, and compa_need. With a relatively large mean (mean=4.22, SD=0.83), most respondents think that innovative CEP possess more functionalities. Furthermore, the ANOVA results show that there is no age group effect regarding the respondents' assessment whether or not innovative CEP possess more functionalities than their predecessor products ($F=0.508$, $p>0.05$). The same holds for evaluating the technological compatibility of innovative CEP with older CEP ($F=0.404$, $p>0.05$) and the importance that these different generations of CEP are compatible with each other ($F=2.33$, $p>0.05$). Here most respondents say that they find technological compatibility important (mean=4.13, SD=1.00) but that this compatibility of innovative CEP with older devices does not really exist in their opinion (mean=2.84, SD=1.02).

Dependent variable	**Total**				**Group means**		
	Means	SD	Min.	Max.	G1	G2	G3
tech_sup	4.0830	0.90161	1	5	4.2685	3.9439	4.0807
funct_sup	4.2213	0.83157	1	5	4.1812	4.2092	4.2733
compa_exist	2.8439	1.01539	1	5	2.9060	2.8112	2.8261
compa_imp	4.1285	1.00163	1	5	4.1544	4.2194	3.9938
compa_small	3.3360	1.22748	1	5	2.6174	3.4031	3.9193
compa_need	3.0731	0.88586	1	5	3.3557	2.9541	2.9565

Table 66: Means and standard deviations overview of the general CEP assessment variables

Dependent variable	**Source**	**Type III sum of squares**	**df**	**Mean square**	**F**	**Sig.**
tech_sup	Group	8.919	2	4.460	5.586	0.004
	Error	401.595	503	0.798		
	Corrected Total	410.514	505			
funct_sup	Group	0.703	2	0.352	0.508	0.602
	Error	348.506	503	0.693		
	Corrected Total	349.209	505			
compa_exist	Group	0.836	2	0.418	0.404	0.668

	Error	519.830	503	1.033		
	Corrected Total	520.666	505			
compa_imp	Group	4.640	2	2.320	2.325	0.099
	Error	502.010	503	0.998		
	Corrected Total	506.650	505			
compa_small	Group	132.582	2	66.291	53.071	0.000
	Error	628.303	503	1.249		
	Corrected Total	760.885	505			
compa_need	Group	16.864	2	8.432	11.178	0.000
	Error	379.430	503	0.754		
	Corrected Total	396.294	505			
Independent variable = age group (G1-G3)						

Table 67: Tests of Between-Subjects Effects

Contrary to the above mentioned aspects of CEP, there are significant age group effects on judging the technical superiority of innovative CEP compared to their predecessors ($F=5.59$, $p<0.05$), whether innovative CEP possess buttons and labels which are too small ($F=53.07$, $p<0.05$), and in how far innovative CEP correspond to their needs ($F=11.18$, $p<0.05$). As explained before, by means of ANOVA significant mean differences for the latter three aspects are detected, but it is unclear, in this first step, between which age groups these differences occur. Therefore, the post-hoc tests are conducted. Due to its inhomogeneous variance the Tamhane T-2 procedure is used for tech_sup, while the Scheffé procedure can be applied for compa_small and compa_need (s. Table 68).

After having performed the post-hoc tests, conclusions can be drawn regarding mean differences of the different age groups (see the results in Table 69): While the youngest age group (G1) rate the technical superiority of innovative CEP significantly higher than middle-aged respondents, no differences can be found between young and old (G1 vs. G3), and between old and middle-aged individuals (G2 vs. G3). It is interesting to see that on the one hand old consumers take a "middle" position and do not differ from the other age groups while on the other hand middle-aged consumers are more critical than young consumers as they rate the superiority of innovative CEP significantly lower.

tech_sup

	(I) Group	(J) Group	Mean difference (I-J)	Std. Error	Sig.	95% Confidence Interval	
						Lower bound	Upper bound
Tamhane	1	2	0.3246*	0.08789	0.001	0.1137	0.5355
		3	0.1877	0.10063	0.178	-0.0540	0.4294
	2	1	-0.3246*	0.08789	0.001	-0.5355	-0.1137
		3	-0.1369	0.10238	0.453	-0.3826	0.1089
	3	1	-0.1877	0.10063	0.178	-0.4294	0.0540
		2	0.1369	0.10238	0.453	-0.1089	0.3826
*= The mean difference is significant at the 0.05 level							

compa_small

	(I) Group	(J) Group	Mean difference (I-J)	Std. Error	Sig.	95% Confidence Interval	
						Lower bound	Upper bound
Scheffé	1	2	-0.7856*	0.12148	0.000	-1.0838	-0.4874
		3	-1.3018*	0.12705	0.000	-1.6137	-0.9899
	2	1	0.7856*	0.12148	0.000	0.4874	1.0838
		3	-0.5162*	0.11888	0.000	-.8080	-0.2243
	3	1	1.3018*	0.12705	0.000	0.9899	1.6137
		2	0.5162*	0.11888	0.000	0.2243	0.8080
*= The mean difference is significant at the 0.05 level							

compa_need

	(I) Group	(J) Group	Mean difference (I-J)	Std. Error	Sig.	95% Confidence Interval	
						Lower bound	Upper bound
Scheffé	1	2	0.4016*	0.09440	0.000	0.1699	0.6334
		3	0.3992*	0.09873	0.000	0.1568	0.6416
	2	1	-0.4016*	0.09440	0.000	-0.6334	-0.1699
		3	-0.0024	0.09238	1.000	-0.2292	0.2244
	3	1	-0.3992*	0.09873	0.000	-0.6416	-0.1568
		2	0.0024	0.09238	1.000	-0.2244	0.2292
*= The mean difference is significant at the 0.05 level							

Table 68: Post-hoc test

	G1 vs. G2	**G1 vs. G3**	**G2 vs. G3**
tech_sup	>	=	=
compa_small	<	<	<
compa_need	>	>	=

Table 69: Post-hoc test summary

The data supports the widespread presumption concerning the button and label size in the context of aging consumers. All three age groups significantly differ with respect to these characteristics. The old age group agrees most strongly with the statement that buttons and labels of innovative CEP are too small (G1 vs. G3 and G2 vs. G3). Even the middle-aged compared to the young respondent group find that innovative CEP's buttons and labels are too small (G1 vs. G2).

Finally, young consumers compared to the other age groups state to a significantly higher rate that innovative CEP correspond to their needs (G1 vs. G2 and G1 vs. G3). In contrast, innovative CEP similarly correspond to the needs of middle-aged and old consumers (G2 vs. G3).

5.7.2.3 Summary

The statistical method of ANOVA helped to detect whether significant differences between the three age groups regarding some characteristics about innovative CEP exist. Regardless of their age, consumers think that innovative CEP possess more functunalities and that compatibility between new and old CEP does not exist although this is important to them. In particular younger consumers perceive a good fit of CEP characteristics with their needs. They have no problems with the ongoing miniaturization process. Opposed to that the oldest age group complains about buttons and labels of innovative CEP which are too small in their opinion.

5.7.3 Impact of Age on CEP Ownership and on Adoption Determinants

The previous chapters evidenced that some age related differences exist. While in Chapter 5.7.1 (p. 236) a negative correlation between age and different types of CEPO was identified, results in Chapter 5.7.2 (pp. 236) showed that some aspects of innovative CEP were assessed differently depending on the age group. In this chapter the hypotheses concerning the effect of age on CEPO and on adoption determinants (H_4 - H_{19}) are tested. This is done with the method of regression analysis which has been explained earlier (s. Ch. 5.4.3, pp. 194). As the effect of age on the single adoption determinants is examined here, *single* regressions are calculated.

5.7.3.1 Assumption Testing

A long list of assumptions for regression analyses has been presented in Chapter 5.4.3.2 (pp. 202). However, not all assumptions need to be tested. While for some of them no specific test to confirm or refute them exists (assumptions 2 and 3),[911] auto-correlated data (assumption 6) is an issue only in longitudinal studies, and multicollinearity (assumption 7) is irrelevant because only the influence of one single independent variable (age) is tested here so that no threatening correlation between the independents can possibly exist. In addition, normally distributed residuals (assumption 8) can be assumed due to a sufficiently large sample at hand.[912] The remaining four assumptions – (a) outliers,[913] (b) correct specification of the model, (c) homoscedasticity, and (d) sample size – are examined in the following.

a) For detecting possible outliers in the data, boxplots of CEPO_w and the adoption determinants were generated for each age group. The univariate outliers are reported in the first two columns of Appendix 7. Then, the multivariate analyses are performed for the same groups. On the one hand the centered leverage values were plotted against the absolute studentized deleted residuals. Bearing on the formula given in Chapter 5.4.3.2 (pp. 202), the following thresholds have been calculated for the different age groups: G1 = CLV>0.146; G2 = CLV>0.112; G3 = CLV>0.134. According to this method, participants who score higher than these critical values, are outliers when at the same time the sdr_abs values are clearly above two.[914] On the other hand the Cook's Distance has been calculated for each determinant. In this study, outliers according to the Cook's Distance are respondents with a D-score larger than the following critical values per age group: G1 = D>0.0288; G2 = D>0.0215; G3 = D>0.0261. The IDs of those respondents with critical values of the two multivariate outlier methods are displayed in columns three and four of Appendix 7.

---

911 The assumption "residuals with a mean of zero" is met when the model is specified correctly. If all systematic influence factors of Y are taken into consideration, the residual error will vary randomly. Positive and negative variations compensate each other so that the mean error is zero. The assumption "mean independence" is met as the subjects are not related to the independent variable.

912 Before searching for outliers, the dataset consists of 511 respondents which is largely more than the required 40 cases.

913 The outlier analysis is performed by taking into consideration the three age groups which do not become relevant until exploring the hypotheses in Ch. 5.7.4 (pp. 247). However, it makes sense to perform the regression analysis of this chapter and Chapter 5.7.4 with the same dataset.

914 In a first step all cases falling more or less in the critical area have been noted. In order to avoid detecting outliers which in reality are not, they are cross checked with the other multivariate outlier method.

As theory portended, there are few matches of univariate and multivariate outliers as univariate outliers do not necessarily have to be multivariate outliers at the same time. And cases that seem "normal" when using univariate outlier measures may have an unusual combination of values and, thus, may be multivariate outliers. The matches of the two multivariate outlier analyses can be found in the last column of Appendix 7. Only cases with peculiar values in both multivariate analyses were removed from the dataset. In G1 and G2 five cases were deleted so that the young age group consists of 145 respondents and the middle aged group of 192 respondents. In the old age group (G3) seven cases were removed to obtain a final group size of 157 respondents.

b) In the context of correct model specification, testing the linearity of the relationship between the independent variable and the dependent variables is of primary importance. As it is tested whether age has a significant (positive or negative) influence on the ownership of CEP and on single adoption determinants, no complex model needs to be tested and no further analyses of correct model specification are necessary.

By plotting age against CEPO and against the adoption determinants respectively, the form of the relationship is inspected. It becomes apparent that the relationships between age and "rel" as well as between age and "soc_pres" tend to be slightly u-shaped on the edges. Further, age does not seem to have a great impact on either CEPO or CEPO_w if consumers are younger than 38. With respect to persons 38 years or older however, the relationship approaches the trend of a straight line quite well. Therefore, the impact of age on CEPO/CEPO_w will additionally be calculated incorporating on the one hand only respondents aged 38 or older and on the other hand respondents aged younger than 38. Similarly, the subjective age plots with CEPO and CEPO_w indicate that age_cog does not have any effect on people perceiving themselves younger than 34. As a consequence, theses regressions are additionally calculated for respondents with a subjective age below 34 and respondents who perceive themselves 34 years of age and older. All other relationships take a form of a linear regression line for all ages which is why linearity can be assumed for them.

c) Next, the model assumption of homoscedasticity is tested. As the focus in this chapter is not on multiple but single regressions, homoscedasticity can be tested graphically. The graphical test is performed by plotting the standardized residuals (ZRESID) against the standardized predicted values (ZPRED). The graphs of the SPSS-output do not show any clear patterns which would point to the existence of heteroscedasticity. The slight funnel tendencies observable in some of the residual plots might, at worst, stand for small deviations from homoscedasticity. They are assumed to be so small that they do not have a considerable impact on the regression equation estimation. Therefore, no interaction effect needs to be modeled or variables need to be transformed. This presumption can be accepted.

d) Finally, a sufficiently large sample is the pre-condition for a generalization of the results from the regression model. After having performed the multivariate outlier analysis, the final number of valid cases adds up to 494 with G1 accounting for 145, G2 for 192, and G3 for 157 respondents, the less stringent requirements are met by far (three to five times as many observations per independent variable). Taking the strictest recommendation (ten to 15 times as many observations per independent variable), only G1 is five observations short. This should have no negative effect on the generalizability of the results. Thus, the last assumption is also achieved. To sum up, none of the assumptions necessary for the regression analysis needed to be rejected – the dataset is suited to perform the calculation of the regression equations.

5.7.3.2 Results of the Single Regression Analyses

Now, as the afore-mentioned requirements for simple regression analyses are met, hypotheses H_4 - H_{19} are tested. They intend to answer the question whether or not age significantly influences CEPO/CEPO_w as well as the single adoption determinants and, if so, whether age has a positive or negative effect on the latter. First of all it can be assumed that age → CEPO and age → CEPO_w relationships including respondents of all ages are slightly underestimated. The discussed split into consumers aged 38 and older shows a strong, negative age influence on CEPO_w (Beta = -0.394; p = 0.000).[915] Consumers up to that age (younger than 38 years) do not show any significant relationship between age and CEPO_w (Beta = 0.008; p = 0.918). The single regressions show that age has a significant, positive effect on cx_prod (Beta = 0.416; p = 0.000), cx_is (Beta = 0.279; p = 0.000), r_ben (Beta = 0.095; p = 0.036), r_psy (Beta = 0.254; p = 0.000), and fin_situ (Beta = 0.128; p = 0.004). Age also has a significant positive influence on social pressure (soc_pres; Beta = 0.154; p = 0.001), although, a negative influence was hypothesized (s. also Table 70).

915 Although the age → CEPO 38+ regression yields a higher Beta value than the result of age → CEPO_w 38+ regression, the difference can be considered as marginal.

Adoption determinants	Influence of age					Cognitive age	
	B	SE B	Beta	Sig.	p < 0,05	age_cog Beta	age_cog Sig.
CEPO_w	-0.162	0.018	-0.373	0.000	yes	-0.360	0.000
CEPO	-0.065	0.007	-0.375	0.000	yes	-0.362	0.000
CEPO_i	-0.017	0.004	-0.177	0.000	yes	-0.183	0.000
CEPO_w 38+	-0.250	0.032	-0.394	0.000	yes	-0.382*	0.000
CEPO 38+	-0.106	0.013	-0.415	0.000	yes	-0.401*	0.000
CEPO_w 38- (18-37)	0.012	0.113	0.008	0.918	no	0.010	0.895
CEPO 38- (18-37)	0.032	0.045	0.055	0.478	no	0.039	0.605
rel	**-0.003**	**0.003**	**-0.045**	**0.321**	**no**	**-0.037**	**0.435**
cx_prod	0.022	0.002	0.416	0.000	yes	0.398	0.000
cx_is	0.015	0.002	0.279	0.000	yes	0.270	0.000
r_ben	0.005	0.003	0.095	0.036	yes	**0.065**	**0.170**
r_psy	0.014	0.002	0.254	0.000	yes	0.280	
soc_pres	0.009	0.002	0.154	0.001	yes	0.154	0.001
int	-0.010	0.003	-0.155	0.001	yes	-0.173	0.000
fin_situ	0.008	0.003	0.128	0.004	yes	0.131	0.005
compa	-0.007	0.003	-0.112	0.013	yes	-0.178	0.000
soc_assist	-0.084	0.013	-0.278	0.000	yes	-0.285	0.000

SE B = standard error of B
*= As for cognitive age, the bend detected in the graphical analysis occurs at the age of 34, the age_cog regressions with CEPO and CEPO_w are executed with one group of respondents with the subjective age of 34 and older and the other group of respondents who perceive themselves between 18 and 33 years of age.

Table 70: Results of the single regressions (influence of age on CEPO/CEPO_w and on the single adoption determinants)

On the other hand, the influence of age is significantly negative for int (Beta = -0.155; p = 0.001), compa (Beta = -0.112; p = 0.013), and soc_assist (Beta = -0.278; p = 0.000). No significant influence was found for rel (Beta = 0.045; p = 0.321). Though r_ben and compa are significantly influenced by age, the effect of age is only small. Opposed to that, especially the perceived product complexity (cx_prod), the perceived in-store complexity (cx_is), and the social assistance (soc_assist) are strongly influenced by age.

Apart from the influence of the objective age, the influence of the perceived age (age_cog) on adoption determinants are also calculated and displayed in the last two columns of Table 70. Surprisingly, it can be recapitulated that the relationships between subjective age and single determinants practically do not differ from chronological age results. Only r_ben (Beta = 0.065; p = 0.170) becomes insignificant for age_cog. But, as noticed already, the chronological age effect is significant but low. The largest difference can be observed for the "compa" variable. While for chronological age the Beta value is -0.112, subjective age better predicts the perceived compatibility of CEP (Beta = -0.178; p = 0.000). Overall, however, the hypothesis H_7 that cognitive age is a better predictor of CEP adoption determinants must be rejected. A summary of the hypotheses tests is given below (s. Table 71).

Notation of hypothesis	Hypothesis was confirmed
H_4	yes
H_5	yes
H_6	yes
H_7	no
H_8	no
H_9	yes
H_{10}	yes
H_{11}	yes
H_{12}	n.a.
H_{13}	yes
H_{14}	yes
H_{15}	yes
H_{16}	yes
H_{17}	n.a.
H_{18}	no
H_{19}	yes

Table 71: Results of the hypotheses tests regarding the impact of age on CEPO/CEPO_w and on the single adoption determinants

5.7.3.3 Summary

Regarding the age effect on the adoption determinants of innovative CEP, all hypotheses were confirmed except for rel and soc_pres. While the relative advantage does not show any age effect, the age effect of social pressure shows in the opposite than expected direction. With higher age the social pressure does not decrease but, instead, increases significantly. All other variables are significantly influenced by age *and* show the hypothesized tendency of relationship. The strongest influence of age can be observed for product complexity, in-store complexity, social assistance, and psychological risk. In contrast to that, age has only a minor influence on benefit risk and compatibility. As regards to the effect of age on the CEPO_w index it is interesting to see that no age related differences exist up to an age of about 40 years. Considering individuals who are 40 years and older, however, the expected strong negative relationship between age and CEPO_w becomes apparent.

Furthermore, the results show that, as expected, it is not sufficient to simply consider an index with the total number of innovative CEP as the different novelty degrees of the CEP are neglected. Age is suitable to predict the possession of innovative CEP (CEP_i) only to a very small extent (Beta = -0.177; p = 0.000) compared to the other indices like CEPO, CEPO_w or CEPO_w 40+. Interestingly, cognitive age cannot generally be described as a better predictor of the adoption determinants, so that H_7 needs to be rejected as well.

5.7.4 Impact of Adoption Determinants on CEP Ownership

In contrast to the previous chapter, which focused on the effect age has on the adoption determinants hypotheses regarding the effect of adoption determinants on the ownership index of innovative CEP (CEPO_w) are tested here (H_{20} - H_{31}). With multiple regression analyses, the overall adequacy of the chosen set of independent variables (RQ 1) as well as the single variables' importance will be assessed (RQ 2) and alternative models will be presented (RQ 3).

5.7.4.1 Assumption Testing

As already discussed in Chapter 5.4.3.2 (pp. 202), not all theoretically derived assumptions of linear regression analyses need to be or can be examined. Moreover, the outlier analysis performed before had at that point already taken into consideration that three groups will be examined later. So, the same dataset can be used and no further outlier analyses are necessary. For that reason, group sizes will remain unchanged and the assumption of adequate sample size is also given. In consequence, three assumptions – (a) the correct specification of the model, (b) homoscedasticity, and the (c) absence of multicollinearity – are discussed subsequently.

(a) As opposed to the regression based hypotheses, where simply the effect of one independent variable on one dependent variable was examined at the same time, now the proposed models are more complex and the correct specification needs to be discussed in more detail. A correct specification cannot be finally proven. However, there are some indicators which may be helpful in avoiding an "underfitting" or "overfitting" of the regression model. First of all an extensive literature review as well as the factor analysis in Chapter 5.6 (pp. 226) led to the final set of ten independent variables which are believed to essentially influence the dependent variable CEPO_w. Variables have (as it was demanded) not been selected randomly but after careful consideration through preliminary research. Therefore, it is the hope of the author that the most relevant determinants are included in the model. Though, it is clear that it is almost impossible to capture all relevant aspects.

The regression analyses performed in this chapter also require that the relationships between independent and dependent variables are linear. Analogously to the previous chapter, a graphical analysis shed light on this issue. It can be summarized that for the large majority of plots there are relatively small deviations from linear relationships. Other forms of relationships can be noticed for rel in G3 (cubic), r_ben in G3 (cubic), r_psy in G2 (cubic), and compa in G3 (inversed u-shaped). But, these deviations from linearity are not very clear, which is why these variables are not transformed and linearity is assumed for all variables of the model.

Assumptions about the coefficients' signs stem from the review of innovation literature and research studies about elderly consumers (s. Ch. 3.5, pp. 87). As described above, the signs of the regression coefficients are important to be reviewed before interpreting the results of the regression analysis. Deviations from the hypothesized tendency of relationship are indicators of model misspecifications. The revision of coefficients will take place in next sub-chapter parallel to the regression equation calculation. The variables in questions (with opposed signs) will be excluded and the model will be calculated again.

(b) Another assumption has postulated that the collected data of determinants needs to be homoscedastic – meaning that the residuals show constant variances. As the regression models contain several independent variables, heteroscedasticity is unveiled with the Glejser test (in contrast to the graphical analysis of the single regressions). It is performed for each of the three age groups. The results of the Glejser tests show non-significant coefficients for G1 ($p=0.20$) and G3 ($p=0.08$). In contrast, the coefficient lies under the requested $p>0.05$ for G2 ($p=0.01$). However, this represents just a moderate violation of homoscedasticity with a minor impact on the G2 regression estimation. As reported earlier, transformation is an option only in situations with strong heteroscedasticity. Therefore, larger negative effects on regression results caused by heteroscedasticity are not expected.

(c) Another requirement that needs to be reviewed before conducting regression analyses is the absence of multicollinearity in the data. Multicollinearity – which stands for an unduly high correlation of the independent variables – can be assessed with the VIF. The VIF-value can easily be added to the standard output of the regression model calculation by requesting "collinearity diagnostics" in SPSS. As the VIF of the single adoption determinants ranges from 1.054 (G1)/1.042 (G2)/1.095 (G3) to a maximum of 1.949 (G1)/1.767 (G2)/1.669 (G3) in the present dataset, the critical value of five is not exceeded (s. Table 72). Accordingly, the independent variables do not correlate too strongly and no multicollinearity problem exists in the dataset.

VIF	G 1	G 2	G 3
rel	1.285	1.268	1.327
cx_prod	1.740	1.614	1.368
cx_is	1.884	1.304	1.201
r_ben	1.308	1.327	1.206
r_psy	1.949	1.767	1.399
soc_pres	1.201	1.186	1.390
int	1.447	1.439	1.669
fin_situ	1.054	1.042	1.232
compa	1.188	1.187	1.291
soc_assist	1.258	1.056	1.095

Table 72: VIF values for G1-G3

5.7.4.2 Results of the Multiple Regression Analyses

Hypotheses H_{20} - H_{31} deal with the influence of adoption determinants on the ownership index of innovative CEP (CEPO_w). After theoretical preliminary work, ten variables had been identified which are believed to exert influence on the dependent variable. It was hypothesized that for all determinants the influence on CEPO_w is significant. RQ 1 dealing with the regressions' goodness of fit is assessed via a comparison of R^2-values. The amount of explained variance is expressed in higher R^2-values. No cues have been found in the literature as to which of the specific determinants might be more important and, thus, have a stronger influence than others (RQ 2). Furthermore, if single determinants should exhibit non-significant influence in the regression equation calculation, no concrete hypotheses about *alternative*models exist (RQ 3). Therefore, a two-fold approach is selected here. First, a regression analysis using the "enter"-method is performed for each age group. This will show which determinants become significant when all independent variables are included in the model simultaneously and, on the other hand, which determinants are redundant and should be eliminated from the model. To avoid the exclusion of an independent variable that may become significant in the absence of another independent variable, the "stepwise" procedure will be applied in a second step (in addition to the enter-method). The stepwise procedure is an automated selection process which usually delivers a very good model.[916] As the stepwise-method is not used alone, and the theoretical plausibility of the results will be considered,[917] this procedure seems most adequate.

For all three regression analyses the relevant parameters are studied in the following order: (1) the F-value to assess whether the effect of one or more independent variables are different from zero and, thus, whether it is acceptable to generalize the obtained regression results for the total population; (2) the coefficient of multiple determination (R^2) to show how much variance is explained by the predictors altogether; and (3) the Beta-values to meaningfully compare the strength of influence of the single independent variables. Moreover (4), as mentioned already, the empirically gathered tendencies of the relationship (positive or negative sign) are compared with the theoretically hypothesized signs to unveil eventual misspecifications of the model and exclude the variables in question.

F-values and R^2

The F-values of 7.413 (G=1), 9.563 (G=1) and 8.005 (G=3) are large enough to reject the null hypothesis that the R^2 is zero in each group. Overall, R^2-values of 0.290 (G = 1), 0.301 (G=2), and 0.431 (G=3) were obtained for the regression equations. Drawing on the classification mentioned earlier, the amounts of variance explained by the adoption determinants represent “good” results (this corresponds to “large” R^2-values). All of the three age group R^2-values fall in the same category (“good”-category ranging from R^2=0.25 to R^2=0.49). Whereas G1 and G2 lie on the lower limit, G3 lies on the upper limit of this category.

Enter Group 1

The set of adoption determinants is able to explain 29.0% of the CEPO_w variance. With interest (int; Beta=0.228; p=0.010) and financial situation (fin_situ; Beta=0.156; p=0.038) only two determinants are significant in the youngest age group. However, a third determinant, the relative advantage (rel; Beta=0.163; p=0.050) gets very close to the cut-off value $p<0.05$ and should therefore not be neglected. As the Beta-values allow comparing the strength of influence of the determinants, it can be concluded that for young consumers the interest in CEP has the strongest influences on CEPO_w, followed by the financial situation and the perceived relative advantage.

Taking a look at the regression coefficients (s.Table 73), it becomes apparent that all three signs of the significant determinants show a tendency of relation-

916 Cf. Fahrmeir, L./Kneib, T./Lang, S. (2007), p. 164.

917 Analogous to the recommendation of Backhaus, K. et al. (2008), p. 101.

ship as it was hypothesized (s. Table 74). So, there is no further evidence of any misspecification of the model for G1.

	Unstandardized coefficients		Standardized coefficients	Sig.	p < 0.05
	B	Std. Error	Beta		
rel	1.233	0.623	0.163	0.050	no
cx_prod	0.374	0.742	0.048	0.616	no
cx_is	-0.898	0.701	-0.128	0.202	no
r_ben	-0.047	0.544	-0.007	0.931	no
r_psy	-1.289	0.709	-0.185	0.071	no
soc_pres	0.804	0.618	0.104	0.195	no
int	1.368	0.526	0.228	0.010	yes
fin_situ	0.844	0.404	0.156	0.038	yes
compa	0.454	0.489	0.074	0.355	no
soc_assist	0.033	0.107	0.025	0.755	no

Table 73: Multiple regression result for G1 (enter-method)

G1 significant independent variables	Hypothesized tendency of relationship	Signs in G 1
rel	+	+
int	+	+
fin_situ	+	+

Table 74: Tendencies of relationship for G1

Enter Group 2

In the "middle-age" group 30.1% of the CEPO_w variance is accounted for by the adoption determinants. Altogether four independent variables significantly influence CEPO_w. While interest (int; Beta = 0.336; p = 0.000) ranks – similarly to the youngest age group – first in the assessment of the most important predictors, in-store complexity (cx_is; Beta = -0.263; p = 0.000) ranks second, followed by the financial situation (fin_situ; Beta = 0.195; p = 0.002) and perceived product complexity (cx_prod; Beta = -0.167; p = 0.036).

	Unstandardized coefficients		Standardized coefficients	Sig.	p < 0.05
	B	Std. Error	Beta		
rel	-0.179	0.500	-0.025	0.721	no
cx_prod	-1.412	0.669	-0.167	0.036	yes
cx_is	-2.216	0.599	-0.263	0.000	yes
r_ben	-0.254	0.560	-0.032	0.651	no
r_psy	0.404	0.640	0.052	0.529	no
soc_pres	0.263	0.589	0.030	0.656	no
int	2.171	0.481	0.336	0.000	yes

fin_situ	1.334	0.433	0.195	0.002	yes
compa	-0.102	0.468	-0.015	0.828	no
soc_assist	0.031	0.088	0.022	0.726	no

Table 75: Multiple regression result for G2 (enter-method)

Here the empirically gathered tendencies of relationship are also in accordance with those postulated in theory (compare the coefficients displayed in Table 75 with the hypothesized signs in Table 76).

G2 significant independent variables	Hypothesized tendency of relationship	Signs in G2
cx_prod	-	-
cx_is	-	-
Int	+	+
fin_situ	+	+

Table 76: Tendencies of relationship for G2

Enter Group 3

In G3, the oldest age group, 43.1% of the CEPO_w variance is shared with the set of the adoption determinants in this sample. A total of five independent variables exert a significant influence on CEPO_w (s. Table 77). Once again, interest is the most important variable (int; Beta=0.535; p = 0.000), this time followed by social pressure (soc_pres; Beta = -0.271; p = 0.000), the received social assistance (soc_assist; Beta = 0.150; p = 0.023), the perceived product complexity (cx_prod; Beta = -0.146; p = 0.047) and finally the financial situation (fin_situ; Beta = 0.140; p = 0.045).

	Unstandardized coefficients		Standardized coefficients		
	B	Std. Error	Beta	Sig.	$p < 0.05$
rel	-0.587	0.464	-0.091	0.208	no
cx_prod	-1.258	0.628	-0.146	0.047	yes
cx_is	-0.094	0.575	-0.011	0.870	no
r_ben	-0.586	0.478	-0.084	0.222	no
r_psy	0.060	0.587	0.008	0.919	no
soc_pres	-1.639	0.445	-0.271	0.000	yes
int	3.039	0.458	0.535	0.000	yes
fin_situ	0.904	0.447	0.140	0.045	yes
compa	-0.637	0.424	-0.107	0.135	no
soc_assist	0.220	0.096	0.150	0.023	yes

Table 77: Multiple regression result for G3 (enter-method)

When comparing the theoretically hypothesized and the empirically gained signs of the predictors in Table 78, it becomes apparent that all signs of the significant determinants are in hypothesized tendency of relationship, except for the social pressure variable (soc_pres). Although it seems that social pressure plays an important role in determining CEPO_w of the oldest consumer group with a respectable Beta of -0.271, the predictor needs to be excluded from the regression equation estimation. A reason for this deviation could lie in a reversed causation of social pressure. It makes little sense to believe that instead of triggering consumers to purchase innovative CEP to keep up with their peer-group, social pressure leads to a refusal of innovative CEP purchases in G3. The effect is more likely reversed for these consumers. As members of the oldest age group possess fewer/less innovative CEP than their younger counterparts, elderly consumers may feel a certain pressure that they actually *should* possess more of these devices. In this situation social pressure would not be a predictor but a consequence of the ownership level of innovative CEP.

G3 significant independent variables	Hypothesized tendency of relationship	Signs in G 3
cx_prod	-	-
soc_pres	+	-
int	+	+
fin_situ	+	+
soc_assist	+	+

Table 78: Tendencies of relationship for G3

Enter Group 3 (modified)

As in the following regression also a wrong relationship tendency could be detected for the "rel"-variale (negative sign instead of positive sign)[918] another regression analysis was performed – this time without the variables soc_pres and rel. The results of the regression equation calculation show a drop of the R^2-value from 0.431 to 0.358. Despite the prominent reduction the R^2-value still corresponds to a "large" R^2-value. The independent variables are still "good" predictors of the dependent variable (s. Table 79).

	Unstandardized coefficients		Standardized coefficients	Sig.	p < 0.05
	B	Std. Error	Beta		
cx_prod	-0.918	0.652	-0.107	0.161	no
cx_is	-0.179	0.604	-0.021	0.767	no
r_ben	-0.653	0.502	-0.094	0.195	no

918 After deletion of soc_pres the relative advantage (rel) has become a significant predictor. But, as it displayed a negative sign in the regression analysis despite an expected positive sign the variable was also deleted.

r_psy	-0.490	0.602	-0.062	0.418	no
int	2.516	0.460	0.443	0.000	yes
fin_situ	1.315	0.455	0.204	0.004	yes
compa	-0.828	0.445	-0.139	0.065	no
soc_assist	0.264	0.100	0.180	0.009	yes

Table 79: Multiple regression result for G3 (enter-method; without soc_pres and rel)

As can be seen in Table 79, only three variables remain with significant influence: the interest in CEP (int; Beta=0.443; p=0.000), the financial situation (fin_situ; Beta=0.204; p=0.004) and, finally, the received social assistance (soc_assist; Beta=0.180; p=0.009).

For the remaining predictors all tendencies of relationship are as they were assumed (s. Table 80).

G3 significant independent variables	**Hypothesized tendency of relationship**	**Signs in G 3**
Int	+	+
fin_situ	+	+
soc_assist	+	+

Table 80: Tendencies of relationship for G3 (without soc_pres and rel)

The next table gives a summary of the multiple regressions when the enter-method is used:

Notation	**Tendency of relationship**	**Hypothesis confirmed**		
		G1	G2	G3
H_{20}	rel $\xrightarrow{+}$ CEPO_w	**yes**	no	wtr
H_{21}	cx_prod $\xrightarrow{-}$ CEPO_w	no	**yes**	no
H_{22}	cx_is $\xrightarrow{-}$ CEPO_w	no	**yes**	no
H_{23}	r_ben $\xrightarrow{-}$ CEPO_w	no	no	no
H_{24}	r_fin $\xrightarrow{-}$ CEPO_w	n.a.	n.a.	n.a.
H_{25}	r_psy $\xrightarrow{-}$ CEPO_w	no	no	no
H_{26}	soc_pres $\xrightarrow{+}$ CEPO_w	no	no	wtr
H_{27}	s_eff $\xrightarrow{+}$ CEPO_w	n.a.	n.a.	n.a.
H_{28}	int $\xrightarrow{+}$ CEPO_w	**yes**	**yes**	**yes**
H_{29}	compa $\xrightarrow{+}$ CEPO_w	no	no	no
H_{30}	fin_situ $\xrightarrow{+}$ CEPO_w	**yes**	**yes**	**yes**
H_{31}	soc_assist $\xrightarrow{+}$ CEPO_w	no	no	**yes**

wtr = wrong tendency of relationship

Table 81: Results of the hypotheses tests regarding the impact of adoption determinants on CEPO_w (G1-G3)

Stepwise / Enter
It has become apparent that not all adoption determinants play a significant role in explaining the level of a consumer's ownership index score of CEP. On the one hand, variables with low correlations with CEPO_w show insignificant results. But, on the other hand, independent variables with high correlations with CEPO_w may also show insignificant results in regression analyses. This might be the case, even when the single predictor taken as only variable in a regression analysis showed significant results before. The formerly significant variable (of the single regression analysis) may become insignificant in the presence of other predictors due to the fact that predictors are rarely completely uncorrelated. As a consequence, variables that have a stronger influence on the dependent variable push other variables out of the model. With the following automated selection process of predictors, alternative models are suggested which leave out insignificant predictors. The stepwise method and, if necessary, the enter-method are used to guarantee that the best combinations of predictors with $p<0.10$ are included and therewith alternative models in explaining CEPO_w are proposed.

Group 1 Stepwise / Enter
After applying the stepwise method of regression equation estimation and the enter-method to integrate fin_situ ($p<0.10$), a model with the following characteristics is obtained: A significant F-value of 12.799 means that the set of independents is unequal to zero. In fact, the four significant predictors explain 26.8% of the CEPO_w variance ($R^2 = 0.268$). In order of their influence the significant predictors are interest (int; Beta = 0.277; p = 0.001), the psychological risk (r_psy; Beta=-0.224; p=0.005), the perceived relative advantage (rel; Beta=0.176; p=0.020), and the financial situation (fin_situ; Beta=0.142; p=0.052).[919] Now, with the exclusion of the interfering non-significant independents, a model is proposed where rel and r_psy are also included (s. Table 82).

	Unstandardized coefficients		Standardized coefficients	Sig.
	B	Std. Error	Beta	
rel	1.333	0.568	0.176	0.020
r_psy	-1.565	0.548	-0.224	0.005

919 As the signs show the hypothesized tendency of relationship, no further variables needed to be excluded.

int	1.662	0.489	0.277	0.001
fin_situ	0.768	0.392	0.142	0.052

Table 82: Multiple regression result for G1 (stepwise- and enter-method)

Group 2 Stepwise / Enter

For the middle-aged group of respondents the stepwise method results in a model with a significant F-value of 19.840 and R^2 of 0.298. Four significant independent variables are also obtained with interest (int; Beta=0.326; p=0.000) being the most important, followed by the perceived in-store complexity (cx_is; Beta=-0.252; p=0.000), the financial situation (fin_situ; Beta=0.188; p=0.003), and the perceived product complexity (cx_prod; Beta=0.176; p=0.020). In terms of the selection of predictors, the stepwise method comes to a result identical to the enter-method which considered all ten independents (s. Table 83). The variables which remain in the model after the automated selection process are those for which significant Beta values were obtained.

	Unstandardized coefficients		**Standardized coefficients**	**Sig.**
	B	**Std. Error**	**Beta**	
int	2.101	0.415	0.326	0.000
cx_is	-2.130	0.566	-0.252	0.000
fin_situ	1.285	0.422	0.188	0.003
cx_prod	-1.228	0.591	-0.145	0.039

Table 83: Multiple regression result for G2 (stepwise- and enter-method)

Group 3 Stepwise / Enter

Applying the stepwise regression method for the oldest age group G3[920] resulted in suggesting two alternative models which differ in the variables r_ben and cx_prod. While both predictors show non-significant influences ($p \geq 0.10$) when included in the same model, they become significant when integrated individually in separate models. Therefore Table 84 presents the results incorporating r_ben, but not cx_prod. This model results in an F-value of 18.674 and R^2 of 0.329. Again, the most important predictor of CEPO_w is interest (int; Beta = 0.448; p = 0.000). But the financial situation (fin_situ; Beta = 0.213; p = 0.003), social assistance (soc_assist; Beta = 0.183; p = 0.007), and benefit risk (r_ben; Beta = -0.133; p = 0.059) are also significant determinants.

920 The variable soc_pres was not integrated into the model due to the wrong sign that has been detected earlier.

	Unstandardized coefficients		Standardized co-efficients	Sig.
	B	Std. Error	Beta	
r_ben	-0.926	0.486	-0.133	0.059
int	2.544	0.380	0.448	0.000
fin_situ	1.370	0.450	0.213	0.003
soc_assist	0.270	0.099	0.183	0.007

Table 84: Multiple regression result for G3 (stepwise- and enter-method; including r_ben)

Table 85 shows the results of the regression analysis including cx_prod and excluding r_ben. The F-value (18.447) and R^2 (0.327) are almost identical. That is why none of the two models is more accurate than the other. The single Beta-values are slightly different with interest (int; Beta = 0.410; p = 0.000), the financial situation (fin_situ; Beta = 0.239; p = 0.001), social assistance (soc_assist; Beta = 0.191; p = 0.005), and instead of r_ben the perceived product complexity (cx_prod; Beta = -0.126; p = 0.085). In contrast to the enter-method that included all independents, the omission of irrelevant variables resulted in significant Beta-values of r_ben and cx_prod, respectively.

	Unstandardized coefficients		Standardized co-efficients	Sig.
	B	Std. Error	Beta	
cx_prod	-1.087	0.627	-0.126	0.085
int	2.329	0.412	0.410	0.000
fin_situ	1.538	0.435	0.239	0.001
soc_assist	0.280	0.099	0.191	0.005

Table 85: Multiple regression result for G3 (stepwise- and enter-method; including cx_prod)

Finally, Table 86 displays the Beta-values of adoption determinants which have a significant influence on CEPO_w. These variables represent the alternative models that can be proposed instead of integrating all the theoretically derived determinants. Fortunately, all R^2-values fall in the "large" R^2-value category. That means that the selected variables represent "good" models despite the omission of several determinants.

	G1		G2		G3 (r_ben)		G3 (cx_prod)	
R^2	0.268		0.298		0.329		0.327	
Order	Variable	Beta	Variable	Beta	Variable	Beta	Variable	Beta
1.	int	0.277	int	0.326	int	0.448	int	0.410
2.	r_psy	-0.224	cx_is	-0.252	fin_situ	0.213	fin_situ	0.239
3.	rel	0.176	fin_situ	0.188	soc_assis t	0.183	soc_assist	0.191
4.	fin_situ	0.142	cx_prod	-0.145	r_ben	-0.133	cx_prod	-0.126

Table 86: Overview of alternative regression models (dependent variable CEPO_w)

5.7.4.3 Summary

The empirical analysis of the adoption determinants' impact on CEPO_w has shown that for no age group all of the hypothesized determinants become significant. Therefore, alternative models were proposed for each age group, leaving out irrelevant variables. In all age groups four determinants were detected as significant influence factors on CEPO_w. One interesting result from the alternative model calculations was that "interest" is the most important determinant of CEPO_w no matter which age group is looked at. "Interest" is an important predictor for CEPO_w, especially in the oldest age group. Further, the financial situation is a significant CEPO_w factor throughout all age groups (second largest Beta value for G3, third largest for G2, and fourth largest for G1). Differences exist for the other significant influence factors. While psychological risk and relative advantage are relevant CEPO_w determinants in the youngest age group, in-store complexity and product complexity are relevant in the middle-age group. Two alternative models for the oldest age group were proposed. Both had social assistance as significant variable, but differed in the fourth determinant being benefit risk in one model and product complexity in the other.

5.7.5 Diverse Research Questions and Analyses

Next to the main hypotheses and research questions, some other research areas were identified that help in understanding the innovation behavior of old and young consumers. First it will be looked at the sources of information that consumers use before a potential purchase of innovative CEP (s. Ch. 5.7.5.1). Then, the role of social assistance is looked at in closer detail (s. Ch. 5.7.5.2) and a descriptive analysis of the situational variables is performed (s. Ch. 5.7.5.3). A closer look is also taken at the issue of cognitive age (s. Ch. 5.7.5.4). Finally, Chapter 5.7.5.5 is aimed to shed light on possible gender differences in CEPO.

5.7.5.1 Information Source Selection

Analyzing the status-quo of information source selection may provide interesting insights into how consumers prepare their purchase decisions (RQ 4). A graphical overview shows the sources selected by the respondents of the three age groups. The information sources are split into personal (s. Figure 33) and non-personal/mass media information sources (s. Figure 34). The charts visualize different information sources and percentages of consumers who would use the respective source prior to purchasing an innovative CEP.[921] The most important source is conversations with family/relatives/friends (72.9%), followed by the internet (68%). Nearly on the same level are the "in-store sales personnel" (55.9%) and test magazines (55.7%). More than 14 percentage points fewer are devoted to newspapers and periodicals (41.1%) and leaflets containing product information (38.9%). Own experiences (32.6%) are on a similar level while other sources are sparely chosen, at least when all age groups are taken together.

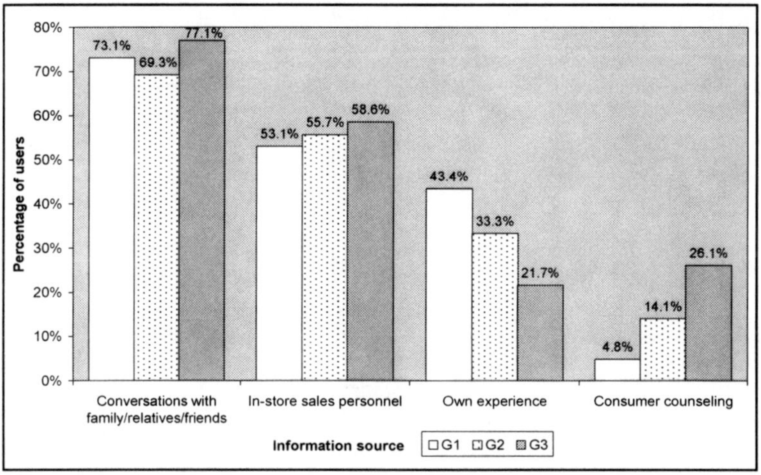

Figure 33: Respondents' selection of personal information sources

Further insight can be gained when *comparing* the relevant information gathering of the three age groups. As regards to personal information channels, no large differences exist between G1, G2 and G3 for the conversations with family/relatives/friends and the in-store sales personnel. Still, it is interesting to see that the oldest age group shows the highest percentage for both (77.1% and 58.6% respectively). In contrast to that, clear differences exist between "own experience" and "consumer counseling." While 43.4% of the youngest consumers heavily rely on own experiences, only 21.7% of the oldest age group members and 33.3% of the middle-aged respondents do equally so. The situation is reversed when taking a look at the use of consumer counseling. Here, 26.1% (G3) stand against 14.1% (G2) and only 4.8% (G1).

921 Multiple answers were allowed.

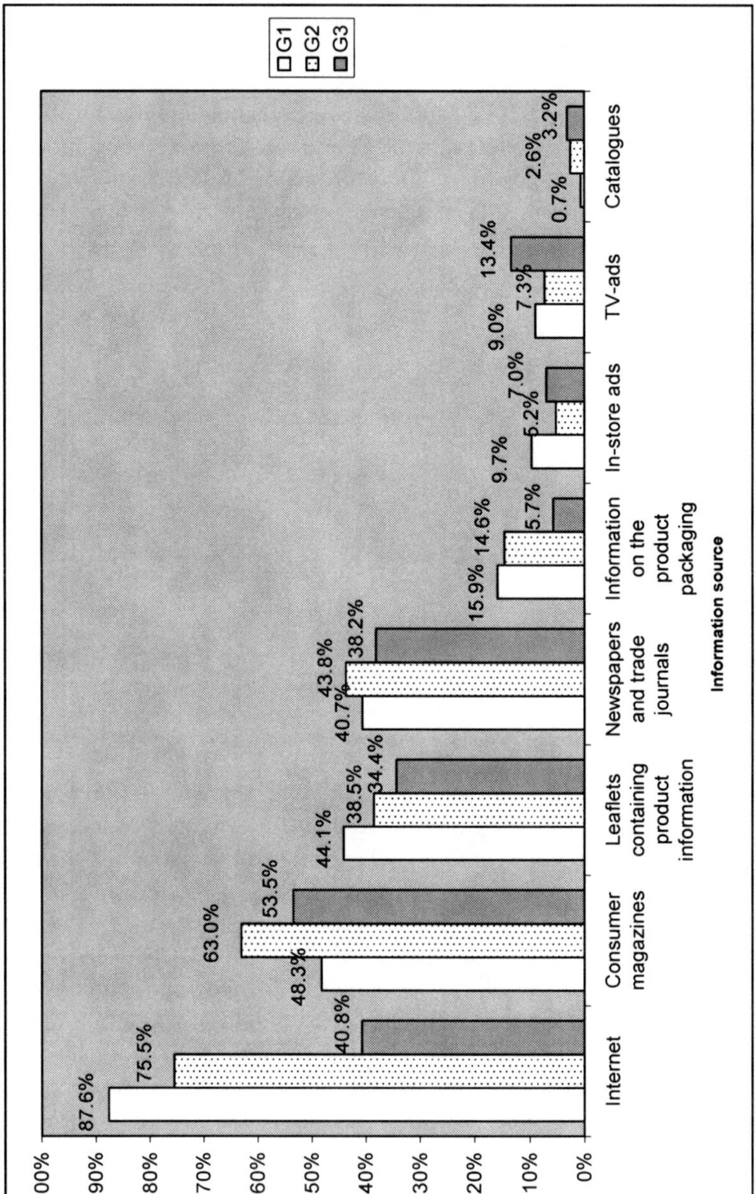

Figure 34: Respondents' selection of non-personal information sources

Concerning non-personal/mass media sources, the most striking differences revolve around the use of the internet. Almost every member of the youngest age group (87.6%) and three-quarter of the middle-aged consumers (75.5%) consult the internet. A respectable 40.8% of the older consumers gather information via the internet. Still, it is a far smaller share than the other two age groups. Next to the internet playing a dominant role (at least for G1 and G2), the second most important non-personal information source across all age groups are test magazines. Middle aged consumers show most interest in this source (63.0%), followed by G3 (53.5%) and G1 (48.3%). Older respondents consult less often existing leaflets containing product information (G3=34.4% vs. G1=44.1% and G2=38.5%) and do not use newspapers and trade journals as often as their younger counterparts (G3=38.2% vs. G1=40.7% and G2=43.8%). The information supplied by the packaging of the product is generally used less often – the percentages of users range from an extreme low 5.7% (G3) to 15.9% (G1), G2 takes a middle position (14.6%). Although on a lower level, it is still interesting to see that TV-ads are important for the elderly. While they are used by few of the younger age groups (9.0% in G1 and 7.3% in G2) a surprising 13.4% of the oldest consumers use this information. In-store ads are of little relevance and catalogues are nearly irrelevant.

It can only be hypothesized about whether the potential consumers would – in a real purchase situation – really use the information sources as indicated. But at least for the oldest age group one further aspect must be taken into consideration. With 28.7% nearly a third of the G3 individuals answered that they would let family members or friends purchase the innovative CEP. This is a large percentage compared to the younger respondents who would act this way only in 8.9% (G2) and in as few as 4.1% (G1) of the cases.

5.7.5.2 Social Assistance

The regression analysis in Chapter 5.7.4.2 (pp. 249) showed that social assistance does not play a significant role in all age groups. However, it could be confirmed that especially elderly consumers benefit from assistance by family members or friends. Older individuals who receive more social assistance score higher on the CEPO_w index. Depending on the set of other variables included in the different regression models, the influence of the social assistance index (soc_assist) varies from Beta 0.183 to Beta 0.191, which is a fairly high effect. To know more precisely whom elderly consumers *can* and *do* ask for advice in a pre-purchase situation, a closer look is taken at the quality and quantity of social contacts of the oldest age group G3. These two aspects (quantity and quality) have been fused into the single index "soc_assist" and are decomposed in the following (RQ 5).

From the point of view of elderly consumers, the persons who would be most helpful when purchasing an innovative CEP are children and friends. Taking together the percentage of individuals who responded with five and four on the five point Likert scale (with five being "very helpful" and one being "not helpful at all"), 60.7% chose "children" and 29.6% "friends" (s. Figure 35 and Figure 36). Grandchildren might be too young for most persons in the age group in order to be helpful in the purchase decision and only 9.3% say that colleagues would be able to be (very) helpful. This assumption can be supported by the high values for the "does not apply" option that more than 50% of the respondents selected for these two groups of potential contact persons. The categories "other relatives," "siblings," "neighbors," and "parents" are practically irrelevant. Parents especially not being a great help is no surprise as the oldest age group is considered here.

A look at the contact frequency with different groups of individuals unveils that social interaction exists most notably with the own children and neighbors (s. Figure 37 and Figure 38). A total of 52.9% of the respondents are in contact with their children at least 2-3 times per week. This is all the more important when remembering that children also represent the most helpful group before a potential purchase of innovative CEP. The second most frequent contact group of G3 members is neighbors. Though 39.7% of the elderly have personal contact with their neighbors 2-3 times per week or more this group is not very relevant in giving support regarding CEP purchases. The situation for colleagues and grandchildren as contact groups is heterogeneous. Though they rank third (9%) and fourth (8.4%) in having contact nearly every day, these two categories of colleagues and grandchildren "do not apply" for about half of the G3 members. On the one hand this is because they are retired and do not maintain contact with former colleagues and on the other hand because these people do not have any grandchildren (yet). More interesting is the "friends" group. Despite a lower portion of "nearly every day"-contacts, slightly more than 70% of elderly consumers interact several times per month or 2-3 times per week. Furthermore, as seen before, friends are valued as a precious source of advice regarding innovative CEP – the second most important after their own children.

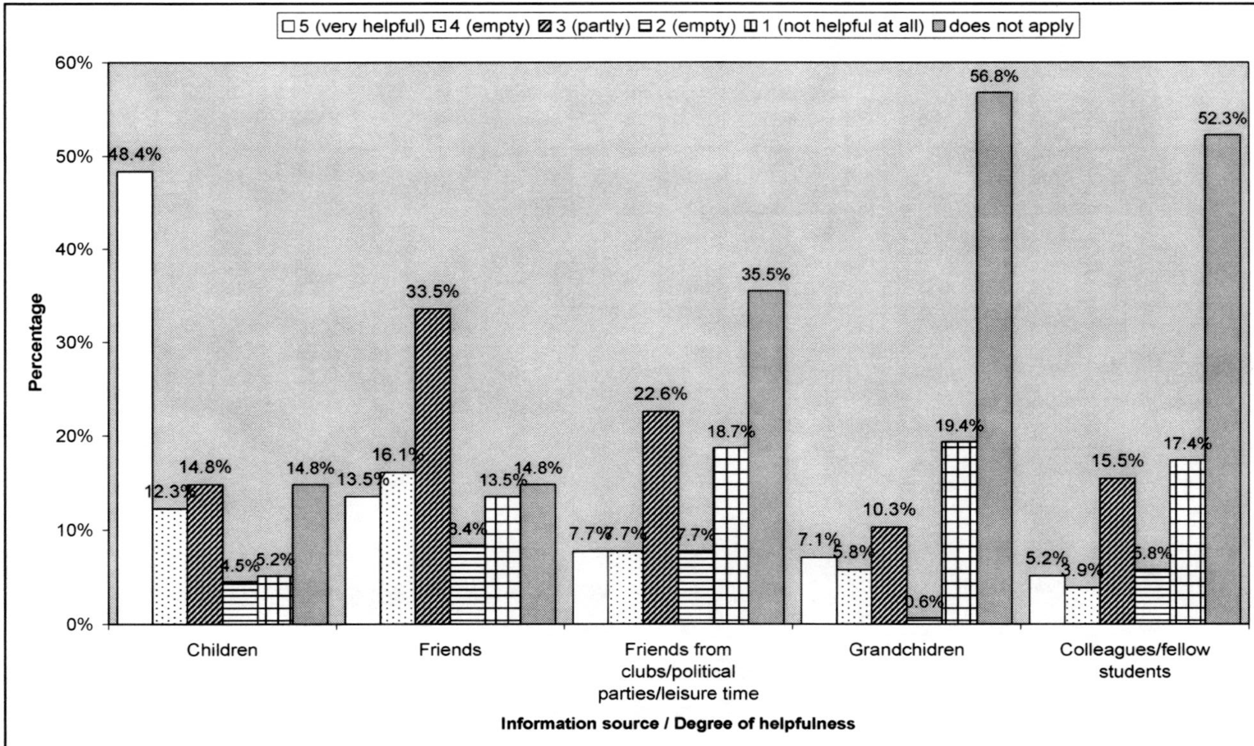

Figure 35: Help potential of social contacts regarding innovative CEP purchases (part 1)

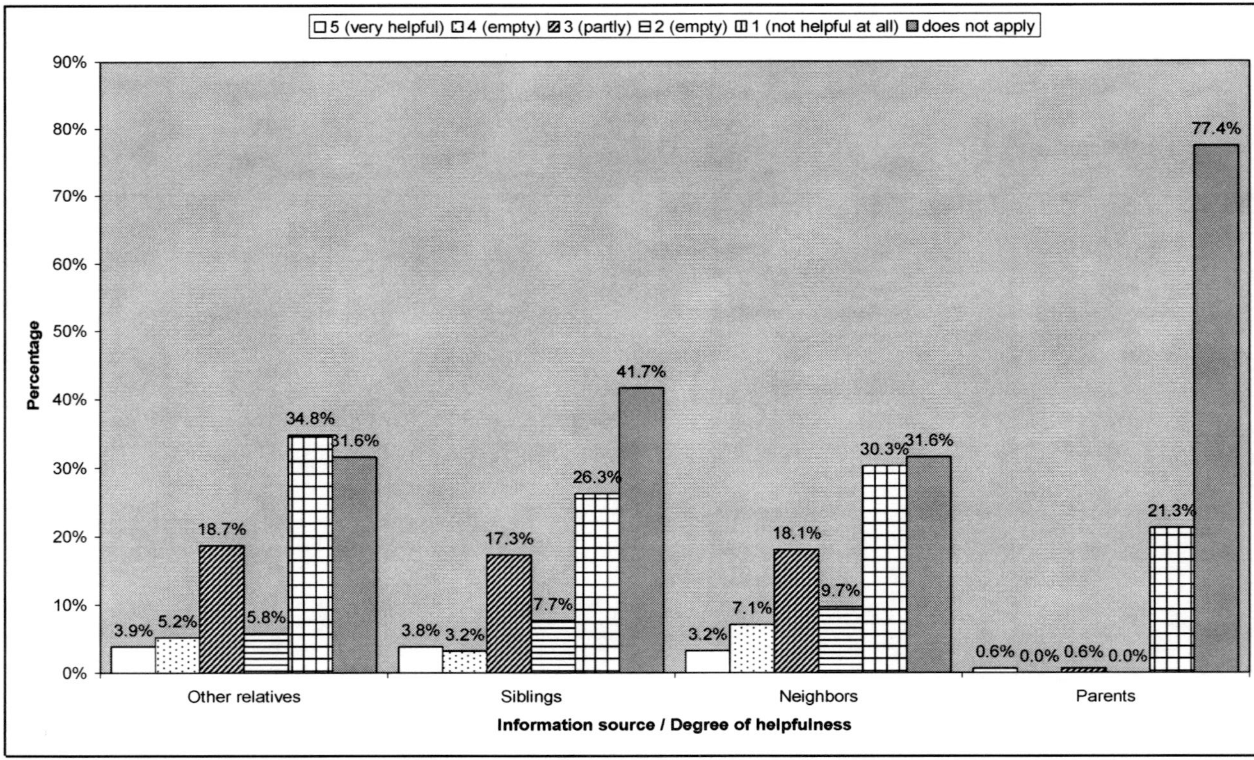

Figure 36: Help potential of social contacts regarding innovative CEP purchases (part 2)

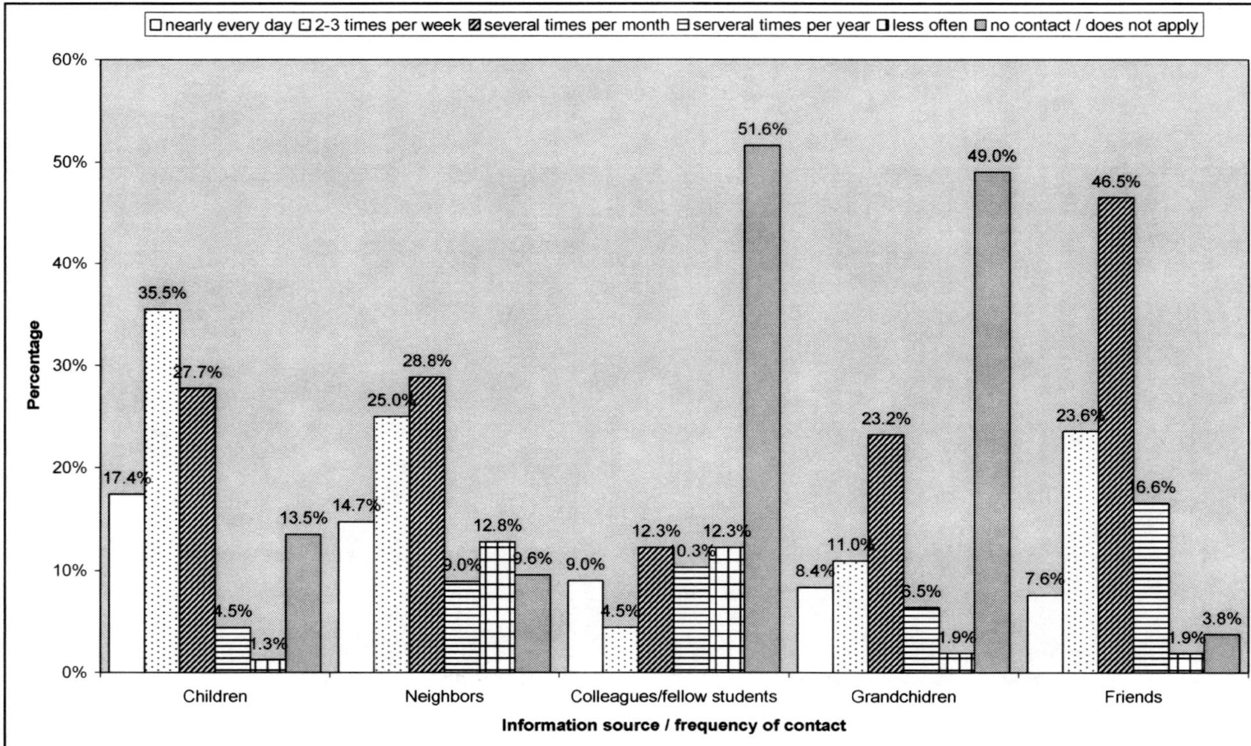

Figure 37: Contact frequency with social contacts (part 1)

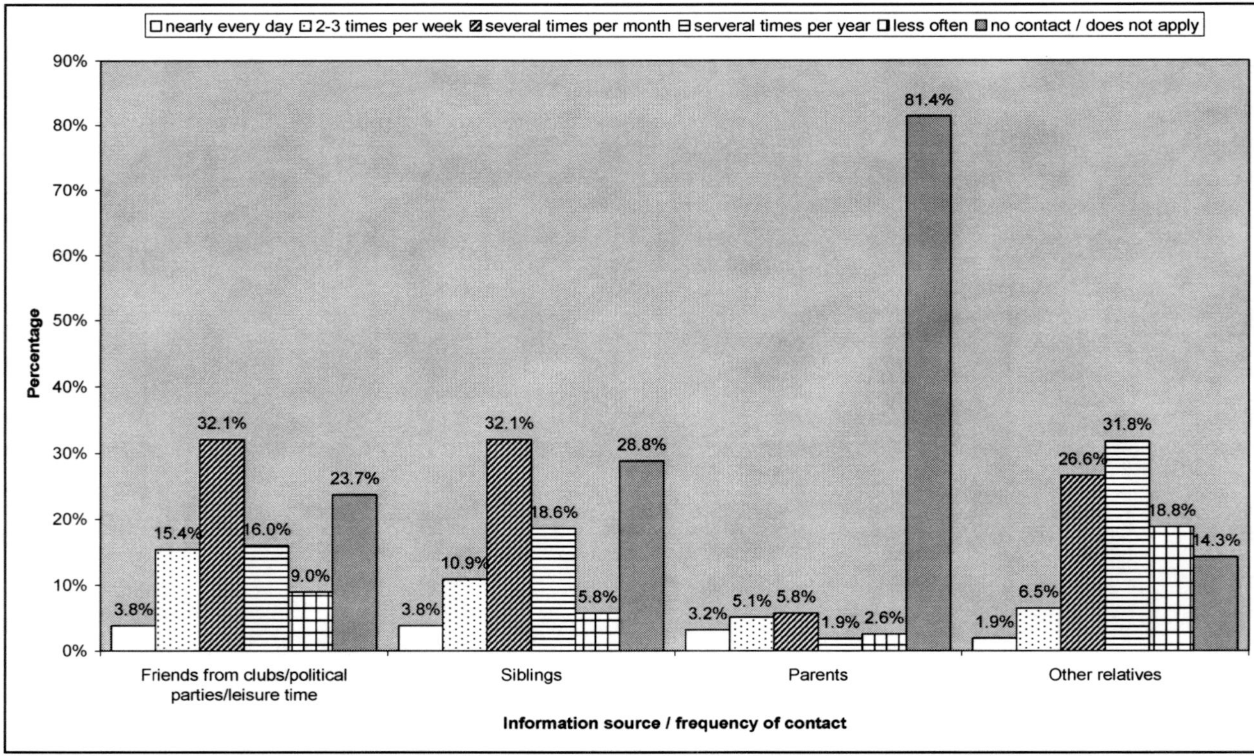

Figure 38: Contact frequency with social contacts (part 2)

It can therefore be summarized that children and friends are the most important persons who supply elderly with the necessary support when it comes to a purchase involving innovative CEP. These two groups are followed by the important groups of colleagues and grandchildren, but only for those individuals where this applies – as more than 50% of the members in G3 do not have any grandchildren or any contact with former colleagues.

5.7.5.3 Situational Variables

To gain further insight into consumer behavior regarding innovative CEP, the issue of situational variables leading to (non-)adoption of innovative CEP was picked up (RQ 6). As their impact on the ownership index (CEPO_w) was expected to be negligible, the situational variables were not integrated as adoption determinants, but seen as variables supplying additional information for improving the offered CEP. While the selected seven items deemed the most important situational variables worked well in the pre-test and were understood by the test persons, the large number of missing values after the first survey wave revealed that the statements and the idea behind it may not have been fully comprehended by many participants.[922] Therefore, the interpretation of the descriptive statistics and the recommendations should be done with care.

Mean	G1	G2	G3	All (G1-G3)
situ_1 (device broke)	2.02	1.91	2.03	1.98
situ_2 (service discontinued)	1.33	1.48	1.68	1.49
situ_3 (more modern device)	2.74	2.16	2.24	2.36
situ_n_1 (financial situation)	2.86	2.40	2.07	2.44
situ_n_2 (product not available)	1.57	1.61	1.37	1.53
situ_n_3 (lack of time)	2.22	2.27	2.04	2.19
situ_n_4 (wait for cheaper price)	2.99	2.86	2.66	2.84

Table 87: Means of the situational variables

The situational variables were measured on a five point Likert scale. Higher means indicate the situation occurred more frequently. Table 87 shows, according to age group, the means of the situational influences/reasons for (non-)purchase of innovative CEP. For all three age groups the most important reason why respondents bought a new, innovative CEP was because they desired to have a more modern device (situ_3). This was important especially for the

922 In order to be consistent and not to influence responses of the remaining questions in the second survey wave that took place, the questionnaire was not changed and the (non-)purchase statements were kept identical.

youngest age group (mean = 2.74) compared to G2 (mean = 2.16) and G3 (mean = 2.24). While the discontinuation of service (situ_2) was the most infrequent situation for all three groups, it represents a far more important reason to purchase a new device for the oldest age group (mean = 1.68) than for the middle-aged (mean = 1.48) and young consumers (mean = 1.33). Regarding the situation that most frequently resulted in *non-purchases* of innovative CEP, all three age groups named one and the same situation most often: the situation to wait for lower prices and thus postponing the purchase (situ_n_4). While out of stock situations (situ_n_2) were the least frequent situation without major age group differences, some interesting differences are observable for the momentary financial situation (situ_n_1). As expected, older consumers are by far less constrained by their financial situation in purchasing novel CEP (mean = 2.07) while opposite to that, this situation occurred far more often for the youngest age group (mean = 2.86) leading to non-purchase of CEP. This supports the results of the hitherto results and the mentioned divergence of buying-willingness and -ability for the group of younger consumers – as it can be assumed, especially for the purchase of expensive CEP innovations.

5.7.5.4 Cognitive Age

Although regression analysis revealed that cognitive age (age_cog) is not a better predictor of CEP adoption determinants than chronological age, a brief look is taken at this variable. From 466 respondents who answered the question of cognitive age, only 18 individuals (3.9%) indicated that they feel older than their actual age. Only two of the persons were older than 50 years – most of the respondents who feel older than their chronological age are between 20 and 35 years old (61.1%). In total, the two perspectives of age for 134 respondents were identical expressing that those persons feel neither older nor younger than they are. A remarkable majority of 314 respondents (67.4%) stated however, that they feel younger than their chronological age. Figure 39 visualizes these differences by plotting chronological age on the x-axis and the difference between cognitive and chronological age on the y-axis.

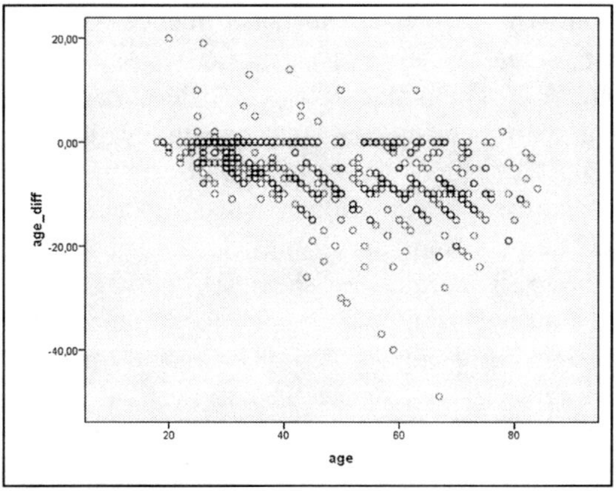

Figure 39: Chronological age and age difference plot

Conducting a correlation analysis between chronological age and the age difference between cognitive age and chronological age results in a high negative correlation (s. Table 88). This means that the older the individuals are, the greater the disparity between the subjective age and the "real" biological age. However, as analyzed in Chapter 5.7.3.2 (pp. 258), the cognitive age of a person did not show a better performance than the chronological age in predicting the adoption determinants.

		age_diff	**age**
age_diff	Pearson Correlation	1.000	-0.444**
	Sig. (2-tailed)		0.000
	N	466	466
age	Pearson Correlation	-.0444**	1.000
	Sig. (2-tailed)	0.000	
	N	466	512
** = Correlation is significant at the 0.01 level (2-tailed).			

Table 88: Correlation analysis between chronological age and age difference

5.7.5.5 Gender Analyses

Throughout the dissertation it has been mentioned that the respondents' gender should not be neglected. Gender differences may exist due to the CEP focus of the study. As gender analyses were not the central focus of the dissertation, no concrete hypotheses were put forward. Some analyses should, however, reveal whether significant gender differences exist (RQ 8).

In order to find out whether gender has an influence on the "age → determinants" relationship, single regressions are performed again. This time, however, two sub-groups are built. The impact of age on determinants and different CEPO indices is calculated for women and men separately. In the *first* analysis of Chapter 5.7.3.2 (pp. 258) only three non-significant regressions were found. Two of the non-significant regressions were CEPO indices for respondents younger than 38. The only non-significant determinant was the relative advantage. The current regression analyses show six non-significant results for the female group and four non-significant results for the male group (s. Table 89). Again there is no significant impact of age observable for the CEPO indices for respondents younger than 38 (for females as well as for males). The non-significant results are, however, also partly due to the smaller number of cases. The most pertinent results are that higher age has a much stronger, negative influence on the CEPO_w index of females (Beta=-0.489; p=0.000) than on the CEPO_w scores of their male counterparts (Beta=-0.289; p=0.000). The impact of age on ownership is even more negative for CEPO_w 38+ (females: Beta=-0.550; p=0.000; males: Beta=-0.372; p=0.000). Interestingly, the age effect regarding the perceived psychological risk is higher for men than for women (females: Beta=0.197; p=0.004; males: Beta=0.285; p=0.000).

	Influence of age					
	Female			**Male**		
Adoption determinants	Beta	Sig.	$p < 0.05$	Beta	Sig.	$p < 0.05$
CEPO_w	-0.489	0.000	yes	-0.289	0.000	yes
CEPO	-0.502	0.000	yes	-0.285	0.000	yes
CEPO_i	-0.241	0.000	yes	-0.121	0.041	yes
CEPO_w 38+	-0.517	0.000	yes	-0.358	0.000	yes
CEPO 38+	-0.550	0.000	yes	-0.372	0.000	yes
CEPO_w 38- (18-37)	**0.130**	**0.329**	**no**	**-0.070**	**0.462**	**no**
CEPO 38- (18-37)	**0.172**	**0.196**	**no**	**-0.019**	**0.843**	**no**
rel	-0.138	0.047	yes	**0.049**	**0.414**	**no**
cx_prod	0.400	0.000	yes	0.422	0.000	yes
cx_is	0.239	0.000	yes	0.296	0.000	yes
r_ben	**0.075**	**0.283**	**no**	**0.101**	**0.090**	**no**
r_psy	0.197	0.004	yes	0.285	0.000	yes
soc_pres	**0.133**	**0.055**	**no**	0.174	0.003	yes
int	-0.175	0.011	yes	-0.121	0.042	yes
fin_situ	**0.125**	**0.071**	**no**	0.139	0.019	yes

compa	**-0.058**	**0.400**	**no**	-0.141	0.018	yes
soc_assist	-0.343	0.000	yes	-0.244	0.000	yes

Table 89: Results of the single regressions after gender split (influence of age on CEPO/CEPO_w and on the single adoption determinants)

When looking at distinctive characteristics, it can further be stated that the social assistance for females decreases much stronger with age (Beta=-0.343; p=0.000) than for the males (Beta=-0.244; p=0.000). When only looking at the oldest age group, however, the social assistance for female and male respondents does not differ much (s. Table 90).[923] Besides, the depicted mean comparison table again shows large differences between the CEPO_w score of female and male respondents belonging to the oldest age group (females: mean=7.44; males: mean=13.15). The same holds true for the discrepancy of the "interest" determinant. In the oldest group of respondents the two means largely differ (females: mean=2.43; males: mean=3.30).

	All		**G1**		**G2**		**G3**	
	Female	**Male**	**Female**	**Male**	**Female**	**Male**	**Female**	**Male**
CEPO_w	12.44	16.24	16.04	17.24	14.18	17.95	7.44	13.15
rel	3.52	3.78	3.90	3.78	3.28	3.72	3.51	3.86
cx_prod	3.61	3.30	3.15	2.89	3.59	3.25	3.99	3.79
cx_is	3.86	3.58	3.62	3.33	3.79	3.54	4.13	3.89
r_ben	3.14	2.95	3.03	2.87	3.17	2.96	3.20	3.03
r_psy	3.78	3.26	3.52	3.11	3.78	3.14	3.99	3.56
soc_pres	1.92	1.94	1.93	1.80	1.71	1.93	2.18	2.10
int	2.58	3.40	2.91	3.45	2.50	3.44	2.43	3.30
fin_situ	4.81	4.95	4.67	4.84	4.72	4.84	5.04	5.22
compa	2.12	2.37	2.37	2.47	1.97	2.48	2.12	2.15
soc_assist	11.11	10.46	13.03	11.52	11.50	10.86	9.13	8.89

Table 90: Mean comparison of adoption determinants and CEPO_w after gender split

For a deeper gender analysis it would make sense to analyze the influence of the determinants on CEPO_w with multiple regression analyses (analogously to the regression analyses in Chapter 5.7.4, pp. 247). However, this is not advisable with the current dataset. The formed sub-groups are too small. Due to the separated analysis of male and female respondents in combination with the three age groups, six sub-groups are obtained. As can be seen in Table 91 only the male middle-aged group (G2) reaches the minimum number of cases of the weaker requirement for regression analyses. Normally each group should consist of at least 100 cases, better 150 cases.

923 This is due to the much higher scores of the social assistance index of women in younger years compared to younger men.

Group	Gender	Number of cases (group size)
G1	Female	54
	Male	91
G2	Female	86
	Male	106
G3	Female	69
	Male	88

Table 91: Group sizes after gender split

6 Conclusion

6.1 Synopsis and Interpretation of the Results

With this study it was aimed to add a small piece to the incomplete puzzle regarding older individuals' consumer behavior. More precisely, deficits were detected in the academic research of older persons' innovation behavior. The analysis of the current market situation and the literature review indicated that younger and older consumers' innovation behavior regarding consumer electronic products (CEP) has not been researched in great detail and, thus, represents an interesting research focus. Therefore, the central objective of the dissertation was to find out whether the ownership of innovative CEP depends on the age of the individual and, if so, what factors can be identified that are responsible for ownership.

With the help of prior gerontological knowledge in combination with theoretical knowledge on general innovation behavior of individuals, the concrete research hypotheses were formulated. At the same time the potential explanatory variables (determinants) of innovation adoption were identified. As no adequate data was available to test the research hypotheses, a survey was set up and more than 500 persons between 18 and 85 years of age were interviewed. The collected data allowed building the final scales in order to measure the theoretically generated determinants. Then, by use of several statistical methods, the hypotheses were tested and research questions were answered.

In a first step, the gathered data indicated a negative relationship between age and different definitions of CEP ownership. That means that ownership hypotheses derived from secondary data were confirmed. The fact that the ownership index of innovative CEP (CEPO_i) shows a weaker negative association with age than the other two indices suggested that older persons possess novelties that are less innovative. After all, distinguishing only between innovations and non-innovations seemed insufficient. If on the one hand all CEP defined as innovations are treated identically and on the other hand all CEP defined as non-innovations are treated equally as well, it is too imprecise and reduces the variance of the "age"-"ownership" relation. The weaker performance of CEPO_i confirms the measure to create a weighted index that takes the different innovativeness scores of the CEP into account.

On the basis of several statements regarding innovative CEP, the three age groups were assessed and a first insight into the topic was gained. There is a consensus regarding the higher number of functionalities of innovative CEP and their poor compatibility with previous generations of products despite the desired attribute of compatibility. For these statements no age differences were found. In contrast, clear differences were found for the design of products. In the opinion of elderly persons innovative CEP have buttons and labels which are too small compared to middle aged and young consumers. Even middle aged consumers complain about this inconvenience and are less satisfied with innovative CEP compared to the youngest consumer group. It is also the middle aged group of persons which seems particularly critical with regard to innovative CEP. This is expressed in low scores of technical superiority. Similarly, a low compatibility of CEP innovations with the consumers' needs was found. Innovative CEP significantly better correspond to the needs of the youngest group of individuals compared to middle aged or old individuals.

As outlined, significant negative relationships between age and CEPO indices were found. These correlational results led to the next reasonable step to find out *why* ownership differences exist. After the extensive literature review, which resulted in a number of possible explanatory variables and numerous items for the operationalization of these variables, the factor analysis finally extracted seven determinants. With the three indices that were created, a total number of ten determinants were examined. The influence of age on these determinants was the subject of hypotheses H_8 - H_{19}. The dependency analysis was also performed for the different ownership indices (H_4 - H_6). This analysis brought a very interesting finding. While the negative impact of age on CEPO, CEPO_i and CEPEO_w was confirmed, it was found that the negative influence is even stronger when only respondents 38 years and older were observed. In turn, no significant impact of age on CEPO and CEPO_w was found for younger persons (18-37 years). It can be assumed that starting with that age persons are more likely to be in the position to express their innovativeness by purchasing greater numbers of innovative CEP. As many innovative CEP are expensive, it is probable that younger consumers cannot afford them and cannot purchase them as they might wish. Regarding the determinants it can be emphasized that especially older consumers perceive high levels of product complexity in situations of innovative CEP purchases. Furthermore, the in-store complexity as well as the psychological discomfort to deal with the details of an innovative CEP purchase is perceived higher at old age. Marketing strategies must, therefore, address this particular point and reduce complexity for example by supporting the trialability of the products and by decision support through qualified service personnel. Also noticeable, but not unexpected, is the decreasing social assistance in innovative CEP purchases as well as the decreasing interest in innovative

CEP at higher age (even though the interest in CEP did not decrease so strongly as could be expected). The required actions of companies to reverse this fact depend on the reasons of the lower interest (e.g. no intended use of the innovations; advantage of technology unclear; fear of the technology etc.). This needs to be analyzed in more detail before recommendations can be given.

Hypotheses H_{20} - H_{31} dealt with the impact of the determinants on the weighted ownership index. Here the regression analyses were performed for the three age groups (18-34; 35-59; 60-85 years). The results of the regression analyses showed that not all determinants are significant factors. By sorting out the irrelevant determinants, alternative models were proposed. The models of all three age groups had in common that “interest” was the most important variable determining CEPO_w. For all age groups also the financial situation played a significant role. This seems reasonable as innovative CEP are usually relatively expensive. However, group differences still exist. For younger consumers the financial situation is not as important compared to other relevant determinants (e.g. due to financial support by parents, or due to fewer fixed household expenses). Contrary to that, relative importance of the financial situation for the oldest person group is higher – meaning that individuals who are financially worse situated in that group also express this by means of fewer/less innovative CEP and vice versa.

In each age group at least one determinant was identified that distinguished the specific group from the others: Only in the youngest consumer group, the psychological risk has a negative impact on ownership of innovative CEP. Thus, those who are reluctant to invest time/money in information research, to deal with the details of a potential purchase, and possibly choose the wrong technological standard, possess fewer/less innovative CEP and vice versa. However, since a positive dependency between age and psychological risk exists, only few of the younger consumers perceive psychological risks.

In the middle aged group the significance of perceived in-store complexity is striking. The strong negative impact can be explained with the huge product range of CEP retailers. In combination with the time limitations that can be expected in this age group (e.g. due to job related obligations, family responsibilities, etc.) the in-store complexity becomes a relevant factor of innovative CEP purchases. More transparency in stores and better services are possible strategies to countervail this tendency.

The variable “social assistance” for innovative CEP purchases becomes relevant only in the oldest age group. The combination of strongly decreasing possibilities with higher age to receive relevant help in the nearer social environment and the importance of this factor for purchases makes this variable an important starting point for product offers and service improvements. Especially for the less integrated older individuals with fewer support by friends or relatives, services offered by retailers are likely to gain importance. These services might be offered in-store in order to support the decision process, but also at home for an eventually necessary installation of the CEP or in case problems during usage arise. The largest challenge in this context will be to adequately substitute the role social assistance plays. People from the social network make older individuals *aware* of CEP innovations and their advantages when they *explain* these advantages. Furthermore, they are credible sources of information as they are independent from producers or retailers.

Even though social assistance is important, the interest in CEP is still by far the strongest determinant of CEPO_w in the oldest age group. As with age the interest decreases it would be interesting to know what reasons are responsible for that.[924] The influence of product complexity is significant (in one of the two alternative models of the oldest age group), but low. As the influence of age on product complexity is very strong, the high complexity perception regarding innovative CEP is universal for older consumers, but does not differentiate between owners of more or less innovative CEP.

Surprisingly, the cognitive age did not perform better than the chronological age regarding the prediction of adoption determinants. Although differences between the perceived and the actual age exist (old persons perceive themselves much younger compared to young individuals), the subjective age does not generally have a stronger influence on the explanatory variables of CEPO. While one interpretation of the weak performance of cognitive age has been the development of "ageless" markets,[925] this cannot be the reason in this study (as chronological age *has* a strong impact on CEPO_w). It seems that today's elderly consumers feel cognitively younger in general – independent from consumption related attitudes (at least regarding innovative CEP).

Regarding gender influences no final conclusion can be drawn. At old age great differences between the CEPO_w scores of females and males exist. Female respondents by far possess fewer/less innovative CEP. Women are also less interested in CEP. No further large differences between the *three age groups* exist regarding the other variables. The stronger influence of age on the psychological risk of males can be interpreted with the higher propensity of men purchasing

924 It has been hyothesized about the reasons above.

925 Cf. Szmigin, I./Carrigan, M. (2000), pp. 519; Schiffman, L. G./Sherman, E. (1991), pp. 187.

CEP devices. Therewith they are more likely to be confronted with situations that they perceive as unpleasant (investing time to gather the most important information, dealing with the details of the purchase, worrying about making wrong decisions regarding the technology standard). Furthermore, a stronger negative impact of age on social assistance can be observed for females. This is due to much higher scores of the social assistance index of younger women compared to young men. The scores of the oldest age group do not differ much between males and females. A major drawback of the gender analysis is that it was not possible to unveil the most important CEPO_w predictors with the help of multiple regressions. Therefore, more research is needed in this area.

If applicable, older consumers trust their children and close friends. For producers or retailers it is difficult to emulate the personal character of the induced information and service strategies. Next to personal information sources, sales personnel and consumer magazines are consulted. Consumer counselling is less important compared to the other sources, but much more important than for younger and middle aged individuals. The internet plays a subordinated role so far. However, the information source selection also needs to be interpreted carefully. Often differences between the sources indicated in the questionnaire and the actually used information sources occur.[926]

6.2 Limitations of the Study and Critical Discussion

In the following the study design and obtained results are critically discussed. Certainly, one limitation of this dissertation is the static snapshot that it gives. The cross-sectional study design, which had to be chosen due to the temporal and financial limitations of a dissertation, can only unveil age *differences*, but no age changes can be investigated. Only longitudinal studies spanning over several decades are suitable to adequately map the dynamic process of aging. Besides, due to cohort effects, it can be expected that the results of the study cannot be transferred to later generations of elderlies without further ado. The future elderlies will have experienced a completely different socialization process than today's senior consumers. They will be much more used to technologies such as mobile communication, computers, the internet, etc. Nevertheless, the relevance of analyzing the innovation behavior and its influencing variables will persist in the future. New CEP and new technologies are going to be developed and introduced to the markets and are going to replace existing solutions.[927] Therefore,

926 Cf. Kroeber-Riel, W./Weinberg, P. (2003), pp. 252.

927 Even during the process of completing the dissertation, rapid technological developments became visible. With a high pace, novel CEP were launched which combine

knowledge on products is going to be outdated and future older consumers might face similar problems with technological novelties like today's elderlies. When today's younger consumers are going to be confronted with a novelty in the future and they are unfamiliar with this novel product or service, they are likely to perceive risks, complexity etc. in the purchase situation.

Another point of concern is the "age" variable: the dominating focus on age and the exact description of "old age." As discussed in the introductory chapter, the definition of "old age" is difficult. Although it seems that a majority of researchers prefers an age threshold of 60 or 65 years, no unanimity in the research community was found. The consideration of the retirement status helped to concretize "old age" and to be more precise about the living conditions of targeted individuals. Though, it can be expected that the so defined consumer group is still very heterogeneous. For these reasons, several authors turned away from an age focus and preferred other personality variables.[928]

Several points of the questionnaire and the operationalization of the theoretical constructs also need to be mentioned here. As already discussed in detail in Chapter 5.4.1.2 (pp. 185), theoretical constructs should be operationalized with at least four items. Due to the broad conceptualization of this study regarding a large number of variables it was not possible to guarantee this minimum number per theoretical construct. In order to prove the stability of the determinants, future research should examine whether the obtained factor structure can be confirmed. Furthermore, two constructs were not found by the factor analysis as expected (financial risk and self-efficacy). Here it would be interesting to find out whether an alternative operationalization would have been more rewarding or whether these two constructs are implemented in the other determinants of the study.

Regarding the CEPO variable it needs to be noticed that the possession of an innovation does not automatically mean that the owner regularly uses the innovation. Continued use, however, is an essential aspect of adoption[929] – only if purchased innovations are continuously used ("full scale use") it is fair to talk about adoption. As a consequence, in future studies it would make sense to survey the use frequency of the respective CEP, too. By additionally investigating the use frequency another drawback can be avoided. Persons with few but inno-

---

more and more functions of previously individual products (as for example the emergence of smart phones or tablet computers).

928 See for example Moschis, G. P. (2003), pp. 521.

929 Low usage intensity could be an indicator for no need of the innovation or dissatisfaction with the product.

vative CEP may score lower on the CEPO index (CEPO_w) compared to individuals who do not throw their old devices away. The persons, who keep their old CEP in the household, even though they do not use them anymore, may score higher on the same index.

One further improvement of the questionnaire encompasses surveying who in a multi-person household purchased the CEP or is responsible for the purchase decision.[930] On the one hand this information allows improving the predictability of the adoption determinants and on the other hand this information is important for the gender analysis. When it is assumed that in multi-person households predominantly men purchase the CEP, the explanatory power of the determinants is reduced for the female survey participants.

It can also be criticized that especially elderly persons might not know all CEP devices listed in the questionnaire. However, in this context this does not seem to be significant. For the objective of the study it was crucial that the respondents know which device they possess and not which other products exist which they do not own.

6.3 Outlook and Further Research

This study achieved that some additional insights into older individuals and their consumer behavior were gained. Yet, the state of knowledge regarding older consumers is still not even remotely satisfactory. Many research areas bear a lot of potential and are hopefully covered by future research studies. With respect to the obtained results of this dissertation, it would be desirable as a first step to have future studies confirm the conceptualized determinants including the operationalization and the empirically gathered scales. Questionnaires will profit from the experience of this study and also avoid some drawbacks such as uncertainty concerning the real use frequency of the specific CEP, the decision maker and purchaser of the innovative CEP, or the operationalization of the situational variables.

With the obtained survey result, namely which determinants are relevant to explain CEPO, it is now possible to develop a structural equation model (SEM). While at the starting point of this dissertation the picture of CEPO influencing variables was diffuse and did not allow constructing an SEM, interesting insight into the consumer behavior was gained. On this basis further research studies may investigate the relevant determinants with this statistical method. Among

930 As only the ownership of CEP is surveyed it remains unclear who purchased the product.

others, the investigation would allow analyzing the existing relationship *between* the determinants by means of an SEM. Therewith the important determinants can be analyzed in more detail which was not possible so far due to the large number of potentially relevant determinants at the outset of this study. The SEM approach would further avoid the inconvenience to create age groups. The age variable can be directly integrated into the SEM. The effect of age on the determinants (which are in turn partly related to each other) and the effect of the determinants on CEPO (CEPO_w) can be analyzed simultaneously.

Another promising future research field is the analysis of eventual differences in the younger and older consumers' reaction to marketing mix variables. Especially on the basis of the obtained results in this study it would be interesting to know *which* marketing variables need to be configured and *how* they need to be configured. This would permit answering for example what ideal measures can be taken on the company side to compensate the lower assistance levels of older individuals, stimulate interest for innovative CEP, and reduce the high levels of perceived complexity – of the products themselves as well as of the assortment in retail stores. But more general aspects would also be of interest: Which product modifications are required? How should the advantages of the innovation best be communicated? What is an adequate price for the innovation and how should the distribution strategy look like? Which services should be offered and how can consumers be further supported before, during, and after the purchase?

Especially with regard to CEP, a more detailed analysis of gender differences would be beneficial to the discussion. For all ages, but especially for old age, a deeper analysis would make sense as with the longer life-expectancy of women, their ratio is continuously going to increase in the future. Due to the reported heterogeneity of older consumers, future studies could additionally assess different approaches to either cluster sub-groups of elderly consumers or, even more farsighted, to detect clusters by considering consumers of all ages and build intergenerational segments.[931] Further an international comparison would enrich the discussion. While the results gained here apply to German consumers of different ages, it would be interesting to replicate this study in other countries in order to see whether there are differences between consumers of other cultural backgrounds.

As mentioned in the introduction of this dissertation, targeting innovations to older consumers bears a tremendous potential for companies. The mere focus on

931 Regarding the necessity of creating segments and the different segmentation criteria see Freter, H. (2009), pp. 399 and Freter, H. (2008), pp. 90.

demographical aspects and older consumers' spending power is, however, misleading. A deeper understanding of their consumer behavior is essential. Making general statements about elderly consumers is very tempting. Yet, there is no sign that older individuals represent a homogenous group. Additionally, non-scientific literature produced a large number of hypotheses about older individuals which resulted in a fuzzy picture and anecdotic knowledge about elderly consumers. Therefore, large scale surveys are very helpful to better grasp the "older consumer" and differences that may exist compared to younger consumers. Research in consumer behavior is complicated by the interplay of age, cohort, and period effects. The older age group is dynamic. For example, CEP knowledge but also needs and expectations of the future elderly will change. The life expectancy will increase and older individuals will stay active for a longer time. It can be expected that tomorrow's elderlies will be more open to today's technologies and innovations. It can, however, only be speculated on how their attitude towards *future* innovations will be. Past and present results about the consumer behavior will be outdated and need to be continuously updated.

Despite the incomplete state of knowledge and the complexity of the older consumer topic, it is important for companies to open up to the specific characteristics of this consumer group. The competitive pressure in the markets will inevitably lead to a better consideration of the older consumer group. In the upcoming years, older buyers will be able to choose between more and more competing products or services considering their specific needs. Thus, only those companies considering the expectations of the steadily growing age group in *more* detail and better than their competitors will be successful. Companies that shut themselves off to these developments will, instead, face problems.

The literature review has shown that still a lot of research advances have to be done because older individuals and more specifically older *consumers* are a complex research field. Profound knowledge on younger and older consumers is necessary and analyses which are tailored to the specific research question are required. Future research can benefit from this study's results. Especially the variables that are strongly influenced by age and the significant determinants of CEPO provide interesting starting points. Models expanding on these results may help companies developing products and services for older *and* younger consumers. These innovations, in turn, may better meet the needs of the consumers compared to today's offers and improve their product and service satisfaction.

Appendix

Appendix 1: Age structure of Germany

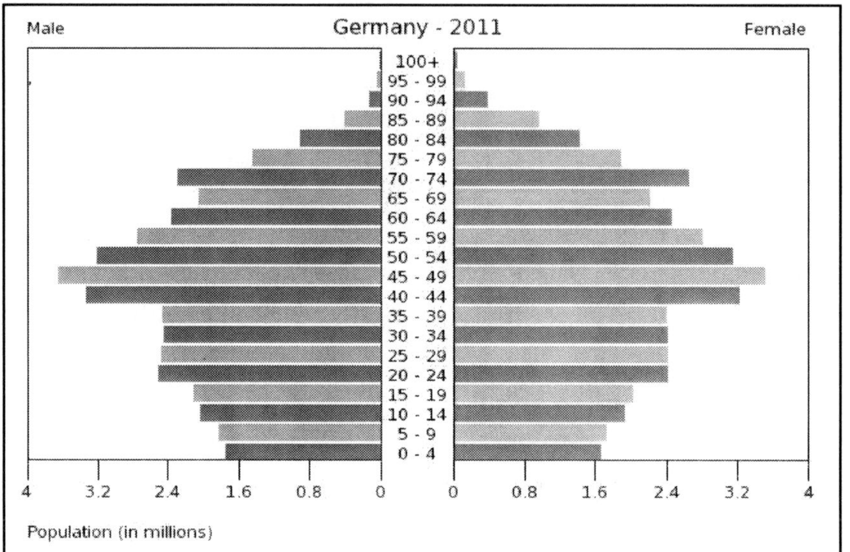

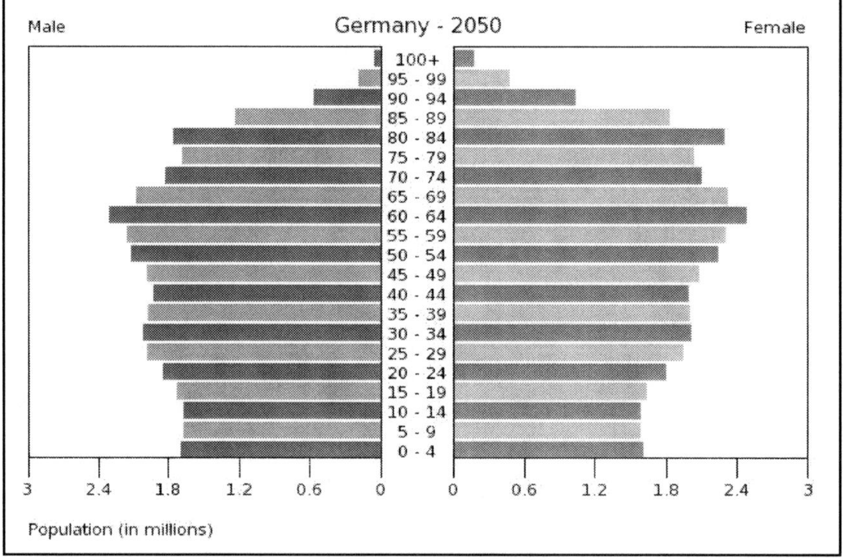

(Source: **U.S. Census Bureau** (2011b), online.)

Appendix 2: Age structure of Japan

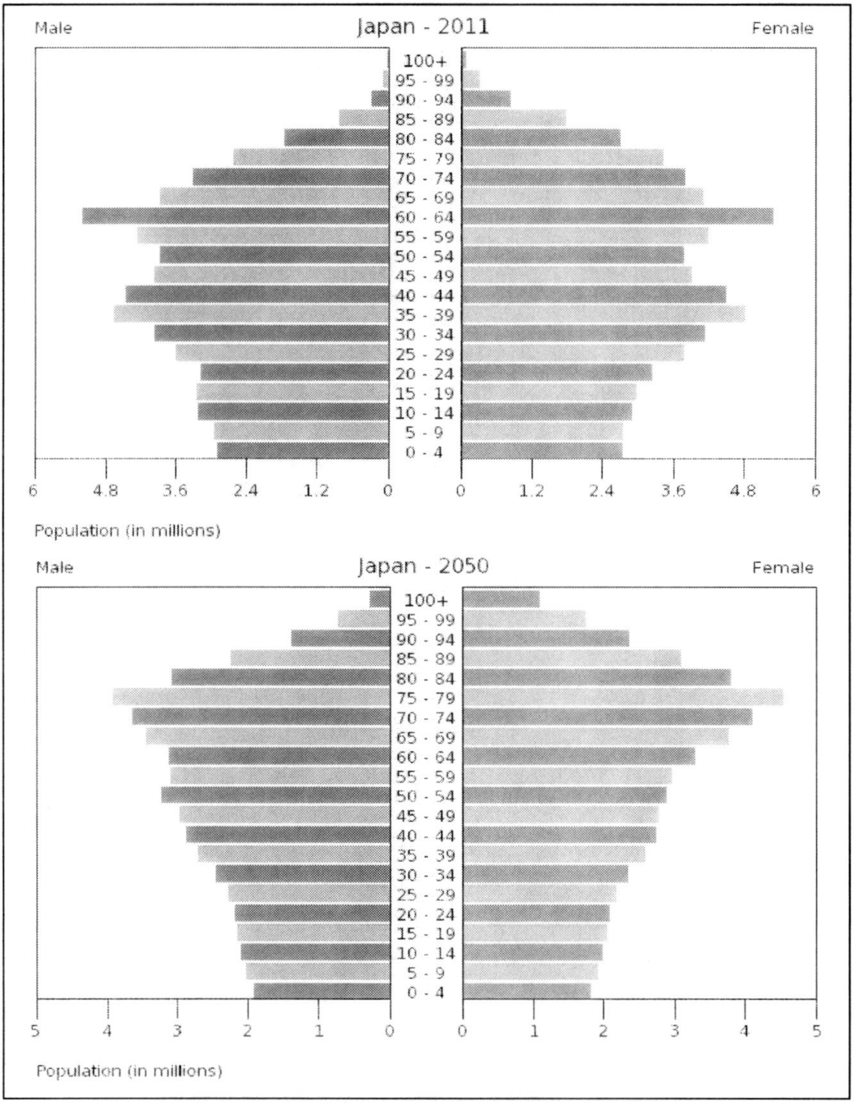

(Source: **U.S. Census Bureau** (2011b), online.)

Appendix 3: Age structure of Afghanistan

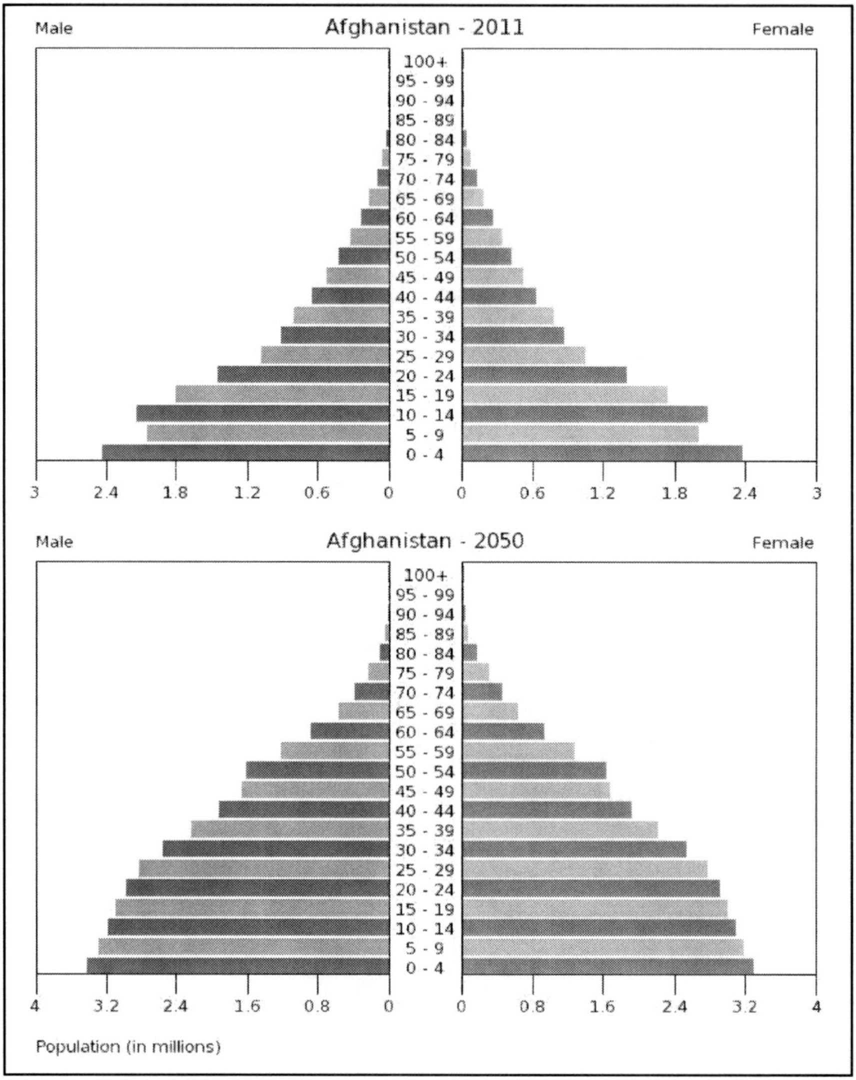

(Source: **U.S. Census Bureau** (2011b), online.)

Appendix 4: Survey questionnaire (in German)

UNIVERSITÄT SIEGEN

Theorie
und Praxis
für Karrieren
von morgen

Meinungsumfrage zur Handhabung von Unterhaltungselektronik-Geräten

Sehr geehrte Damen und Herren,

herzlichen Dank für Ihr Interesse an dieser Umfrage. Im Rahmen meiner Forschungsarbeit an der Universität Siegen, möchte ich Sie nach Ihrer Meinung zu Produkten aus dem Bereich "Unterhaltungselektronik" fragen. Da **kein besonderes Vorwissen nötig** ist, ist jeder herzlich eingeladen teilzunehmen.

Die Umfrage dient der Forschung und hat keinen kommerziellen Hintergrund. Es werden **keine Namen oder Adressdaten erfragt** – die Anonymität wird somit gewährleistet!

SOS-Kinderdörfer ÄRZTE OHNE GRENZEN

Machen Sie mit und unterstützen Sie dabei die SOS-Kinderdörfer, Ärzte ohne Grenzen, oder den Deutschen Tierschutzbund. Denn **für jeden vollständig ausgefüllten Fragebogen wird ein Betrag von 2 € an die von Ihnen gewählte Hilfsorganisation gespendet** (Genaueres dazu finden Sie auf der letzten Seite des Fragebogens).

Vielen Dank für Ihre Unterstützung.

Dipl. Kfm. Armin Tank

Anmerkungen zum Ausfüllen des Fragebogens:

Die Beantwortung der Fragen dauert ca. 10-12 Minuten. Es gibt weder „richtige" noch „falsche" Antworten. Kreuzen Sie einfach spontan das an, was Ihrer Meinung nach am ehesten auf Sie zutrifft!

Wenn im Folgenden von **„innovativen Geräten der Unterhaltungselektronik"** die Rede ist, so sind damit neuartige Produkte wie z.B.:

- *Flachbildfernseher (Plasma-/LCD-TV)*
- *Neueste Videokamera-/Camcorder-Generation (mit Festplatte/Speicherchip)*
- *Digitale Videorekorder (Festplattenrekorder)*

gemeint.

Sie brauchen aber kein Vorwissen bezüglich dieser Produkte!

Bitte umblättern...

Bitte geben Sie für jede Aussage an, in welchem Maße Sie zustimmen bzw. ablehnen.
Einfach ankreuzen oder umkreisen.
Je mehr Punkte Sie für eine Frage vergeben, desto höher ist Ihre Zustimmung.

Ich finde, dass innovative Unterhaltungselektronik-Geräte im Vergleich zu ihren Vorgängern...	stimme gar nicht zu		teils/ teils		stimme voll zu
...technisch überlegen sind.	1	2	3	4	5
...mehr Funktionen bieten.	1	2	3	4	5
...ein gutes Preis-Leistungs-Verhältnis besitzen.	1	2	3	4	5

Ich finde es wichtig, dass neu gekaufte Unterhaltungselektronik-Geräte technisch überlegen sind.	1	2	3	4	5
Ich finde es wichtig, dass neu gekaufte Unterhaltungselektronik-Geräte mehr Funktionen besitzen.	1	2	3	4	5
Ich finde, dass der Besitz innovativer Unterhaltungselektronik das Ansehen bei Freunden oder anderen Leuten steigert.	1	2	3	4	5

	sehr unvorteil-haft		weder noch		sehr vorteil-haft
Wie vorteilhaft finden Sie im Allgemeinen innovative Unterhaltungselektronik-Geräte im Vergleich zu ihren Vorgängern?	1	2	3	4	5

Ihre Meinung zur Verständlich- und Bedienbarkeit	stimme gar nicht zu		teils/ teils		stimme voll zu
Ich finde, dass innovative Unterhaltungselektronik-Geräte leicht zu verstehen sind.	1	2	3	4	5
Ich finde, dass innovative Unterhaltungselektronik-Geräte einfach zu bedienen sind.	1	2	3	4	5
Ich finde, dass die Anzahl der Funktionen von innovativen Unterhaltungselektronik-Geräten verwirrend ist.	1	2	3	4	5
Ihre Meinung zur Angebotssituation im Geschäft					
Ich finde, dass die Auswahl an Unterhaltungselektronik-Geräten in den Geschäften zu groß ist.	1	2	3	4	5
Ich finde, dass es schwer ist, sich zwischen ähnlichen Produktvarianten von Unterhaltungselektronik-Geräten zu entscheiden.	1	2	3	4	5
Ich finde, dass die Produktinformationen von Unterhaltungselektronik-Geräten in Geschäften schwer verständlich sind (z.B. Prospekte, Verpackungen, Verkaufsschilder).	1	2	3	4	5

	gar nicht komplex		weder noch		sehr komplex
Wie komplex finden Sie im Allgemeinen die Kaufentscheidung von innovativen Unterhaltungselektronik-Geräten?	1	2	3	4	5

Bitte umblättern...

Stellen Sie sich nun die Situation vor, dass Sie ein innovatives Unterhaltungselektronik-Gerät kaufen würden.

Wenn ich ein innovatives Unterhaltungselektronik-Gerät kaufen würde, hätte ich die Befürchtung, dass...	**stimme gar nicht zu**		**teils/ teils**		**stimme voll zu**
...die Preise für das Gerät schon bald nach dem Kauf fallen.	1	2	3	4	5
...das Gerät sein Geld nicht wert ist.	1	2	3	4	5
...das Gerät nicht den Nutzen erbringt, den ich erwartet hatte.	1	2	3	4	5
...das Gerät schon bald wieder veraltet ist.	1	2	3	4	5
...mich einige Freunde für einen Angeber halten.	1	2	3	4	5
...das Gerät aufgrund von „Kinderkrankheiten“ nicht zuverlässig funktioniert.	1	2	3	4	5
...ich viel Zeit aufbringen müsste, um die Bedienung des Geräts zu erlernen.	1	2	3	4	5
...ich viel Zeit aufbringen müsste, um vor dem Kauf die wichtigsten Informationen zu sammeln.	1	2	3	4	5
...ich mich bei den verschiedenen Technologie-Standards falsch entscheide.	1	2	3	4	5
Ich fände es unangenehm, mich mit den Details eines Kaufs innovativer Unterhaltungselektronik auseinandersetzen zu müssen.	1	2	3	4	5
Ich traue mir zu, ein innovatives Unterhaltungselektronik-Gerät zu kaufen, auch ohne Unterstützung im Kaufprozess.	1	2	3	4	5
Ich würde mich wohl dabei fühlen, ein innovatives Unterhaltungselektronik-Gerät alleine zu bedienen.	1	2	3	4	5
Wenn ich keine innovativen Unterhaltungselektronik-Geräte kaufen würde, hätte ich die Befürchtung, dass mich einige Freunde für unfortschrittlich halten.	1	2	3	4	5

	gar nicht riskant		**weder noch**		**sehr riskant**
Wie riskant finden Sie im Allgemeinen den Kauf von innovativen Unterhaltungselektronik-Geräten?	1	2	3	4	5

	stimme gar nicht zu		**teils/ teils**		**stimme voll zu**
Ich finde, dass innovative Unterhaltungselektronik-Geräte mit älteren Geräten kombinierbar (d.h. kompatibel) sind.	1	2	3	4	5
Ich finde es wichtig, dass innovative Geräte mit älteren Geräten kombinierbar (d.h. kompatibel) sind.	1	2	3	4	5
Ich finde, dass das Wissen über **ältere** Unterhaltungs-elektronik Geräte auch auf innovative Geräte anwendbar ist.	1	2	3	4	5
Ich finde, dass mein Wissen über **ältere** Unterhaltungselektronik-Geräte gut ist.	1	2	3	4	5
Ich finde, dass innovative Unterhaltungselektronik-Geräte zu kleine Tasten und Beschriftungen besitzen.	1	2	3	4	5
Ich finde, dass innovative Unterhaltungselektronik-Geräte meinen Bedürfnissen entsprechen.	1	2	3	4	5

Bitte umblättern...

Allgemeines Interesse und Erfahrungen mit Unterhaltungselektronik-Geräten

	stimme gar nicht zu		teils/ teils		stimme voll zu
Ich interessiere mich im Allgemeinen für Geräte der Unterhaltungselektronik.	1	2	3	4	5
Ich finde es spannend, mich über Neuheiten im Bereich der Unterhaltungselektronik zu informieren.	1	2	3	4	5

	sehr niedrig		weder noch		sehr hoch
Wie hoch ist Ihre Bereitschaft, Geld für innovative Unterhaltungselektronik auszugeben?	1	2	3	4	5

Haben Sie in den letzten 2-3 Jahren ein oder mehrere neue Unterhaltungselektronik-Geräte gekauft?

❑ ja	❑ nein

↓

Wenn ja, bewerten Sie bitte auch folgende Aussage:	stimme gar nicht zu		teils/ teils		stimme voll zu
Ich habe mit diesem Gerät/diesen Geräten überwiegend gute Erfahrungen gemacht.	1	2	3	4	5

Bitte nennen Sie die Geräte, mit denen Sie ggf. schlechte ______________________

Erfahrungen in den letzten 2-3 Jahren gemacht haben. ______________________

Wie oft ist es in den letzten 12 Monaten vorgekommen, dass Sie ein innovatives Unterhaltungselektronik-Gerät gekauft haben...?	gar nicht vorge-kommen		manchmal		oft vorge-kommen
...weil Ihr altes Gerät kaputt gegangen ist.	1	2	3	4	5
...weil es keinen Service mehr für Ihr altes Gerät gab.	1	2	3	4	5
...weil Sie ein moderneres Gerät haben wollten.	1	2	3	4	5
Wie oft ist es in den letzten 12 Monaten vorgekommen, dass Sie ein innovatives Unterhaltungselektronik-Gerät nicht gekauft haben...?					
...auf Grund Ihrer momentanen finanziellen Situation.	1	2	3	4	5
...da das gewünschte Produkt nicht erhältlich war.	1	2	3	4	5
...auf Grund fehlender Zeit, sich mit einem möglichen Kauf zu beschäftigen.	1	2	3	4	5
...weil Sie warten wollten, bis die Geräte günstiger werden.	1	2	3	4	5

Bitte umblättern...

Wie häufig haben Sie in der Regel persönlichen oder telefonischen Kontakt mit folgenden Personen?	seltener	mehrmals im Jahr	mehrmals im Monat	2-3 mal pro Woche	fast jeden Tag	gar nicht/ trifft nicht zu
Eltern	1	2	3	4	5	0
Geschwister	1	2	3	4	5	0
Kinder	1	2	3	4	5	0
Enkel	1	2	3	4	5	0
Sonstige Verwandte	1	2	3	4	5	0
Freunde	1	2	3	4	5	0
Nachbarn	1	2	3	4	5	0
Arbeitskollegen/Kommilitonen	1	2	3	4	5	0
Bekannte aus Vereinen/Partei/Freizeit	1	2	3	4	5	0

Wie hilfreich wären Ihnen folgende Personen beim Kauf eines innovativen Geräts aus der Unterhaltungselektronik?	überhaupt nicht hilfreich		teils/ teils		sehr hilfreich	trifft nicht zu
Eltern	1	2	3	4	5	0
Geschwister	1	2	3	4	5	0
Kinder	1	2	3	4	5	0
Enkel	1	2	3	4	5	0
Sonstige Verwandte	1	2	3	4	5	0
Freunde	1	2	3	4	5	0
Nachbarn	1	2	3	4	5	0
Arbeitskollegen/Kommilitonen	1	2	3	4	5	0
Bekannte aus Vereinen/Partei/Freizeit	1	2	3	4	5	0

Bitte umblättern...

Ich besitze folgende Geräte... (Mehrfachnennungen möglich) Bitte zutreffendes ankreuzen.			
❑	Flachbildfernseher (Plasma-/LCD-TV)	❑	Handy mit Kamera
❑	Flachbildschirm (Computer)	❑	Handy ohne Kamera
❑	Computer	❑	Digitaler Fotoapparat (mind. 8 Megapixel)
❑	Laptop/Notebook	❑	Digitaler Fotoapparat (weniger als 8 Megapixel)
❑	Beamer	❑	Digitaler Bilderrahmen
❑	Heimkinosystem (Dolby Digital, DTS, 5.1,..)	❑	Camcorder mit Festplatte oder Speicherchip
❑	Navigationssystem	❑	Anderer Camcorder
❑	DVD-Player (Einzelgerät, d.h. kein Computer-Laufwerk)	❑	iPhone
❑	DVD-Recorder (Einzelgerät)	❑	iPod
❑	Festplattenrekorder (Digitaler Videorekorder)	❑	MP3-Player (Einzelgerät)
❑	Blu-Ray Disc Player (Einzelgerät)	❑	Mobiler Organizer
❑	HD-DVD Player (Einzelgerät)	❑	Mobiler Videoplayer
❑	Videokassettenrekorder	❑	Wii oder Sony Playstation 3
❑	Kombi-Gerät (DVD-Player+Videokassettenrekorder)	❑	Nintendo DS oder Sony Playstation Portable
❑	**Keins der Geräte in dieser Liste**		

Welche Hauptinformationsquellen würden Sie vor dem Kauf eines innovativen Unterhaltungselektronik-Geräts nutzen?

Bitte kreuzen Sie aus der folgenden Liste **maximal 5** Informationsquellen an.	
❑	Werbeprospekte
❑	Gespräche mit Familie/Verwandten/Freunden
❑	Eigene Erfahrung
❑	Werbung im Geschäft
❑	Verkaufspersonal im Geschäft
❑	Informationen auf der Verpackung
❑	Internet
❑	TV-Werbung
❑	Zeitungen/Fachzeitschriften
❑	Testzeitschriften
❑	Verbraucherberatung
❑	(Versand-)Kataloge
❑	**Ich würde Familienmitglieder (z.B. Kinder, Enkel,..) oder Freunde das Gerät für mich kaufen lassen.**

Bitte umblättern...

Zum Schluss noch einige Angaben zu Ihrer Person:

Welchen Bildungsabschluss haben Sie?	
❑	Ohne Abschluss
❑	Volks-/Hauptschule
❑	Mittlere Reife
❑	Abitur
❑	Universität/Fachhochschule
❑	Sonstiges: ____________

Wie ist Ihr aktueller Beschäftigungsstatus?	
❑	Ausbildung
❑	Student
❑	Angestellter/Beamter
❑	Freiberufler/Selbständiger
❑	Umschulung
❑	Rentner/Pensionär
❑	Hausfrau/Hausmann
❑	z.Zt. ohne Beschäftigung
❑	Sonstiges: ____________

Wie alt sind Sie?	_____ Jahre.
Wie alt fühlen Sie sich?	_____ Jahre.

Geschlecht

❑ weiblich	❑ männlich

Wie ist Ihr Familienstand?

❑ Ledig	❑ Verheiratet	❑ Geschieden	❑ Verwitwet

Wie viele Personen leben <u>außer Ihnen</u> ständig in Ihrem Haushalt?	_____ Personen.

	sehr negativ	**negativ**	**eher negativ**	**weder noch**	**eher positiv**	**positiv**	**sehr positiv**
Wie würden Sie Ihre allgemeine finanzielle Situation beschreiben?	1	2	3	4	5	6	7

****

Entscheiden Sie sich hier für die ***<u>Hilfsorganisation Ihrer Wahl</u>***!
Bitte wählen Sie diejenige Organisation aus, an welche die vorgesehene Spende gehen soll.

❑	SOS-Kinderdörfer	SOS-Kinderdörfer
❑	Ärzte ohne Grenzen - Médecins Sans Frontières	ÄRZTE OHNE GRENZEN
❑	Deutscher Tierschutzbund e.V.	DEUTSCHER TIERSCHUTZBUND E.V.
❑	Keine Spende.	

Auch im Namen der Organisationen möchte ich mich herzlich für Ihre Teilnahme bedanken!

****

Appendix 5: Correlation matrix of the scale items

	rel_ tech	rel_ funct	rel_ coper	cx_ und_r	cx_ use_r	cx_ conf	cx_ choice	cx_ sim	cx_ isinf	perf_ expec	perf_ troub	fin_ worth	fin_ price	fin_ obs	psy_ detail	psy_ info	psy_ wrong	psy_ opera	pres_ unprog	pres_ prest	s_eff_ buy	s_eff_ oper	int_ ce
rel_tech	1,000	0.508**	0.214**																				
rel_funct	0.508**	1,000	0.183**																				
rel_coper	0.214**	0.183**	1,000																				
cx_und_r				1,000	0.782**	0.577**			0.387**						0.321**			0.478**			-0.309**		
cx_use_r				0.782**	1,000	0.556**			0.308**									0.485**					
cx_conf				0.577**	0.556**	1,000	0.328**	0.373**	0.406**						0.336**		0.347**	0.520**					
cx_choice						0.328**	1,000	0.501**	0.325**														
cx_sim						0.373**	0.501**	1,000	0.493**							0.319**		0.307**					
cx_isinf				0.387**	0.308**	0.406**	0.325**	0.493**	1,000									0.338**					
perf_expec										1,000	0.343**	0.627**		0.342**			0.348**	0.312**					
perf_troub										0.343**	1,000	0.337**					0.327**						
fin_worth										0.627**	0.337**	1,000	0.322**	0.310**			0.352**						
fin_price												0.322**	1,000	0.458**									
fin_obs										0.342**		0.310**	0.458**	1,000		0.307**							
psy_detail				0.321**		0.336**									1,000	0.413**	0.424**	0.452**			-0.396**		-0.374**
psy_info								0.319**						0.307**	0.413**	1,000	0.450**	0.523**					
psy_wrong						0.347**				0.348**	0.327**	0.352**			0.424**	0.450**	1,000	0.437**			-0.323**		
psy_opera				0.478**	0.485**	0.520**		0.307**	0.338**	0.312**					0.452**	0.523**	0.437**	1,000					
pres_unprog																			1,000	0.458**			
pres_prest																			0.458**	1,000			
s_eff_buy				-0.309**											-0.396**		-0.323**				1,000	0.317**	0.329**
s_eff_oper																					0.317**	1,000	
int_ce															-0.374**						0.329**		1,000
int_inno															-0.434**						0.360**		0.750**

**. Correlation is significant at the 0.01 level

*. Correlation is significant at the 0.05 level

Appendix 6: Descriptive statistics of the final scales (per age group)

G=1	Mean	Median	Mode	SD	Var.	Skew	Kurtosis	Min.	Max.
rel	3.83	4.00	3.50	0.823	0.678	-0.314	-0.612	2	5
cx_prod	2.98	3.00	3.00	0.797	0.636	0.091	0.449	1	5
cx_is	3.42	3.33	3.67	0.885	0.785	-0.460	0.048	1	5
r_ben	2.95	3.00	3.00	0.948	0.900	0.069	-0.532	1	5
r_psy	3.23	3.33	3.33	0.889	0.792	-0.120	-0.195	1	5
soc_pres	1.86	2.00	1.00	0.813	0.662	0.779	-0.078	1	4
int	3.25	3.00	3.00	1.060	1.124	-0.162	-0.568	1	5
fin_situ	4.79	5.00	4.00[a]	1.156	1.338	-0.448	0.344	1	7
compa	2.47	2.40	2.40	1.039	1.080	0.474	-0.170	0.4	5
soc_assist	11.9	11.80	0.00[a]	4.901	24.025	-0.067	0.670	0	27

G=2	Mean	Median	Mode	SD	Var.	Skew	Kurtosis	Min.	Max.
rel	3.51	3.50	4.00	1.045	1.092	-0.453	-0.380	1	5
cx_prod	3.40	3.33	3.00	0.882	0.778	-0.015	-0.354	1	5
cx_is	3.66	3.67	4.33	0.878	0.771	-0.454	-0.392	1	5
r_ben	3.05	3.00	2.50[a]	0.966	0.934	-0.024	-0.439	1	5
r_psy	3.42	3.33	3.00	0.952	0.907	-0.113	-0.690	1	5
soc_pres	1.83	1.50	1.00	0.871	0.760	0.891	-0.122	1	4.5
int	3.01	3.00	3.50	1.144	1.309	-0.109	-0.891	1	5
fin_situ	4.79	5.00	5.00	1.096	1.202	-0.590	0.274	1	7
compa	2.27	2.40	2.40	1.090	1.190	0.560	-0.080	0.4	5
soc_assist	11.2	11.50	0.00	5.453	29.739	-0.049	-0.225	0	28.4

G=3	Mean	Median	Mode	SD	Var.	Skew	Kurtosis	Min.	Max.
rel	3.71	4.00	4.00	1.022	1.046	-0.737	0.068	1	5
cx_prod	3.89	4.00	5.00	0.765	0.587	-0.190	-0.483	1.67	5
cx_is	3.98	4.00	4.33	0.788	0.621	-0.809	0.303	1.67	5
r_ben	3.10	3.00	3.00	0.933	0.871	-0.187	-0.203	1	5
r_psy	3.71	3.67	3.67	0.843	0.711	-0.449	-0.198	1.33	5
soc_pres	2.12	2.00	1.00	1.082	1.173	0.851	-0.056	1	5
int	2.93	3.00	2.50	1.189	1.414	0.131	-0.901	1	5
fin_situ	5.13	5.00	5.00	1.033	1.069	-0.553	0.440	2	7
compa	2.14	2.00	2.40	1.120	1.256	0.602	0.324	0.2	5
soc_assist	8.99	8.80	4.00	4.447	19.779	0.386	0.009	0	21.6

Appendix 7: Results of the univariate and multivariate outlier analysis

Group	Outlier (CEPO_w boxplot)	Outlier (adoption determinants boxplots)	Outlier (graphical residual scores)	Cook's Distance	Multivariate matches
G1	228	191	169	6	169
	257	202	176	168	181
		213	181	169	228
		229	209	181	257
		246	214	204	299
		289	228	228	
		295	257	257	
			295	289	
			299	299	
G2	344	329	316	316	316
	360	468	360	326	360
	430		376	328	388
	465		386	360	391
	468		388	367	464
	485		390	388	
			391	391	
			464	449	
			465	452	
			488	460	
				464	
				468	
G3		509	501	501	501
		510	514	526	526
		515	526	541	581
		541	551	581	591
		551	555	591	616
		553	558	616	651
		554	566	651	652
		555	579	652	
		557	581		
		558	591		
	-	563	599		
		571	602		
		572	616		
		579	617		
		582	651		
		584	652		
		598	657		
		638			
		648			
		650			
		651			

Cook's Distance values are ranked depending their size (starting with the largest values)

Bibliography

Aaker, D. A./Bagozzi, R. P. (1979): Unobservable Variables in Structural Equation Models with Application in Industrial Setting, in: Journal of Marketing Research, 16 (May), pp. 147-158.

Aaker, D. A./Kumar, V./Day, G. S. (2001): Marketing Research, 7. ed., New York (NY) [et al.].

Abraham, S. C. S./Hayward, G. (1984): Understanding Discontinuance: Towards a More Realistic Model of Technological Innovation and Industrial Adoption in Britain, in: Technovation, Vol. 2, No. 3, June, pp. 209-231.

Akhter, S. H. (2003): Digital Divide and Purchase Intention: Why Demographic Psychology Matters, in: Journal of Economic Psychology, Vol. 24, Issue 3, June, pp. 321-327.

Alba, J. W./Hutchinson, J. W. (1987): Dimensions of Consumer Expertise, in: Journal of Consumer Research, Vol. 13, Issue 4, March, pp. 411-454.

Allen, M. P. (1997): Understanding Regression Analysis, New York (NY).

Allison, P. D. (1999): Multiple Regression. A primer, Thousand Oaks (Calif.) [et al.].

Anderson, D. R./Sweeney, D. J./Williams, T. A. (2009): Statistics for Business and Economics, 10. ed., Mason (Ohio).

Antil, J. H. (1988): New Product or Service Adoption: When Does it happen?, in: Journal of Consumer Marketing, Vol. 2, No. 4, Fall, pp. 5-16.

Arndt, J. (1967a): Role of Product-Related Conversations in the Diffusion of a New Product, in: Journal of Marketing Research, Vol. 4, Issue 3, Aug, pp. 291-295.

Arndt, J. (1967b): Word of Mouth Advertising and Informal Communication, in: Cox, D. F. (Ed.): Risk Taking and Information Handling in Consumer Behavior, Boston, pp. 188-239.

Arndt, J. (1968): Word-of-mouth Advertising and Perceived Risk, in: Kassarjian, H. H./Robertson, T. S. (Eds.): Perspectives in Consumer Behavior, Glenview (Ill.), pp. 330-336.

Assael, H. (1992): Consumer Behavior and Marketing Action, 4. ed., Boston.

Atchley, R. C. (1989): A Continuity Theory of Normal Aging, in: The Gerontologist, Vol. 29, No. 2, pp. 183-190.

Auken, S. v./Barry, T. E./Anderson, R. L. (1993): Observations Toward the Internal Validation of Cognitive Age Measures in Advertising Research, Journal of Advertising Research, May/June 1993.

Aulinger, R. (1970): Die Produktinnovation im Bereich der Konsumgüter, Diss., Wien 1969.

Bäcker, G. (2008): Altersarmut als soziales Problem der Zukunft?, in: Deutsche Rentenversicherung, Bd. 63, Heft 4, pp. 357-367.

Backes, G. M./Clemens, W. (1998): Lebensphase Alter. Eine Einführung in die sozialwissenschaftliche Alternsforschung, Weinheim, München.

Backhaus, K. et al. (2008): Multivariate Analysemethoden. Eine anwendungsorientierte Einführung, 12. ed., Berlin [et al.].

Bäckman, L./Small, B. J./Wahlin, A. (2001): Aging and Memory. Cognitive and Biological Perspectives, in: Birren, J. E./Schaie, W. K. (Eds.): Handbook of the Psychology of Aging, 5. ed., San Diego [et al.], pp. 349-377.

Bagozzi, R. P./Lee, K.-H. (1999): Consumer Resistance to, and Acceptance of, Innovations, in: Advances in Consumer Research, Vol. 26, Issue 1, pp. 218-225.

Bähr-Seppelfricke, U. (1999): Diffusion neuer Produkte. Der Einfluß von Produkteigenschaften, Diss., Wiesbaden 1999.

Baltes, P. B. (1990): Entwicklungspsychologie der Lebensspanne: Theoretische Leitsätze, in: Psychologische Rundschau, Vol. 41, pp. 1-24.

Baltes, P. B./Baltes, M. M. (1994): Gerontologie: Begriff, Herausforderung und Brennpunkte, in: Baltes, P. B./Mittelstraß, J./Staudinger, U. M. (Eds.): Alter und Altern: Ein interdisziplinärer Studientext zur Gerontologie. - Sonderausgabe des 1992 erschienenen 5. Forschungsberichts der Akademie der Wissenschaft zu Berlin, Berlin, pp. 1-34.

Bandura, A. (1986): Social Foundations of Thought and Action. A Social Cognitive Theory, Englewood Cliffs (NJ).

Bandura, A./Adams, N. E./Beyer, J. (1977): Cognitive Processes Mediating Behavioral Change, in: Journal of Personality and Social Psychology, Vol. 35, No. 3, pp. 125-139.

Bänsch, A. (2002): Käuferverhalten, 9. ed., München; Wien.

Barak, B./Gould, S. (1985): Alternative Age Measures: A Research Agenda, in: Advances in Consumer Research, Vol. 12, Issue 1, pp. 53-58.

Barak, B./Schiffman, L. G. (1981): Cognitive Age: A Nonchronological Age Variable, in: Advances in Consumer Research, Vol. 8, Issue 1, pp. 602-606.

Barry, T. E. (1987): The Development of the Hierarchy of Effects: An Historical Perspective, in: Current Issues & Research in Advertising, Vol. 10, No. 2, pp. 251-295.

Bartos, R. (1980): Over 49: The Invisible Consumer Market, in: Harvard Business Review, No. 58 (1), Feb, pp. 140-148.

Bass, F. M. (1969): A New Product Growth for Model Consumer Durables, in: Management Science, Vol. 15, No. 5, Jan, pp. 215-227.

Bass, F./Jain, D./Krishnan, T. (2000): Diffusion Models, in: Mahajan, V./Muller, E./Wind, Y. (Eds.): New Product Diffusion Models, Boston; Dordrecht; London, pp. 99-122.

Bauer, H. H./Sauer, N. E./Becker, S. (2003): Risikowahrnehmung und Kaufverhalten im Internet, in: Markteting ZFP (Zeitschrift für Forschung und Praxis), Heft 3, 3. Quartal, pp. 183-199.

Bauer, R. A. (1960): Consumer Behavior as Risk Taking, in: Hancock, R. S. (Ed.): Dynamic Marketing for a Changing World, Proceedings of the 43rd National Conference of the American Marketing Association, Chicago (Ill.), pp. 389-398.

Beal, G. M./Rogers, E. M./Bohlen, J. M. (1957): Validity of the Concept of Stages in the Adoption Process, in: Rural Sociology, Vol. 22, No. 2, June, pp. 166-168.

Bearden, W. O./Etzel, M. J. (1982): Reference Group Influence on Product and Brand Purchase Decisions, in: Journal of Consumer Research, Vol. 9, Sep, pp. 183-194.

Behrend, C. (2000): Arbeit im Alter, in: Wahl, H.-W./Tesch-Römer, C. (Eds.): Angewandte Gerontologie in Schlüsselbegriffen, Stuttgart; Berlin; Köln, pp. 402-406.

Belk, R. W. (1974): An Exploratory Assessment of Situational Effects in Buyer Behavior, in: Journal of Marketing Research, Vol. 11, Issue 2, May, pp. 156-163.

Belk, R. W. (1975): Situational Variables and Consumer Behavior, in: Journal of Consumer Research, Vol. 2, Issue 3, Dec, pp. 157-164.

Bender, R./Ziegler, A./Lange, S. (2007): Varianzanalyse - Artikel Nr. 21 der Statistik-Serie in der DMW, in: Deutsche Medizinische Wochenschrift, 132, pp. e57-e60.

Berekoven, L./Eckert, W./Ellenrieder, P. (2009): Marktforschung. Methodische Grundlagen und praktische Anwendung, 12. ed., Wiesbaden.

Bethke-Brenken, I./Brenken, G. (2010): Aufbruch in den Ruhestand. Anleitung zum Gestalten und Genießen, München; Basel.

Bettman, J. R. (1973): Perceived Risk and Its Components: A Model and Empirical Test, in: Journal of Marketing Research, Vol. 10, Issue 2, May, pp. 184-190.

Bettman, J. R. (1979): An Information Processing Theory of Consumer Choice, Reading (Mass.) [et al.].

Birren, J. E./Birren, B. A. (1996): Autobiography: Exploring the Self and Encouraging Development, in: Birren, J. E. et al. (Eds.): Aging and Biography. Explorations in Adult Development, New York (NY), pp. 283-299.

Birren, J. E./Schroots, J. J. (2001): Histrory of Geropsychology, in: Birren, J. E./Schaie, W. K. (Eds.): Handbook of the Psychology of Aging, 5. ed., San Diego [et al.], pp. 3-28.

Black, K. (2010): Business Statistics: Contemporary Decision Making, 6. ed., Hoboken (NJ).

Blackwell, R. D./Miniard, P. W./Engel, J. F. (2001): Consumer Behavior, 9. ed., Fort Wort [et al.].

Bloch, P. H. (1981): An Exploration into the Scaling of Consumers' Involvement with a Product Class, in: Advances in Consumer Research, Vol. 8, Issue 1, pp. 61-65.

Bloch, P. H./Sherrell, D. L./Ridgway, N. M. (1986): Consumer Search: An Extended Framework, in: Journal of Consumer Research, Vol. 13, Issue 1, June, pp. 119-126.

BMAS (2008): Ergänzender Bericht der Bundesregierung zum Rentenversicherungsbericht 2008 gemäß § 154 Abs. 2 SGB VI (Alterssicherungsbericht 2008). Bundesministerium für Arbeit und Soziales, Berlin.

BMFSFJ (2006): Hilfe- und Pflegebedürftige in Alteneinrichtungen 2005. Schnellbericht zur Repräsentativerhebung im Forschungsprojekt "Möglichkeiten und Grenzen selbständiger Lebensführung in Einrichtungen" (MuG IV). Bundesministerium für Familie, Senioren, Frauen und Jugend, München.

Böcker, F. (1991): Verhaltenswissenschaftlich fundierte Innovationspolitik, in: Ordelheide, D./Rudolph, B./Büsselmann, E. (Eds.): Betriebswirtschaftslehre und ökonomische Theorie, Stuttgart, pp. 371-392.

Böcker, F. (1996): Marketing, 6. ed., Stuttgart.

Böcker, F./Gierl, H. (1987): Determinanten der Diffusion neuer industrieller Produkte, in: Zeitschrift für Betriebswirtschaft (ZfB), 57. Jg., Heft 7, pp. 684-698.

Bohlen, J. M. et al.(1968): Adopters of New Farm Ideas, in: Kassarjian, H. H./Robertson, T. S. (Eds.): Perspectives in Consumer Behavior, Glenview (Ill.), pp. 351-361.

Böhler, H. (2004): Marktforschung, 3. ed., Stuttgart.

Bortz, J. (2005): Statistik für Human- und Sozialwissenschaftler, 6. ed., Berlin [et al.].

Bortz, J./Schuster, C. (2010): Statistik für Human- und Sozialwissenschaftler, 7. ed., Berlin; Heidelberg.

Bourne, F. S. (1968): Group Influence in Marketing, in: Kassarjian, H. H./Robertson, T. S. (Eds.): Perspectives in Consumer Behavior, Glenview (Ill.), pp. 289-296.

Bovair, S./Kieras, D. E./Polson, P. G. (1990): The Acquisition and Performance of Text-Editing Skill: A Cognitive Complexity Analysis, in: Human-Computer Interaction, Vol. 5 Issue 1, pp. 1-48.

Bowe, F. (1988): Why Seniors Don't Use Technology, in: Technology Review, August-September, pp. 34-40.

Breitschopf, B./Haller, I./Grupp, H. (2005): Bedeutung von Innovationen für die Wettbewerbsfähigkeit. Ausgewählte Indikatoren zur Messung der Wettbewerbsfähigkeit am Beispiel Deutschlands, in: Albers, S./Gassmann, O. (Eds.): Handbuch Technologie- und Innovationsmanagement. Strategie - Umsetzung - Controlling, Wiesbaden, pp. 43-60.

Brewer, M. B. (2000): Research Design and Issues of Validity, in: Reis, H. T./Charles M. Judd (Eds.): Handbook of Research Methods in Social and Personality Psychology, New York (NY) [et al.], pp. 3-16.

Brockhoff, K. (1999): Produktpolitik, 4. ed., Stuttgart.

Brockhoff, K. (2002): Produktinnovation, in: Albers, S. (Ed.): Handbuch Produktmanagement: Strategieentwicklung - Produktplanung - Organisation - Kontrolle, 2. ed., Wiesbaden , pp. 25-54.

Brooker, G. (1984): An Assessment of an Expanded Measure of Perceived Risk, in: Advances in Consumer Research, Vol. 11, Issue 1, pp. 439-441.

Brose, P. (1982): Planung, Bewertung und Kontrolle technologischer Innovationen, Berlin 1982.

Brosius, F. (2006): SPSS 14, Heidelberg.

Brown, J. J./Reingen, P. H. (1987): Social Ties and Word-of-Mouth Referral Behavior, in: Journal of Consumer Research, Vol. 14, Issue 3, Dec, pp. 350-362.

Brown, S. A./Venkatesh, V. (2005): Model of Adoption of Technology in Households: A Baseline Model Test and Extension Incorporating Household Life Cycle, in: MIS Quarterly, Vol. 29, Issue 3, Sep, pp. 399-426.

Brünner, B. O. (1997): Die Zielgruppe Senioren. Eine interdisziplinäre Analyse der älteren Konsumenten, Diss., Frankfurt (a. M.) 1997.

Bühl, A. (2008): SPSS 16. Einführung in die moderne Datenanalyse, 11. ed., München [et al.].

Bühner, M. (2011): Einführung in die Test- und Fragebogenkonstruktion, 3. ed., München [et al.].

Bundesinstitut für Bevölkerungsforschung (2011a): Medianalter, http://www.bib-demografie.de/cln_099/nn_1645598/SharedDocs/Glossareintraege/DE/M/medianalter.html, 28.9.2011.

Bundesinstitut für Bevölkerungsforschung (2011b): Medianalter in Deutschland, 1950 bis 20260, http://www.bib-demografie.de/cln_099/nn_1645598/DE/DatenundBefunde/02/Abbildungen/a__02__16__medianalter__d__1950__2060.html, 28.9.2011.

BVV (2007): Bundesverband Audiovisuelle Medien - BVV Business Report (2006/2007), Hamburg.

Byrne, B. M. (2001): Structural Equitation Modeling with AMOS. Basic Concepts, Applications, and Programming, Mahwah (NJ).

Campbell, R. R. (1966): A Suggested Paradigm of the Individual Adoption Process, in: Rural Sociology, Vol. 31, No. 4, Dec, pp. 458-466.

Capgemini (2005): Wirtschaftsstimmung. Eine Umfrage unter deutschen Führungskräften, Berlin.

Carmichael, A. (2002): Older Users of Digital TV - Challenges and Opportunities, in: Consumer Policy Review, Vol. 12, Issue 6, Nov/Dec, pp. 217-222.

Carr, D./Bodnar-Deren, S. (2009): Gender, Aging and Widowhood , in: Uhlenberg, P. (Ed.): International Handbook of Population Aging, Dordrecht [et al.], pp. 705-728.

Cattell, R. B. (1979): Personality and Learning Theory, New York (NY).

Cattell, R. B. (1987): Intelligence: Its Structure, Growth, and Action, Amsterdam; New York (NY).

Cavallotti, C. A. P./Cerulli, L. (2008): Age-related Changes of the Human Eye, Totowa (NJ).

Cavanaugh, J. C./Blanchard-Fields, F. (2002): Adult Development and Aging, 4. ed., Belmont.

Chernev, A. (2003): When More Is Less and Less Is More: The Role of Ideal Point Availability and Assortment in Consumer Choice, in: Journal of Consumer Research, Vol. 30, Issue 2, Sep, pp. 170-183.

Child, I. (1968): Personality in Culture, in: Borgatta, E. F./Lambert, W. W. (Eds.): Handbook of Personality Theory and Research, Chicago, pp. 82-145.

Christophersen, T. (2007): Usability im Online-Shopping. Entwicklung eines Fragebogeninstrumentes (ufosV2) unter Berücksichtigung formativer und reflektiver Messmodelle, Kiel 2006.

Christophersen, T./Grape, C. (2006): Die Erfassung latenter Konstrukte mit Hilfe formativer und reflektiver Messmodelle, in: Albers, S. et al. (Eds.): Methodik der empirischen Forschung, Wiesbaden, pp. 115-132.

Clemens, W./Naegele, G. (2004): Lebenslagen im Alter, in: Kruse, A./Martin, M. (Eds.): Enzyklopädie der Gerontologie. Alternsprozesse in multidisziplinärer Sicht, Bern [et al.], pp. 387-402.

Cleve, J. (2011): Media Markt: Wie blöd ist das Konzept wirklich?, Regensburg.

Clow, K. E./Baack, D. (2005): Hierarchy of Effects Model, in: Concise Encyclopedia of Advertising, New York (NY), London, Oxford, pp. 90-91.

Cohen, J. (1988): Statistical Power Analysis for the Behavioral Sciences, 2. ed., Hillsdale (NJ) [et al.].

Cohen, J. et al. (2003): Applied Multiple Regression/Correlation Analysis for the Behavioral Sciences, 3. ed., Mahwah (NJ); London.

Compeau, D. R./Higgins, C. A. (1995): Computer Self-Efficacy: Development of a Measure and Initial Test, in: MIS Quarterly, Vol. 19, Issue 2, June, pp. 189-211.

Cooper, R. G. (1979): The Dimensions of Industrial New Product Success and Failure, in: Journal of Marketing, Vol. 43, Summer 1979, pp. 93-103.

Cooper, R. G. (2000): Product Innovation and Technology Strategy, in: Research Technology Management, Vol. 43, Issue 1, Jan/Feb 2000, pp. 38-41.

Costa, P. T./McCrae, R. R. (1988): Personality in Adulthood: A Six-year Longitudinal Study of Self-reports and Spouse Ratings on the NEO Personality Inventory, in: Journal of Personality and Social Psychology, Vol 54, No. 5, May, pp. 853-863 .

Costa, P. T./McCrae, R. R. (1994): Set Like Plaster? Evidence for the Stability of Adult Personality, in: Heatherton, T. F./Weinberger, J. L. (Eds.): Can Personality Change?, 4. ed., Washington, D.C., pp. 21-40.

Costello, A. B./Osborne, J. W. (2005): Best Practices in Exploratory Factor Analysis: Four Recommendations for Getting the Most From Your Analysis, in: Practical Assessment, Research & Evaluation, Vol. 10, No. 7, July, pp. 1-9.

Cote, J. A./Buckley, R. (1988): Measurement Error and Theory Testing in Consumer Research: An Illustration of the Importance of Construct Validation, in: Journal of Consumer Research, Vol. 14, No. 4, March, pp. 579-582.

Cox, D. F. (1967): Risk Handling in Consumer Behavior - An Intensive Study of Two Cases, in: Cox, D. F. (Ed.): Risk Taking and Information Handling in Consumer Behavior, Boston, pp. 34-81.

Craik, F. I. M./Salthouse, T. A. (2000): The Handbook of Aging and Cognition, 2. ed., Mahwah.

Cravens, D. W./Shipp, S. H. (1991): Market-Driven Strategies for Competitive Advantage, in: Business Horizons, Vol. 34, Issue 1, Jan/Feb91, pp. 53-61.

Crawford, C. M. (1979): New Product Failure Rates - Facts and Fallacies, in: Research Management, Vol. 22, No.5, Sep, pp. 9-13.

Crews, J. E./Whittington, F. J. (2000): Vision Loss in an Aging Society. A Multidisciplinary Perspective, New York (NY).

Csillag, A. L. (2005): Atlas of the Sensory Organs. Functional and Clinical Anatomy, Totowa (NJ).

Cumming, E./Henry, W. E. (1961-1979): Growing Old, New York (NY).

Cunningham, M. T. (1969): The Application of Product Life Cycles to Corporate Strategy: Some Research Findings, in: British Journal of Marketing, Vol. 3, Issue 1, Spring, pp. 32-44.

Cunningham, S. M. (1967a): Perceived Risk as a Factor in Informal Consumer Communications, in: Cox, D. F. (Ed.): Risk Taking and Information Handling in Consumer Behavior, Boston, pp. 265-288.

Cunningham, S. M. (1967b): The Major Dimensions of Perceived Risk, in: Cox, D. F. (Ed.): Risk Taking and Information Handling in Consumer Behavior, Boston, pp. 82-108.

Dandekar, T. (1999): Warum altern wir? Biologische Aspekte des Älterwerdens, in: Niederfranke, A. /Naegele, G./Frahm, E. (Eds.): Funkkolleg Altern. Band 1: Die vielen Gesichter des Alterns, Opladen, Wiesbaden, pp. 239-276.

Danko, W. D./Schaninger, C. M. (1990): Attitudinal and Leisure Activity Differences Across Modernized Household Life Cycle Categories, in: Advances in Consumer Research, Vol. 17, pp. 886-894.

Danner, D. B./Schröder, H. C. (1994): Biologie des Alterns (Ontogenese und Evolution), in: Baltes, P. B./Mittelstraß, J./Staudinger, U. M. (Eds.): Alter und Altern: Ein interdisziplinärer Studientext zur Gerontologie. - Sonderausgabe des 1992 erschienenen 5. Forschungsberichts der Akademie der Wissenschaft zu Berlin, Berlin, pp. 95-123.

Davis, F. D. (1989): Perceived Usefulness, Perceived Ease of Use, and User Acceptance of Information Technology, in: MIS Quarterly, Vol. 13, Issue 3, Sep, pp. 318-340.

De Ayala Rafael Jaime (2009): The Theory and Practice of Item Response Theory, New York (NY).

Derrick, F. W./Lehfeld, A. K. (1980): The Family Life Cycle: An Alternative Approach, in: The Journal of Consumer Research, Vol. 7, Sep, pp. 214-217.

Dewar, R./Dutton, J. (1986): The Adoption of Radical and Incremental Innovations: An Empirical Analysis, in: Management Science, Vol. 32, Issue 11, Nov 1986, pp. 1422-1433.

Dhebar, A. (1995): Complementarity, Compatibility, and Product Change: Breaking with the Past?, in: Journal of Product Innovation Management, Vol. 12, Issue 2, March, pp. 136-152.

Dholakia, U. M. (1997): An Investigation of the Relationship Between Perceived Risk and Product Involvement, in: Advances in Consumer Research, Vol. 24, Issue 1, pp. 159-167.

Diaz-Bone, R. (2006): Statistik für Soziologen, Konstanz.

Dickerson, M. D./Gentry, J. W. (1983): Characteristics of Adopters and Non-Adopters of Home Computers, in: Journal of Consumer Research, Vol. 10, Issue 2, Sep, pp. 225-235.

Diehl, M. (1999): Self-Development in Adulthood and Aging: The Role of Critical Life Events, in: Ryff, C. D./Marshall, V. W. (Eds.): The Self and Society in Aging Processes, New York (NY), pp. 150-183.

Diener, E./Oishi, S./Lucas, R. E. (2003): Personality, Culture, and Subjective Well-being: Emotional and Cognitive Evaluations of Life, in: Annual Review of Psychology, 54(1), pp. 403-425.

Dietz, J.-W. (1989): Gründung innovativer Unternehmen, Wiesbaden 1989.

Ding-Greiner, C./Lang, E. (2004): Alternsprozesse und Krankheitsprozesse - Grundlagen, in: Kruse, A./Martin, M. (Eds.): Enzyklopädie der Gerontologie. Alternsprozesse in multidisziplinärer Sicht, Bern [et al.], pp. 182-206.

Disselkamp, M. (2005): Innovationsmanagement. Instrumente und Methoden zur Umsetzung im Unternehmen, Wiesbaden.

Dix, A. et al. (2004): Human-Computer Interaction, 3. ed., Harlow.

Dobbelstein, T./Pflaum, M. (2007): Generation 45+. Eine lebensgefühlbasierte Marktsegmentierung für den Gesundheitsmarkt, Saarbrücken.

Dorbritz, J./Gärtner, K. (1995): Die demographische Bedeutung des Familienstandes. Schriftenreihe des Bundesministerium für Familie, Senioren, Frauen und Jugend. Band 44, Stuttgart, Berlin; Köln.

Dowling, G. R. (1986): Perceived Risk: The Concept and Its Measurement, in: Psychology & Marketing, Vol. 3, Issue 3, Fall, pp. 193-210.

Dziuban, C. D./Shirkey, E. C. (1974): When is a Correlation Matrix Appropriate for Factor Analysis? Some Decision Rules, in: Psychological Bulletin, 81, No. 6, pp. 358-361.

Eastman, J. K./Iyer, R. (2005): The Impact of Cognitive Age on Internet Use of the Elderly: An Introduction to the Public Policy Implications, in: International Journal of Consumer Studies, Vol. 29, Issue 2, March, pp. 125-136.

Ehrlich, D. et al. (1957): Postdecision Exposure to Relevant Information, in: Journal of Abnormal and Social Psychology, Vol. 54, pp. 98-102.

Eichler, H.-V. (1991): Besonderheiten der Geschenksituation und ihre Auswirkungen auf das Konsumentenverhalten, Frankfurt (a. M.) [et al.] 1990.

Eitner, C. (2008): Die Reaktionsfähigkeit des deutschen Einzelhandels auf den demographischen Wandel. Eine qualitative und quantitative Analyse unter zielgruppen- und netzwerkspezifischen Gesichtspunkten, Diss., Bochum 2008.

Engstler, H. et al. (2004): Die Zeitverwendung älterer Menschen, in: Statistisches Bundesamt (Ed.): Alltag in Deutschland. Analysen zur Zeitverwendung, Wiesbaden, pp. 216-246.

Enslin, A.-P. (2003): Generationen 50 plus - die Ausgegrenzten der mobilen Informationsgesellschaft. Text- und Rezeptionsanalysen aktueller Mediennutzung, Diss., Marburg (Diss in Gießen) 2003.

Eppler, M. J./Mengis, J. (2004): The Concept of Information Overload: A Review of Literature from Organization Science, Accounting, Marketing, MIS, and Related Disciplines, in: The Information Society, Vol. 20, Issue 5, Nov/Dec, pp. 325-344.

Erber, J. T. (2010): Aging and Older Adulthood, 2. ed., Chichester.

Ettlie, J. E./Vellenga, D. B. (1979): The Adoption Time Period for Some Transportation Innovations, in: Management Science, Vol. 25, Issue 5, May, pp. 429-443.

Fahrmeir, L./Kneib, T./Lang, S. (2007): Regression. Modelle, Methoden und Anwendungen, Berlin; Heidelberg.

Fantapié Altobelli, C. (1991): Die Diffusion neuer Kommunikationstechniken in der Bundesrepublik Deutschland. Erklärung, Prognose und marketingpolitische Implikationen, Diss., Heidelberg 1990.

Farrell, J./Shapiro, C. (1989): Optimal Contracts with Lock-In, in: American Economic Review, Vol. 79, No. 1, March, pp. 51-68.

Federsel-Lieb, C. (1992): Kommunikationspolitik im Seniorenmarkt, Diss., Bayreuth 1992.

Fehlau, E. G. (2011): 30 Minuten Erfolgsfaktor 50+, Offenbach.

Festinger, L. (1957): A Theory of Cognitive Dissonance, Evanston (Ill.) [et al.].

Fidell, L. S./Tabachnick, B. G. (2003): Preparatory Data Analysis, in: Weiner, I. B./Schinka, J. A./Velicer, W. F. (Eds.): Handbook of Psychology. Vol. 2, Research Methods in Psychology, Hoboken (NJ), pp. 115-142.

Field, A. P. (2009): Discovering statistics using SPSS, 3. ed., London [et al.].

Filipp, S.-H. (1990): Kritische Lebensereignisse, 2. ed., München.

Filipp, S.-H. (1999): Lebenserfahrung und Lebenssinn. Biographische Aspekte des Alterns, in: Niederfranke, A./Naegele, G./Frahm, E. (Eds.): Funkkolleg Altern. Band 1: Die vielen Gesichter des Alterns, Opladen, Wiesbaden, pp. 101-136.

Fischer, M. (2001): Produktlebenszyklus und Wettbewerbsdynamik. Grundlagen für die ökonomische Bewertung von Markteintrittsstrategien, Wiesbaden 2001.

Fisher, R. J./Price, L. L. (1992): An Investigation into the Social Context of Early Adoption Behavior, in: Journal of Consumer Research, Vol. 19, Issue 3, Dec, pp. 477-486.

Fitzgerald Bone, P. (1992): Determinants of Word-of-Mouth Communications During Product Consumption, in: Advances in Consumer Research, Vol. 19, Issue 1 pp. 579-583.

Fliegel, F. C./Kivlin, J. E. (1966): Attributes of Innovations as Factors in Diffusion, in: The American Journal of Sociology, Vol. 72, No. 3, Nov, pp. 235-248.

FOCUS Magazin Verlag GmbH (2006): Communication Networks 10.0 - Basisdaten für Kommunikationsstrategien in der modernen Wissensgesellschaft, München.

FOCUS Magazin Verlag GmbH (2007): Communication Networks 11.0 - Basisdaten für Kommunikationsstrategien in der modernen Wissensgesellschaft, München.

Fox, J. (1991): Regression Diagnostics, Newbury Park, (Calif.) [et al.].

Fox, J. (1997): Applied Regression Analysis, Linear Models, and Related Methods, Thousand Oaks [et al.].

Fox, J. (2005): Linear Models: Problems, in: Kempf-Leonard, K. (Ed.): Encyclopedia of Social Measurement, Vol. 2, Amsterdam, pp. 515-522.

Fox, J. (2008): Applied Regression Analysis and Generalized Linear Models, 2. ed., Los Angeles [et al.].

Fox, M. C./Roscoe Jr., A. M./Feigenbaum, A. M. (1984): A Longitudinal Analysis of Consumer Behavior in the Elderly Population , in: Advances in Consumer Research, Vol. 11, Issue 1, pp. 563-568.

Franz, C. E./White, K. M. (1985): Individuation and Attachment in Personality Development: Extending Erikson's Theory, in: Journal of Personality, Vol. 53, Issue 2, June 1985, pp. 224-256.

Frederickson, H. G./Rowe, L. A./Boise, W. B. (1974): Organizational Innovation: Current Research and Evolving Concepts, in: Public Administration Review, Vol. 34, Issue 3, May/Jun74, pp. 284-293.

Freter, H. (1974): Marktsegmentierungsmerkmale, in: Marketing-Enzyklopädie. Das Marketingwissen unserer Zeit in 3 Bänden. Band 2: Käuferverhalten - Produktmanagement, München, pp. 15-31.

Freter, H. (1982): Effizientes Marketing in mittelständischen Unternehmen - Ansätze zum Abbau bestehender Defizite, in: Schöttle, K. M. (Ed.): Jahrbuch des Marketing, Essen, pp. 113-124.

Freter, H. (1983): Marketing-Strategien im Mittelstand, in: Gabele, E. (Ed.): Märkte, Mitarbeiter, Management. Erfolgreiche Führung kleiner und mittlerer Unternehmen I, Bamberg, pp. 23-46.

Freter, H. (1987): Marketing, klassisches Instrumentarium: Nur mit Vernetzung ist das Optimum zu erreichen, Gabler's Magazin. Betriebswirtschaft für Manager, No. 9, pp. 24-26.

Freter, H. (2001a): Marktsegmentierung, in: Diller, H. (Ed.): Vahlens Großes Marketing Lexikon. Band 2, 2. ed., München, pp. 1069-1074.

Freter, H. (2001b): Marktsegmentierungsmerkmale, in: Diller, H. (ed.): Vahlens Großes Marketing Lexikon. Band 2, 2. ed., München, pp. 1074-1076.

Freter, H. (2004): Marketing. Die Einführung mit Übungen, München.

Freter, H. (2008): Markt- und Kundensegmentierung. Kundenorientierte Markterfassung und -bearbeitung, 2. ed., Stuttgart.

Freter, H. (2009): Identifikation und Analyse von Zielgruppen, in: Bruhn, M./ Esch, F.-R./Langner, T. (Ed.): Handbuch Kommunikation. Grundlagen, innovative Ansätze, praktische Umsetzungen, Wiesbaden, pp. 397-411.

Freter, H./Keßler, R./Wangen, E. (1980): Marketing mit kleinem Budget, Schriftenreihe des BF/M (Betriebswirtschaftliche Forschungszentrum für Fragen der mittelständischen Wirtschaft e. V. an der Universität Bayreuth), Band 1, pp. 43-60.

Fromm, S. (2008): Multiple lineare Regressionsanalyse, in: Baur, N./Fromm, S. (Eds.): Datenanalyse mit SPSS für Fortgeschrittene. Ein Arbeitsbuch, 2. ed., Wiesbaden, pp. 345-369.

Gabler (2004): Innovation, in: Gabler Wirtschafts Lexikon, 16. ed., Wiesbaden, pp. 1496-1497.

Garcia, R./Calantone, R. (2002): A Critical Look at Technological Innovation Typology and Innovativeness Terminology: A Literature Review, in: Journal of Product Innovation Management, Vol. 19, Issue 2, March, pp. 110-132.

Garson, G. D. (2011a): "Multiple Regression," from Statnotes: Topics in Multivariate Analysis. Retrieved 09.05.2011 from http://faculty.chass.ncsu.-edu/garson/PA765/regress.htm#assume.

Garson, G. D. (2011b): "Univariate GLM, ANOVA, and ANCOVA," from Statnotes: Topics in Multivariate Analysis. Retrieved 11.03.2011 from http://faculty.chass.ncsu.edu/garson/PA765/anova.htm#assume.

Gassmann, O./Reepmeyer, G. (2006): Wachstumsmarkt Alter. Innovationen für die Zielgruppe 50+, München; Wien.

Gatignon, H./Robertson, T. S. (1985): A Propositional Inventory for New Diffusion Research, in: Journal of Consumer Research, Vol. 11, Issue 4: March, pp. 849-867.

Gatignon, H./Robertson, T. S. (1986): Integration of Consumer Diffusion Theory and Diffusion Models: New Research Directions, in: Mahajan, V./Wind, Y. (Eds.): Innovation Diffusion Models of New Product Acceptance; Vol. 5 of Series on Econometrics and Management Sciences, Cambridge (Mass.), pp. 37-59.

Gatignon, H./Robertson, T. S. (1989): Technology Diffusion: An Empirical Test of Competitive Effects, in: Journal of Marketing, Vol. 53, Issue 1, Jan, pp. 36-49.

Gatignon, H./Robertson, T. S. (1991): Innovative Decision Processes, in: Robertson, T. S./Kassarjian, H. H. (Eds.): Handbook of Consumer Behavior, Englewood Cliffs, pp. 316-348.

Gefromm, U. (2010): Die Entwicklung des Konsumverhaltens japanischer Senioren im Luxusgütersegment. Soziodemographische Rahmendaten und Lebensstilelemente, Diss., Berlin (Diss in Hamburg) 2009.

Gerbing, D. W./Anderson, J. C. (1988): An Updated Paradigm for Scale Development Incorporating Unidimensionality and its Assessment, in: Journal of Marketing Research, Vol. 25, May 1988, pp. 186-192.

Gergen, K. J./Back, K. W. (1966): Communication in the Interview and the Disengaged Respondent, in: Public Opinion Quarterly, Vol. 30, Issue 3, Fall, pp. 385-398.

Gerhards, M./Klingler, W. (2006): Mediennutzung in der Zukunft. Traditionelle Nutzungsmuster und innovative Zielgruppen, in: Media Perspektiven (Ed.): Ausgabe 2, pp. 75-90.

Geroski, P./Machin, S./Reenen, J. V. (1993): The Profitability of Innovating Firms, in: RAND Journal of Economics, Vol. 24, Issue 2, Summer93, pp. 198-211.

Gerpott, T. J. (2005): Prognose des Markterfolgs von Produktinnovationen, in: Albers, S./Gassmann, O. (Eds.): Handbuch Technologie- und Innovationsmanagement. Strategie - Umsetzung - Controlling, Wiesbaden, pp. 437-454.

GfK (1993): Die neuen Alten, Nürnberg.

GfK (2006): Consumer Electronics Market Europe (EC) 2006, Nürnberg.

GfK (2008): Press Release 2008-08-27. Innovation Drives Growth in Technical Goods, Nürnberg.

Gfu/BVT/GfK (2006): Consumer Electronics Marktindex Deutschland (CEMIX).

Gfu/GfK (2009): Der Markt für Consumer Electronics - Deutschland 2009, Frankfurt (a. M.); Nürnberg.

Gierl, H. (1987): Die Erklärung der Diffusion technischer Produkte, Berlin 1987.

Gilly, M. C./Enis, B. M. (1982): Recycling the Family Life Cycle: A proposal for Redefinitions, in: Advances in Consumer Research, Vol. 9, pp. 271-276.

Gilly, M. C./Zeithaml, V. A. (1985): The Elderly Consumer and Adoption of Technologies, in: Journal of Consumer Research, Vol. 12, Issue 3; Dec, pp. 353-357.

Glejser, H. (1969): A New Test for Heteroskedasticity, in: Journal of American Statistical Association, Vol. 64, No. 325, March, pp. 316-323.

Goebel, J./Grabka, M. M. (2011): Zur Entwicklung der Altersarmut in Deutschland, in: DIW Wochenbericht, Nr. 25, 23. Juni 2011, pp. 3-16.

Golder, P. N./Tellis, G. J. (1997): Will It Ever Fly? Modeling the Takeoff of New Consumer Durables, in: Marketing Science, Vol. 16; No. 3, pp. 256-270.

Goldsmith, R. E./Flynn, L. R./Goldsmith, E. B. (2003): Innovative Consumers and Market Mavens, in: Journal of Marketing Theory and Practice, Vol. 11, Issue 4, Fall 2003, pp. 54-64.

Goldsmith, R. E./Goldsmith, E. B. (1996): An Empirical Study of Overlap of Innovativeness, in: Psychological Reports, Part 2, Vol. 79, Issue 3, Dec, pp. 1113-1114.

Goldsmith, R. E./Hofacker, C. F. (1991): Measuring Consumer Innovativeness, in: Journal of the Academy of Marketing Science, Vol. 19, Issue 3, Summer, pp. 209-221.

Gorsuch, R. L. (1997): Exploratory Factor Analysis. It's Role in Item Analysis, in: Journal of Personality Assessment, Vol. 68 Issue 3, pp. 532-560.

Gorsuch, R. L. (2003): Factor Analysis, in: Weiner, I. B./Schinka, J. A./Velicer, W. F. (Eds.): Handbook of Psychology. Vol. 2, Research Methods in Psychology, Hoboken (NJ), pp. 143-164.

Göthlich, S. E. (2006): Zum Umgang mit fehlenden Daten in großzahligen empirischen Erhebungen, in: Albers, S. et al. (Eds.): Methodik der empirischen Forschung, Wiesbaden, pp. 133-150.

Gourville, J. T./Soman, D. (2005): Overchoice and Assortment Type: When and Why Variety Backfires, in: Journal of Marketing Management, Vol. 24, Issue 3, Summer, pp. 382-395.

Graham, J. W./Cumsille, P. E./Elek-Fisk, E. (2003): Methods for Handling Missing Data, in: Weiner, I. B./Schinka, J. A./Velicer, W. F. (Eds.): Handbook of Psychology. Vol. 2, Research Methods in Psychology, Hoboken (NJ), pp. 87-114.

Greenleaf, E. A./Lehmann, D. R. (1991): Causes of Delay in Consumer Decision Making: An Exploratory Study, in: Advances in Consumer Research, Vol. 18, Issue 1, pp. 470-475.

Greenleaf, E. A./Lehmann, D. R. (1995): Reasons for Substantial Delay in Consumer Decision Making, in: Journal of Consumer Research, Vol. 22, Issue 2, Sep, pp. 186-199.

Greenstein, S. M. (1993): Did Installed Base Give an Incumbent Any (Measurable) Advantages in Federal Computer Procurement?, in: RAND Journal of Economics, Vol. 24, No. 1, Spring, pp. 19-39.

Gregan-Paxton, J./ John, D. R. (1997): Consumer Learning by Analogy: A Model of Internal Knowledge Transfer, in: Journal of Consumer Research, Vol. 24, Issue 3, Dec, pp. 266-284.

Guadagnoli, E./Velicer, W. F. (1988): Relation of Sample Size to the Stability of Component Patterns, in: Psychological Bulletin, Vol. 103, No. 2, pp. 265-275.

Gunter, B. (1998): Understanding the Older Consumer. The Grey Market, London; New York (NY).

Guttman, L. (1953): Image Theory for the Structure of Quantitative Variates, in: Psychometrika 18 pp. 277-296.

Gwinner, K. P./Stephens, N. (2001): Testing the Implied Mediational Role of Cognitive Age, in: Psychology & Marketing, Vol. 18, Issue 10, Oct, pp. 1031-1048.

Haan, N./Millsap, R./Hartka, E. (1986): As Time Goes By: Change and Stability in Personality Over Fifty Years, in: Psychology and Aging, Vol 1, No. 3, pp. 220-232.

Haedrich, G./Tomczak, T. (1996): Produktpolitik, Stuttgart.

Haimann, R. (2005): Alt! Wie die wichtigste Konsumentengruppe der Zukunft die Wirtschaft verändert, Frankfurt (a. M.).

Hampson, S. E. (1992): The Construction of Personality. An Introduction, 2. ed., London; New York (NY).

Hansen, F. (1972): Consumer Choice Behavior. A Cognitive Theory, Diss., New York (NY); London 1972.

Hardie, B. G. S./Robertson, T. S./Ross, Jr. W. T. (1996): Technology Adoption: Amplifying Versus Simplifying Innovations, in: Marketing Letters, Vol. 7, Issue 4, Oct, pp. 355-369.

Hartig, J./Frey, A./Jude, N. (2007): Validität, in: Moosbrugger, H./Kelava, A. (Eds.): Testtheorie und Fragebogenkonstruktion, Heidelberg, pp. 135-163.

Härtl-Kasulke, C. (1998): Marketing für Zielgruppen ab 50. Kommunikationsstrategien für 50plus und Senioren, Neuwied.

Hauschildt, J. (2004): Innovationsmanagement, 3. ed., München.

Havighurst, R. J. et al. (1970): Adjustment to Retirement. A Cross-national Study, 2. ed., Assen.

Hawkins, D. I./Best, R. J./Coney, K. A. (2004): Consumer Behavior. Building Marketing Strategy, 9. ed., Boston [et al.].

Heinze, R. G./Naegele, G./Schneiders, K. (2011): Wirtschaftliche Potentiale des Alters, Stuttgart.

Helm, R. (2001): Planung und Vermarktung von Innovationen: Die Präferenz für verschiedene Innovationsumfänge unter Berücksichtigung des optimalen Stimulationsniveaus und marktbezogener Einflussfaktoren, Stuttgart.

Hens, M. (2004): Professionelle Ladengestaltung für 50plus-Kunden. Verkaufsförderung am POS für die Zielgruppe der Zukunft Hens, BBE-Unternehmensberatung GmbH (Ed.): Köln.

Herbig, P. A./Palumbo, F. (1994): The Effect of Culture, in: Technological Forecasting and Social Change, Vol. 46, No. 1, pp. 71-101.

Herrmann, A. K. (2003): Kaufverhalten bei Innovationen auf dem Lebensmittelmarkt. Entwicklung eines kausalanalytischen Messinstruments für functional food, Diss., Wiesbaden 2003.

Herrmann, A./Huber, F. (2000): Kundenorientierte Produktgestaltung - Ziele und Aufgaben, in: Herrmann, A. et al. (Eds.): Kundenorientierte Produktgestaltung, München, pp. 3-18.

Herrmann, A./Landwehr, J. R. (2008): Varianzanalyse, in: Herrmann, A./Homburg, C./Klarmann, M. (Eds.): Handbuch Marktforschung. Methoden - Anwendungen - Praxisbeispiele, 3. ed., Wiesbaden, pp. 579-606.

Hildebrandt, L. (1998): Kausalanalytische Validierung in der Marketingforschung, in: Hildebrandt, L./Homburg, C. (Eds.): Die Kausalanalyse. Ein Instrument der empirischen betriebswirtschaftlichen Forschung, Stuttgart, pp. 85-110.

Hippel, E. v. (1992): Adapting Market Research to the Rapid Evolution of Needs for New Products and Services, in: Sloan School of Management, Massachusetts Institute of Technology, Working paper No. 3374-92-BPS, Jan, pp. 1-23.

Hirschberg, N. (1978): A Correct Treatment of Traits, in: London, H. (Ed.): Personality: A New Look at Metatheories, Washington, pp. 45-68.

Hirschman, E. C. (1980): Innovativeness, Novelty Seeking, and Consumer Creativity, in: Journal of Consumer Research, Vol. 7, Issue 3, Dec, pp. 283-295.

Hirunyawipada, T./ Paswan, A. K. (2006): Consumer Innovativeness and Perceived Risk: Implications for High Technology Product Adoption, in: Journal of Consumer Marketing, Vol. 23, No. 4, pp. 182-198.

Hock, E.-M./Bader, B. (2001): Kauf- und Konsumverhalten der 55plus-Generation, St. Gallen.

Hoeffler, S. (2003): Measuring Preferences for Really New Products, in: Journal of Marketing Research, Vol. 40, Issue 4, Nov. 2003, pp. 406-420.

Hoeffler, S. (2002): Conceptualizing Newness and Positioning Really New Products, in: Advances in Consumer Research, Vol. 29, Issue 1, pp. 267.

Hoffmann, E./Menning, S./Schelhase, T. (2009): Demografische Perspektiven zum Altern und zum Alter, in: Böhm, K./Tesch-Römer, C./Ziese, T. (Eds.): Gesundheit und Krankheit im Alter. Beiträge zur Gesundheitsberichterstattung des Bundes, Berlin, pp. 21-30.

Hohl, N. A. D./Naskrent, J. (2009): Involvement. Forschungsstand und Neukonzeption, Siegen.

Holak, S. L. (1988): Determinants of Innovative Durables Adoption. An Empirical Study with Implications for Early Product Screening, in: Journal of Product Innovation Management, Vol. 5, Issue 1, March, pp. 50-69.

Holak, S. L./Lehmann, D. R. (1990): Purchase Intentions and the Dimensions of Innovation: An Exploratory Model, in: Journal of Product Innovation Management, Vol. 7, No. 1, March, pp. 59-73.

Holak, S. L./Lehmann, D. R./Sultan, F. (1987): The Role of Expectations in the Adoption of Innovative Consumer Durables: Some Preliminary Evidence, in: Journal of Retailing, Vol. 63, Issue 3, Fall, pp. 243-259.

Holm, K. (1970): Zuverlässigkeit von Skalen und Indizes, in: Kölner Zeitschrift für Soziologie und Sozialpsychologie, 22, pp. 356-386.

Holmes, T. H./Rahe, R. H. (1967): The Social Readjustment Scale, in: Journal of Psychosomatic Research, Vol. 11, Issue 2, Aug 1967, pp. 213-218.

Homburg, C. et al. (2008): Methoden der Datenanalyse im Überblick, in: Herrmann, A./Homburg, C./Klarmann, M. (Eds.): Handbuch Marktforschung. Methoden - Anwendungen - Praxisbeispiele, 3. ed., Wiesbaden, pp. 151-173.

Homburg, C./Giering, A. (1996): Konzeptualisierung und Operationalisierung komplexer Konstrukte. Ein Leitfaden für die Marketingforschung, in: Marketing Zeitschrift für Forschung und Praxis, 1. Quartal, 18, pp. 5-24.

Homburg, C./Krohmer, H. (2003): Marketingmanagement. Strategie - Instrumente - Umsetzung - Unternehmensführung, Wiesbaden.

Hooyman, N. R./Kiyak, H. A. (1999): Social Gerontology. A Multidisciplinary Perspective, 5. ed., Boston; Mass [et al.].

Hooyman, N. R./Kiyak, H. A. (2008): Social Gerontology. A Multidisciplinary Perspective, 8. ed., Boston; Mass [et al.].

Horton, R. L. (1976): The Structure of Perceived Risk: Some Further Progress, in: Journal of the Academy of Marketing Science, Vol. 4, No. 4, pp. 694-706.

Hoyer, W. D./MacInnis, D. J. (2004): Consumer Behavior, 3. ed., Boston; New York (NY).

Hoyer, W. J./Rybash, J. M./Roodin, P. A. (1999): Adult Development and Aging, 4. ed., Boston [et al.].

Hübler, O. (2005): Einführung in die empirische Wirtschaftsforschung. Probleme, Methoden und Anwendungen, 10. ed., München; Wien.

Hunke, G. (2011): Best Practice Modelle im 55plus Marketing. Bewährte Konzepte für den Dialog mit Senioren, Wiesbaden.

Hupp, O. (1999): Seniorenmarketing: Analyse des Informations- und Entscheidungsverhalten älterer Konsumenten unter besonderer Berücksichtigung des Involvements und der Marktsegmentierung, Diss., Saarbrücken.

Huxhold, O./Mahne, K./Naumann, D. (2010): Soziale Integration, in: Motel-Klingebiel, A./Wurm, S./Tesch-Römer, C. (Eds.): Altern im Wandel. Befunde des Deutschen Alterssurveys (DEAS), Stuttgart, pp. 215-233.

IBM Global Business Services (2007): Konvergenz oder Divergenz? Erwartungen und Präferenzen der Konsumenten an die Telekommunikations- und Medienangebote von morgen, Stuttgart.

Institut der deutschen Wirtschaft (2004): Insolvenzen. Von Rekord zu Rekord, in: Informationsdienst des Instituts der deutschen Wirtschaft, Jg. 30; No. 4, 22. Jan, pp. 8.

Institut der deutschen Wirtschaft (2006): Unternehmensinsolvenzen. Ein kleiner Lichtblick, in: Informationsdienst des Instituts der deutschen Wirtschaft, Jg. 32, No. 8, 23. Feb, pp. 2.

Institut der deutschen Wirtschaft (2011): Unerwartet frühe Wende, in: Informationsdienst des Instituts der deutschen Wirtschaft, Jg. 37, No. 4, 27. Jan, pp. 8.

Jäckel, M./Wollscheid, S. (2004): Mediennutzung im Tagesverlauf: Ausweitung des Angebots und Strukturen der Zeitverwendung, in: Statistisches Bundesamt (Ed.): Alltag in Deutschland. Analysen zur Zeitverwendung, Wiesbaden, pp. 373-411.

Jacoby, J. (1971): Personality and Innovation Proneness, in: Journal of Marketing Research, Vol. 8, No. 2, May, pp. 244-247.

Jacoby, J./Kaplan, L. B. (1972): The Components of Perceived Risk, in: Venkatesan, M. (Ed.): Proceedings of the Third Annual Conference of the Association for Consumer Research, College Park, pp. 382-393.

Jain, K./Srinivasan, N. (1990): An Empirical Assessment of Multiple Operationalizations of Involvement, in: Advances in Consumer Research, Vol. 17, Issue 1, pp. 594-602.

Jamison, J. R. (2007): Healthcare for an Aging Population. Meeting the Challenge, Edinburgh [et al.].

Janssen, J./Laatz, W. (2007): Statistische Datenanalyse mit SPSS für Windows. Eine anwendungsorientierte Einführung in das Basissystem und das Modul exakte Tests, 6. ed., Berlin; Heidelberg; New York (NY).

Jayanti, R. K./McManamon, M. K./Whipple, T. W. (2004): The Effects of Aging on Brand Attitude Measurement, in: Journal of Consumer Marketing, Vol. 21, Issue 4, pp. 264-273.

John, D. R./Cole, C. A. (1986): Age Differences in Information Processing: Understanding Deficits in Young and Elderly Consumers, in: Journal of Consumer Research, Vol. 13, Issue 3, Dec, pp. 297-315.

Jones, C. J./Meredith, W. (1996): Patterns of Personality Change Across the Life Span, in: Psychology and Aging, Vol. 11, No. 1, pp. 57-65.

Jugel, S. (1991): Ansatzpunkte einer Marketingkonzeption für technologische Innovationen, Stuttgart 1991.

Kaas, K. P. (1973): Diffusion und Marketing. Das Konsumentenverhalten bei der Einführung neuer Produkte, Diss., Stuttgart.

Kaiser, H. F./Rice, J. (1974): Little Jiffy, Mark IV, in: Educational and Psychological Measurement, 34, pp. 111-117.

Kakkar, P./Lutz, R. J. (1981): Situational Influence on Consumer Behavior: A Review, in: Kassarjian, H. H./Robertson, T. S. (Eds.): Perspectives in Consumer Behavior, 3. ed., Glenview (Ill.) [et al.], pp. 204-214.

Kalish, S. (1985): A New Product Adoption Model with Price, Advertising, and Uncertainty, in: Management Science, Vol. 31, Issue 12, Dec, pp. 1569-1585.

Kamakura, W. A./Balasubramanian, S. K. (1988): Long-term View of the Diffusion of Durables, in: International Journal of Research in Marketing, Vol. 5, Issue 1, March, pp. 1-13.

Kamenz, U. (1987): Einkommensverwendung im Familienlebenszyklus, in: Helmstädter, E. (Ed.): Schriften zur Textilwirtschft, Frankfurt (a. M.) [et al.].

Kapferer, J.-N./Laurent, G. (1985): Consumer's Involvement Profile: New Empirical Results, in: Advances in Consumer Research, Vol. 12, Issue 1, pp. 290-295.

Kaplan, L. B./Szybillo, G. J./Jacoby, J. (1974): Components of Perceived Risk in Product Purchase: A Cross-Validation, Journal of Applied Psychology, Vol. 59, No. 3, pp. 287-291.

Karahanna, E./Agarwal, R./Angst, C. M. (2006): Reconceptualizing Compatibility Beliefs in Technology Acceptance Research, in: MIS Quarterly, Vol. 30, Issue 4, Dec, pp. 781-804.

Karahanna, E./Straub, D. W./Chervany, N. L. (1999): Information Technology Adoption Across Time: A Cross-Sectional Comparison of Pre-

Adoption and Post-Adoption Beliefs, in: MIS Quarterly, Vol. 23, Issue 2, June, pp. 183-213.

Kassarjian, H. H. (1971): Personality and Consumer Behavior: A Review, in: Journal of Marketing Research, Vol. 8, Issue 4, Nov, pp. 409-418.

Kassarjian, H. H. (1981): Low Involvement: A Second Look, in: Advances in Consumer Research, Vol. 8, Issue 1, pp. 31-34.

Kastenbaum, R. et al. (1972): The Ages of Me: Toward Personal and Interpersonal Definitions of Functional Aging, in: Aging and Human Development, No. 3, pp. 197-211.

Katona, G. (1964): The Mass Consumption Society, New York (NY) [et al.].

Katona, G./Mueller, E. (1955): A Study of Purchase Decisions, in: Clark, L. H. (Ed.): Consumer Behavior. The Dynamics of Consumer Reaction, New York (NY), pp. 30-87.

Katz, E. (1961): The Social Itinerary of Technical Change: Two Studies on the Diffusion of Innovation, in: Human Organization, Vol. 20, No. 2, Summer, pp. 70-82 .

Katz, E. (1968): The Two-step Flow of Communication, in: Kassarjian, H. H./Robertson, T. S. (Eds.): Perspectives in Consumer Behavior, Glenview (Ill.), pp. 337-350.

Kaufmann, A. (2000): Seniorenbanking. Strategische Ansatzpunkte einer zielgerichteten Bankpolitik für ältere Menschen, Diss., Frankfurt (a. M.) [et al.].

Kaupp, P. (1997): Ältere im Schatten der Werbung? Eine Literaturstudie zur Seniorendarstellung in der Werbung, Berlin.

Keit, T. Z./Reynolds, C. R. (2003): Measurement and Design Issues in Child Assessment Research, in: Reynolds, C. R./Kamphaus, R. W. (Eds.): Handbook of Psychological and Educational Assessment of Children. Intelligence, Aptitude, and Achievment, 2. ed., New York (NY), pp. 79-111.

Kelly, G. A. (1963): A Theory of Personality, New York (NY) [et al.].

Ketcham, C. J./Stelmach, G. E. (2001): Age-Related Declines in Motor Control, in: Birren, J. E./Schaie, W. K. (Eds.): Handbook of the Psychology of Aging, 5. ed., San Diego [et al.], pp. 313-348.

Kieser, A. (1974): Management der Produktinnovation: Strategie, Planung und Organisation, in: Management International Review, Vol. 14, pp. 3-22.

King, C. W./Summers, J. O. (1970): Overlap of Opinion Leadership Across Consumer Product Categories, in: Journal of Marketing Research, Vol. 7, Issue 1, Feb, pp. 43-50.

Kirk, R. E. (1995): Experimental Design. Procedures for the Behavioral Sciences, 3. ed., Pacific Grove (Calif.).

Kirsch, J. (2003): Senioren - Marktsegment der Zukunft, in: Rabe, C. (Ed.): Zukunftsperspektiven des Marketing - Paradigmenwechsel und Neuakzentuierungen. Festschrift anläßlich der Emeritierung von Prof. Dr. Dr. h.c. Hans Hörschgen, Berlin, pp. 181-196.

Kogler, A. (1991): Investitionen in Produkt- und Prozessinnovationen. Eine evolutionsökonomische Analyse der Struktur und Entwicklung des Innovationsmix zwischen Produkt- und Prozessinnovationen, Frankfurt (a. M.) [et al.].

Kohlbacher, F./Herstatt, C. (2011): The Silver Market Phenomenon. Marketing and Innovation in the Aging Society, 2. ed., Heidelberg [et al.].

Kolland, F. (2000): Freizeit, in: Wahl, H.-W./Tesch-Römer, C. (Eds.): Angewandte Gerontologie in Schlüsselbegriffen, Stuttgart; Berlin; Köln, pp. 178-183.

Kölzer, B. (1995): Senioren als Zielgruppe. Kundenorientierung im Handel, Diss., Wiesbaden.

Koppelmann, U. (2001): Produktmarketing. Entscheidungsgrundlagen für Produktmanager, 6. ed., Berlin [et al.].

Kotzbauer, N. (1992): Erfolgsfaktoren neuer Produkte. Der Einfluss der Innovationshöhe auf den Erfolg technischer Produkte, Diss., Frankkfurt (a. M.) [et al.].

Krack-Roberg, E./Grobecker, C. (2011): Bevölkerungsentwicklung 2009, in: Statistisches Bundesamt (Ed.): Wirtschaft und Statistik, Mai 2011, Wiesbaden, pp. 419-433.

Krieb, C./Reidl, A. (2001): Seniorenmarketing: So erreichen sie die Zielgruppe der Zukunft, Landsberg/Lech.

Kroeber-Riel, W./Weinberg, P. (2003): Konsumentenverhalten, 8. ed., München.

Kruse, A./Martin, M. (2004): Enzyklopädie der Gerontologie. Alternsprozesse in multidisziplinärer Sicht, Bern [et al.].

Kühn, K./Porst, R. (1999): Befragung alter und sehr alter Menschen: Besonderheiten, Schwierigkeiten und methodische Konsequenzen. Ein Literaturbericht, Mannheim.

Kuntze, S. (2011): Altern wie ein Gentleman. Zwischen Müßiggang und Engagement, München.

Kuß, A./Tomczak, T. (2004): Käuferverhalten. Eine marketingorientierte Einführung, 3. ed., Stuttgart.

LaBay, D. G./Kinnear, T. C. (1981): Exploring the Consumer Decision Process in the Adoption of Solar Energy Systems, in: Journal of Consumer Research, Vol. 8, Issue 3, Dec, pp. 271-278.

Landschulze, S. (2010): Marketing Xplus - Marketing für Senioren. Der Einfluss des Alters auf die Präferenzen für Neuerungen und Abwechslung im Konsum von Fast moving consumer goods, Diss., Hamburg.

Lang, F. R. (2000): Soziale Beziehungen im Alter: Ergebnisse der empirischen Forschung, in: Wahl, H.-W./Tesch-Römer, C. (Eds.): Angewandte Gerontologie in Schlüsselbegriffen, Stuttgart; Berlin; Köln, pp. 142-147.

Lang, F. R./Carstensen, L. L. (1998): Social Relationships and Adaptation in Later Life, in: Edelstein, B. (Ed.): Comprehensive Clinical Psychology. Vol. 7: Geropsychology, Oxford, pp. 55-72.

Lansing, J. B./Kish, L. (1957): Family Life Cycle as an Independent Variable, in: American Social Review, Vol. 22, No. 5, Oct, pp. 512-519.

Lau, G. T./Ng, S. (2001): Individual and Situational Factors Influencing Negative Word-of-Mouth Behaviour, in: Canadian Journal of Administrative Sciences, Vol. 18, Issue 3, Sep, pp. 163-178.

Laumann, C. (2007): Marketing für 'Junge Alte.' So erreicht man die Generation 50plus, Saarbrücken.

Laurinkari, J. (1988): Alte Menschen als Konsumenten, in: Jahrbuch der Absatz- und Verbrauchsforschung, 34. Jg., Nr. 2, pp. 154-163.

Lavidge, R. J./Steiner, G. A. (1961): A Model for Predictive Measurements of Advertising Effectiveness, in: Journal of Marketing, Vol. 25, Issue 6, Oct, pp. 59-62.

Lazarus, R. S./Folkman, S. (1984): Stress, Appraisal, and Coping, New York (NY).

Lehr, U. (2007): Psychologie des Alterns, 11. ed., Wiebelsheim.

Leker, J. (2005): F&E Controlling, in: Albers, S./Gassmann, O. (Ed.): Handbuch Technologie- und Innovationsmanagement. Strategie - Umsetzung - Controlling, Wiesbaden, pp. 567-584.

Leonard-Barton, D. (1985): Experts as Negative Opinion Leaders in the Diffusion of a Technological Innovation, in: Journal of Consumer Research, Vol. 11, Issue 4, March, pp. 914-926.

Leventhal, R. (1997): Aging Consumers and their Effects on the Marketplace, in: Journal of Consumer Marketing, Vol. 14, Issue 4/5, pp. 276-281.

Levitt, T. (1965): Exploit the Product Life Cycle, in: Harvard Business Review, Vol. 43, Issue 6, Nov/Dec , pp. 81-94.

Lewis, H. G. (1997): Seniorenmarketing: die besten Werbe- und Verkaufskonzepte, Landsberg/Lech.

Lichtenstein, D. R./Ridgway, N. M./Netemeyer, R. G. (1993): Price Perceptions and Consumer Shopping Behavior: A Field Study, in: Journal of Marketing Research, Vol. 30, Issue 2, May, pp. 234-245.

Lienert, G. A./Raatz, U. (1998): Testaufbau und Testanalyse, 6. ed., Weinheim.

Litfin, T. (2000): Adoptionsfaktoren. Empirische Analyse am Beispiel eines innovativen Telekommunikationsdienstes, Diss., Wiesbaden 1999.

Loomis, C. P. (1936): The Study of the Life Cycle of Families, in: Rural Sociology, 1, pp. 180-199.

Loomis, C. P./Hamilton, C. H. (1936): Family Life Cycle Analysis, in: Social Forces, 15, pp. 225-231.

Lunsford, D. A./Burnett, M. (1992): Marketing Product Innovations to the Elderly: Understanding The Barriers To Adoption, in: Journal of Consumer Marketing, Vol. 9, Issue 4, Fall, pp. 53-63.

Lutschewitz, H./Kutschker, M. (1977): Die Diffusion von innovativen Investitionsgütern. Theoretische Konzeption und empirische Befunde, München.

MacCallum, R. C. et al.(1999): Sample Size in Factor Analysis, in: Psychological Methods, 4 (1), March, pp. 84-99.

Mahajan, V./Muller, E./Bass, F. M. (1990): New Product Diffusion Models in Marketing: A Review and Directions for Research, in: Journal of Marketing, Vol. 54, Issue 1, Jan, pp. 1-26.

Mahajan, V./Muller, E./Kerin, R. A. (1984): Introduction Strategy for New Products with Positive and Negative Word-of-Mouth, in: Management Science, Vol. 30, No. 12, Dec, pp. 1389-1404.

Mahajan, V./Muller, E./Wind, Y. (2000): New Product Diffusion Models, Boston; Dordrecht; London.

Mahajan, V./Peterson, R. A. (1985): Models for Innovation Diffusion, Beverly Hills; London; New Dehli.

Mahajan, V./Sharma, S./Buzell, R. D. (1993): Assessing the Impact of Competitive Entry on Market. Expansion and Incumbent Sales, in: Journal of Marketing, Vol. 57, July, pp. 39-52.

Malhotra, N. K. (2004): Marketing Research. An Applied Orientation, 4. internat. ed., Upper Saddle River (NJ).

Manning, K. C./Bearden, W. O./Madden, T. J. (1995): Consumer Innovativeness and the Adoption Process, in: Journal of Consumer Psychology, Vol. 4, Issue 4, pp. 329-345.

Martens, J. (2003): Statistische Datenanalyse mit SPSS für Windows, 2. ed., München.

Martin, M./Kliegel, M. (2005): Psychologische Grundlagen der Gerontologie, Stuttgart.

Martinez, E./Polo, Y./Flavian, C. (1998): The Acceptance and Diffusion of New Consumer Durables: Differences Between First and Last Adopters, in: Journal of Consumer Marketing, Vol. 15, Issue 4 pp. 319-342.

Masoro, E. J. (1999): Challenges of Biological Aging, New York (NY).

Mathur, A. (1999): Adoption of Technological Innovations by the Elderly: A Consumer Socialization Perspective, in: Journal of Marketing Management, Vol. 9, Issue 3, Winter, pp. 21-35.

Mathur, A./Moschis, G. P. (2005): Antecedents of Cognitive Age: A Replication and Extension, in: Psychology & Marketing, Vol. 22, Issue 12, Dec, pp. 969-994.

McCrae, R. R./Costa, P. T. (2003): Personality in Adulthood, 2. ed., New York (NY).

McDonald, H./Corkindale, D./Sharp, B. (2003): Behavioral versus Demographic Predictors of Early Adoption: A Critical Analysis and Comparative Test, in: Journal of Marketing Theory & Practice, Vol. 11, Issue 3, Summer 2003, pp. 84-97.

Medialine (2001): Der Markt der Unterhaltungselektronik. Daten, Fakten, Trends.

Mediendaten Südwest (2007): Geräteausstattung BRD Gesamt (Mediendaten Südwest), http://www.mediendaten.de/relaunch/medienausstattung/geraeteausstattung.php, 14.9.2007.

Meffert, H. (1976): Die Durchsetzung von Innovationen in der Unternehmung und am Markt, in: Zeitschrift für Betriebswirtschaft (ZfB), 46. Jg., Nr. 2, pp. 77-100.

Meffert, H. (1985): Marketing und neue Medien, Stuttgart.

Meffert, H. (2000): Marketing. Grundlagen marktorientierter Unternehmensführung, 9. ed., Wiesbaden.

Meißner, W. (1989): Innovation und Organisation. Die Initiierung von Innovationsprozessen in Organisationen, Stuttgart.

Melenhorst, A.-S./ Rogers, W. A./Caylor, E. C. (2001): The Use of Communication Technologies by Older Adults: Exploring the Benefits from the Users Perspective, in: Proceedings of the Human Factors and Ergonomics Society 45th Annual Meeting, Oct. 8-12, pp. 221-225.

Menzel, H. (1981): Interpersonal and Unplanned Communications: Indispensable or Obsolete?, in: Roberts, E. B. et al. (Eds.): Biomedical Innovation, Cambridge (Mass.); London, pp. 155-163.

Meyer-Hentschel, G./Meyer-Hentschel, H. (2010): Jahrbuch Senioren-Marketing 2010/2011. Strategien und Innovationen, Frankfurt (a. M.).

Meyer-Hentschel, H. (1990): Produkt- und Ladengestaltung im Seniorenmarkt. Ein verhaltenswissenschaftlicher Ansatz, Diss., Saarbrücken.

Meyer-Hentschel, H./Meyer-Hentschel, G. (2004): Seniorenmarketing. Generationsgerechte Entwicklung und Vermarktung von Produkten und Dienstleistungen, Göttingen.

Midgley, D. F. (1977): Innovation and New Product Marketing, London.

Midgley, D. F. (1987): A Meta-Analysis of the Diffusion of Innovations Literature, in: Advances in Consumer Research, Vol. 14, Issue 1, pp. 204-207.

Midgley, D. F./Dowling, G. R. (1978): Innovativeness: The Concept and Its Measurement, in: The Journal of Consumer Research, Vol. 4, No. 4, March, pp. 229-242.

Miegel, M. (2002): Die deformierte Gesellschaft. Wie die Deutschen ihre Wirklichkeit verdrängen, Berlin; München.

Miller, G. A./Chapman, J. P. (2001): Misunderstanding Analysis of Covariance, in: Journal of Abnormal Psychology, 110, No. 1, pp. 40-48.

Miller, K. E./Ginter, J. L. (1979): An Investigation of Situational Variation in Brand Choice Behavior and Attitude, in: Journal of Marketing Research, Vol. 16, Issue 1, Feb, pp. 111-123.

Mitchell, V.-W. (1999): Consumer Perceived Risk: Conceptualisations and Models, in: European Journal of Marketing, Vol. 33, Issue: 1/2, pp. 163-195.

Mitchell, V.-W./Boustani, P. (1993): Market Development Using New Products and New Customers: A Role for Perceived Risk, in: European Journal of Marketing, Vol. 27, No. 2, pp. 17-32.

Mitchell, V.-W./Boustani, P. (1994): A Preliminary Investigation into Pre- and Post-Purchase Risk Perception and Reduction, in: European Journal of Marketing, Vol. 28, No. 1, pp. 56-71.

Mitchell, V.-W./Papavassiliou, V. (1999): Marketing Causes and Implications of Consumer Confusion, in: Journal of Product & Brand Management, Vol. 8, Issue 4, pp. 319-339.

Mitchell, V.-W./Walsh, G./Mo Yamin (2005): Towards a Conceptual Model of Consumer Confusion, in: Advances in Consumer Research, Vol. 32, Issue 1, pp. 143-150.

Mittal, B./Lee, M.-S. (1988): Separating Brand-Choice Involvement from Product Involvement Via Consumer Involvement Profiles, in: Advances in Consumer Research, Vol. 15, Issue 1, pp. 43-49.

Möhlen, M. (2007): Automobilmarketing. Entwicklung eines Kommunikationskonzeptes für Premiumanbieter auf dem deutschen Automobilmarkt im Hinblick auf die Zielgruppe Senioren 50plus, Diss., München [et al.].

Mollenkopf, H. et al. (2004): Räumlich-soziale Umwelten älterer Menschen: Die ökogerontologische Perspektive, in: Kruse, A./Martin, M. (Eds.): Enzyklopädie der Gerontologie. Alternsprozesse in multidisziplinärer Sicht, Bern [et al.], pp. 343-361.

Montoya-Weiss, M. M./Calantone, R. (1994): Determinants of New Product Performance. A Review and Meta-Analysis, in: Journal of Product Innovation Management, Vol. 11, Issue 5; Nov, pp. 397-417.

Moore, G. C./Benbasat, I. (1991): Development of an Instrument to Measure the Perceptions of Adopting an Information Technology Innovation, in: Information Systems Research, Vol. 2, Issue 3, Sep, pp. 192-222.

Moreau, C. P./Lehmann, D. R./Markman, A. B. (2001): Entrenched Knowledge Structures and Consumer Response to New Products, in: Journal of Marketing Research, Vol. 38, Issue 1, Feb, pp. 14-29.

Moschis, G. P. (1992): Marketing to Older Consumers. A Handbook of Information for Strategy Development, Westport; London.

Moschis, G. P. (1994): Marketing Strategies for the Mature Market, Westport.

Moschis, G. P. (2003): Marketing to Older Adults: An Updated Overview of Present Knowledge and Practice, in: Journal of Consumer Marketing, Vol. 20, No. 6, pp. 516-525.

Moschis, G. P./Mathur, A. (1993): How They're Acting Their Age, in: Marketing Management, Vol. 2, Issue 2 pp. 41-50.

Moschis, G. P./Mathur, A. (2006): Older Consumer Responses to Marketing Stimuli: The Power of Subjective Age, in: Journal of Advertising Research, Vol. 46, Issue 3, Sep, pp. 339-346.

Motel-Klingebiel, A./Simonson, J./Romeu-Gordo, L. (2011): Materielle Sicherung Älterer: Befunde des Deutschen Alterssurveys (DEAS), in: Informationsdienst Altersfragen, Bd. 38, Heft 2, pp. 11-17.

Mukherjee, A./Hoyer, W. D. (2001): The Effect of Novel Attributes on Product Evaluation, in: Journal of Consumer Research, Vol. 28, Issue 3, Dec, pp. 462-472.

Müller-Hagedorn, L. (1984): Die Erklärung von Käuferverhalten mit Hilfe des Lebenszykluskonzeptes, in: Wirtschaftswissenschaftliches Studium, Heft 11, pp. 561-569.

Müller-Hagedorn, L. (1986): Das Konsumentenverhalten: Grundlagen für die Marktforschung, Wiesbaden.

Murphy, P. E./Staples, W. A. (1979): A Modernized Family Life Cycle, in: Journal of Consumer Research, Vol. 6, No. 1, June, pp. 12-22.

Nabih, M. I./Bloem, S. G./Poiesz, T. B. C. (1997): Conceptual Issues in the Study of Innovation Adoption Behavior, in: Advances in Consumer Research, Vol. 24, Issue 1, pp. 190-196.

Nachtigall, C./Wirtz, M. (2006): Wahrscheinlichkeitsrechnung und Inferenzstatistik. Statistische Methoden für Psychologen Teil 2, 4. ed., Weinheim, München.

Naegele, G. (2000): Finanzielle Absicherung und Armut im Alter, in: Wahl, H.-W./Tesch-Römer, C. (Eds.): Angewandte Gerontologie in Schlüsselbegriffen, Stuttgart; Berlin; Köln, pp. 393-401.

Naumann, D./Romeu-Gordo, L. (2010): Gesellschaftliche Partizipation: Erwerbstätigkeit, Ehrenamt und Bildung, in: Motel-Klingebiel, A./Wurm, S./Tesch-Römer, C. (Eds.): Altern im Wandel. Befunde des Deutschen Alterssurveys (DEAS), Stuttgart, pp. 118-141.

Ness, O. (2007): Tourismusmarketing für Senioren. Auswirkungen des demographischen Wandels auf die Tourismusbranche, Saarbrücken.

Neßlböck, R. (2008): Entdeckung der Generation 60+ im Tourismus, Saarbrücken.

Newton, R. R./Rudestam, K. E. (1999): Your Statistical Consultant. Answers to Your Data Analysis Questions, Thousand Oaks (Calif.); London; New Delhi.

Niedbal, M. (2005): Vorankündigung von Produktinnovationen. Eine marktprozesstheoretische Analyse der Käufer- und Wettbewerbsreaktionen, Diss., Wiesbaden.

Noll, H.-H./Schöb, A. (2002): Lebensqualität im Alter, in: Deutsches Zentrum für Altersfragen (Ed.): Expertisen zum vierten Altenbericht der Bundesregierung. Band 1. Das hohe Alter - Konzepte, Forschungsfelder, Lebensqualität, Hannover, pp. 229-314.

Nunnally, J. C./Bernstein, I. H. (1994): Psychometric Theory, 3. ed., New York (NY) [et al.].

Olbrich, E. (1995): Normative Übergänge im menschlichen Lebenslauf: Entwicklungskrisen oder Herausforderungen, in: Filipp, S.-H. (Ed.): Kritische Lebensereignisse, 3. ed., Weinheim, pp. 123-138.

Opaschowski, H. (2004): Der Generationenpakt. Das soziale Netz der Zukunft, Darmstadt.

Osl, P. (2010): Dienstleistungen für Independent Living. Kundenanforderungen, Potenzialbewertung und Handlungsempfehlungen, Berlin.

Ostlund, L. E. (1972): Identifying Early Buyers, in: Journal of Advertising Research, Vol. 12, Issue 2, Apr, pp. 25-30.

Ostlund, L. E. (1973): Factor Analysis Applied to Predictors of Innovative Behavior, in: Decision Sciences, Vol. 4, Issue 1, Jan, pp. 92-108.

Ostlund, L. E. (1974): Perceived Innovation Attributes as Predictors of Innovativeness, in: Journal of Consumer Research, Vol. 1, Issue 2, pp. 23-29.

Otten, D. (2008): Die 50+ Studie. Wie die jungen Alten die Gesellschaft revolutionieren, Reinbek/Hamburg.

Pagano, R. R. (2009): Understanding Statistics in the Behavioral Sciences, 9. ed., Belmont, CA.

Palda, K. S. (1966): The Hypothesis of a Hierarchy of Effects: A Partial Evaluation, in: Journal of Marketing Research, Vol. 3, Issue 1, Feb, pp. 13-24.

Panne, F. (1977): Das Risiko im Kaufentscheidungsprozess des Konsumenten. Die Beiträge risikotheoretischer Ansätze zur Erklärung des Kaufentscheidungsverhaltens des Konsumenten, Diss., Zürich; Frankfurt (a. M.).

Panne, G. v. d./Beers, C. v./Kleinknecht, A. (2003): Success and Failure of Innovation: A Literature Review, in: International Journal of Innovation Management, Vol. 7, No. 3, Sep, pp. 309-338.

Papalia, D. E. (2007): Adult Development and Aging, 3. ed., Boston [et al.].

Pechtl, H. (1991): Innovatoren und Imitatoren im Adoptionsprozess von technischen Neuerungen, Diss., Bergisch Gladbach [et al.].

Pepels, W. (1995): Käuferverhalten und Marktforschung. Eine praxisorientierte Einführung, Stuttgart.

Pepels, W. (1999): Innovationsmanagement, Berlin.

Pepels, W. (2003): Produktmanagement. Produktinnovation, Markenpolitik, Programmplanung, Prozessorganisation, 4. ed., München; Wien.

Peschmann, M. (2005): Senioren im Automobilmarketing. Konzeptionelle Grundlagen und empirische Analyse zum Kaufverhalten jüngerer und älterer Automobilkäufer und ihre Implikationen für eine zielgruppenadäquate marktorientierte Unternehmensführung, Diss., Rostock.

Peskes, M. (2001): Senioren im Internet. Theoretisch fundierte Ansatzpunkte zur zielgerichteten Ansprache der Senioren-Community-Szene, Diss., Kassel.

Pett, M. A./Lackey, N. R./Sullivan, J. J. (2003): Making Sense of Factor Analysis. The Use of Factor Analysis for Instrument Development in Health Care Research, Thousand Oaks (Calif.).

Pfaff, D. (1990): Marktstudie zur Unterhaltungselektronik. Eine empirisch gestützte Untersuchung des Kaufverhaltens bei Radio- und Fernsehartikeln unter besonderer Berücksichtigung von Hersteller- und Händlerimages der Siegener Bevölkerung, 2. ed., Siegen.

Phillips, L. W./Sternthal, B. (1977): Age Differences in Information Processing: A Perspective on the Aged Consumer, in: Journal of Marketing Research, Vol. 14, Issue 4, Nov, pp. 444-457.

Pindyck, R. S./Rubinfeld, D. L. (1998): Econometric Models and Economic Forecasts, 4. ed., Boston [et al.].

Pohl, A. (1996): Leapfrogging bei technologischen Innovationen. Ein Erklärungsansatz auf Basis der Theorie des wahrgenommenen Risikos, Diss., Wiesbaden 1996.

Pohl, A. (2004): Preiszufriedenheit bei Innovationen. Nachfrageorientierte Analyse am Beispiel der Tourismus- und Airlinebranche, Wiesbaden.

Pohlmann, S. (2002): Ageing as a Global Phenomenon, in: Pohlmann, S. (Ed.): Facing an Ageing World - Recommendations and Perspectives, Regensburg, pp. 1-12.

Polanyi, M. (1985): Implizites Wissen, Frankfurt (a. M.).

Pompe, H.-G. (2007): Marktmacht 50plus. Wie Sie Best Ager als Kunden gewinnen und begeistern, Wiesbaden.

Porter, C. E./Donthu, N. (2006): Using the Technology Acceptance Model to Explain How Attitudes Determine Internet Usage: The Role of Perceived Access Barriers and Demographics, in: Journal of Business Research, Vol. 59, Issue 9, Sep, pp. 999-1007.

Pötzsch, O. (2007): Geburten in Deutschland, in: Statistisches Bundesamt (Ed.), Wiesbaden, pp. 1-35.

Price, L. L./Feick, L. F. (1984): The Role of Interpersonal Sources in External Search: An Informational Perspective, in: Advances in Consumer Research, Vol. 11, Issue 1, pp. 250-255.

Ram, S. (1987): A Model of Innovation Resistance, in: Advances in Consumer Research, Vol. 14, Issue 1, pp. 208-212.

Ram, S./Sheth, J. N. (1989): Consumer Resistance to Innovations: The Marketing Problem and Its Solutions, in: Journal of Consumer Marketing, Vol. 6, Issue 2, pp. 5-14.

Ray, M. L. (1973): Marketing Communication and the Hierarchy of Effects, in: Clarke, P. (Ed.): New Models for Mass Communication Research, Beverly Hills, pp. 147-176.

Reichart, S. V. (2002): Kundenorientierung im Innovationsprozess. Die erfolgreiche Integration von Kunden in den frühen Phasen der Produktentwicklung, Wiesbaden 2001.

Reidl, A. (2007): Seniorenmarketing. Mit älteren Zielgruppen neue Märkte erschließen, 2. ed., Landsberg (a. L.).

Reimann, H. (1974): Kommunikations-Systeme. Umrisse einer Soziologie der Vermittlungs- und Mitteilungsprozesse, 2. ed., Tübingen.

Reitzler, R. (2001): Versicherungen für Senioren. Perspektiven für das Zielgruppen-Marketing, Diss., Wiesbaden.

Robertson, T. S. (1967): Consumer Innovators: The Key to New Product Success, in: California Management Review, Vol. 10, Issue 2, Winter, pp. 23-30.

Robertson, T. S. (1968): Purchase Sequence Responses: Innovators vs. Non-Innovators, in: Journal of Advertising Research, Vol. 8, No.1, March, pp. 47-52.

Robertson, T. S. (1971): Innovative Behavior and Communication, New York (NY) [et al.].

Robertson, T. S. (1976): Low-Commitment Consumer Behavior, in: Journal of Advertising Research, Vol. 16, Issue 2, Apr, pp. 19-24.

Robertson, T. S./Myers, J. H. (1969): Personality Correlates of Opinion Leadership and Innovative Buying Behavior, in: Journal of Marketing Research, Vol. 6, Issue 2, May, pp. 164-168.

Robertson, T. S./Zielinski, J./Ward, S. (1984): Consumer Behavior, Glenview (Ill.) [et al.].

Robinson, B./Lakhani, C. (1975): Dynamic Price Models for New-Product Planning, in: Management Science, Vol. 21, Issue 10, June, pp. 1113-1122.

Robnett, R. H./Chop, W. C. (2010): Gerontology for the Health Care Professional, 2. ed., Sudbury, MA [et al.].

Roehrich, G./Valette-Florence, P./Ferrandi, J.-M. (2002): An Exploration of the Relationships Between Innate Innovativeness and Domain Specific Innovativeness, in: Cahier de Recherche du CERAG, Série Recherche, n° 2002-11, pp. 1-20 .

Rogers, E. M. (1962): Diffusion of Innovations, New York (NY) [et al.].

Rogers, E. M. (1983): Diffusion of Innovations, 3. ed., New York (NY) [et al.].

Rogers, E. M. (1995): Diffusion of Innovations, 4. ed., New York (NY) [et al.].

Rogers, E. M. (2003): Diffusion of Innovations, 5. ed., New York (NY) [et al.].

Rogers, E. M./Shoemaker, F. F. (1971): Communication of Innovations. A Cross-Cultural Approach, 2. ed., New York (NY); Toronto.

Rogers, W. A./Fisk, A. D. (2001): Understanding the Role of Attention in Cognitive Aging Research, in: Birren, J. E./Schaie, W. K. (Eds.): Handbook of the Psychology of Aging, 5. ed., San Diego [et al.], pp. 267-287.

Rokeach, M. (1960): The Open and Closed Mind: Investigations into the Nature of Belief Systems and Personality Systems, New York (NY).

Roselius, T. (1971): Consumer Rankings of Risk Reduction Methods, in: Journal of Marketing, Vol. 35, Issue 1, Jan, pp. 56-61.

Rosenberg, N. (1976): On Technological Expectations, in: Economic Journal, Vol. 86, Issue 343, Sep, pp. 523-535.

Rosenmayr, L. (1976): Schwerpunkte der Soziologie des Alters, in: König, R./Rosenmayr, L. (Eds.): Familie: Alter, Handbuch der empirischen Sozialforschung; Bd. 7, 2. ed., München; Stuttgart, pp. 218-406.

Ross, I. (1982): Information Processing and the Older Consumer: Marketing and Public Policy Implications, in: Advances in Consumer Research, Vol. 9, Issue 1, pp. 31-39.

Rößing, A. (2008): Senioren als Zielgruppe des Handels, Bremen.

Runkel, M. (2004): Environmental and Resource Policy for Consumer Durables, Berlin; Heidelberg.

Ruth, J.-E./Coleman, P. (1996): Personality and Aging: Coping and Management of the Self in Later Life, in: Birren, J. E./Schaie, W. K. (Eds.): Handbook of the Psychology of Aging, 4. ed., San Diego [et al.], pp. 308-322.

Ryle, G. (1963): The Concept of Mind, Repr., London.

Sääksjärvi, M. (2003): Consumer Adoption of Technological Innovations, in: European Journal of Innovation Management, Vol. 6, No. 2, pp. 90-100.

Sääksvuori, A./Immonen, A. (2005): Product Lifecycle Management, 2. ed., Berlin; Heidelberg.

Sabisch, H. (1991): Produktinnovationen, Stuttgart.

Scariano, S. M./Davenport, J. M. (1987): The Effects of Violation of Independence Assumptions in the One-way ANOVA, in: The American Statistician, Vol. 41, 2, pp. 123-129.

Schachtschabel, D. O. (2004): Humanbiologie des Alterns, in: Kruse, A./Martin, M. (Eds.): Enzyklopädie der Gerontologie. Alternsprozesse in multidisziplinärer Sicht, Bern [et al.], pp. 167-181.

Schaie, W. K./Hofer, S. M. (2001): Longitudinal Studies in Aging Research, in: Birren, J. E./Schaie, W. K. (Eds.): Handbook of the Psychology of Aging, 5. ed., San Diego [et al.], pp. 53-77.

Schaninger, C. M./ Danko, W. D. (1993): A Conceptual and Empirical Comparison of Alternative Household Life Cycle Models, in: Journal of Consumer Research, Vol. 19, No. 4, March, pp. 580-594.

Schendera, C. F. G. (2007): Datenqualität mit SPSS, München; Wien.

Schermelleh-Engel, K./Werner, C. (2007): Methoden der Realibilitätsbestimmung, in: Moosbrugger, H./Kelava, A. (Eds.): Testtheorie und Fragebogenkonstruktion, Heidelberg, pp. 113-133.

Schewe, C. D. (1988): Marketing to our Aging Population: Responding to Physiological Changes, in: Journal of Consumer Marketing, Vol. 5, No. 3, pp. 61-73.

Schiffman, L. G./Kanuk, L. L. (2004): Consumer Behavior, 8. ed., Upper Saddler River (NJ).

Schiffman, L. G./Sherman, E. (1991): Value Orientations of New-Age Elderly: The Coming of an Ageless Market, in: Journal of Business Research, Vol. 22, Issue 2, March, pp. 187-194.

Schiffman, S. (2009): Sensory Impairment: Taste and Smell Impairments with Aging, in: Bales, C. W./Ritchie, C. S. (Eds.): Handbook of Clinical Nutrition and Aging, 2. ed., New York (NY), pp. 77-97.

Schirrmacher, F. (2004): Das Methusalem-Komplott, München.

Schmalen, H./Pechtl, H. (1992): Technische Neuerungen in Kleinbetrieben. Eine empirische Untersuchung zur Einführung von elektronischer Datenverarbeitung in Handwerksbetrieben, Stuttgart.

Schmalen, H./Pechtl, H. (1996): Die Rolle der Innovationseigenschaften als Determinanten im Adoptionsverhalten, in: Zeitschrift für betriebswirtschaftliche Forschung, 48. Jg., Nr. 9, pp. 816-836.

Schmalen, H./Xander, H. (2002): Produkteinführung und Diffusion, in: Albers, S. (Ed.): Handbuch Produktmanagement: Strategieentwicklung - Produktplanung - Organisation - Kontrolle, 2. ed., Wiesbaden, pp. 439-468.

Schmidt, I. (1997): Ältere Mitbürger als Zielgruppe in Marktforschung und Werbung, in: Planung & Analyse, 2, pp. 36-39.

Schmidt, M./Hollensen, S. (2006): Marketing Research. An International Approach, Harlow.

Schnell, R./Hill, P. B./Esser, E. (2008): Methoden der empirischen Sozialforschung, 8. ed., München; Wien.

Schönert, W. (1997): Senioren-Werbung: Die Mühen werden sich lohnen, in: Marketing Journal, Nr. 2, 30, pp. 114-117.

Schorb, B. (2009): Medien und höheres Lebensalter. Theorie - Forschung - Praxis, Wiesbaden.

Schulz, B. (2005): Senioren als Bankkunden. Ein Beratungs- und Betreuungskonzept für Finanzdienstleister, Diss., Wiesbaden [et al.].

Schulz, R. (1972): Kaufentscheidungsprozesse des Konsumenten; Diss. (Prozessmodelle komplexer Kaufentscheidungen des Konsumenten und ihr Aussagewert für das Marketing), Wiesbaden.

Schumpeter, J. A. (1912): Theorie der wirtschaftlichen Entwicklung, Leipzig.

Schumpeter, J. A. (1947): The Creative Response in Economic History, in: Journal of Economic History, Vol. 7, No. 2, Nov, pp. 149-159.

Schuster, H.-W. (1994): Prestigegeleitetes Konsumentenverhalten. Teil 1: Typologie des Distinktionskonsums, in: GfK - Jahrbuch der Absatz- und Verbrauchsforschung, Jg. 40 Nr. 2, pp. 108-121.

Schweiger, G./Mazanec, J./Wiegele, O. (1976): Das Modell des "erlebten Risikos"("perceived risk"): Struktur und Operationalisierungskonzepte, in: Der Markt - Zeitschrift für Absatzwirtschaft und Marketing, Nr. 60, pp. 93-102.

Schweizer, M./Kotouc, A. J./Wagner, T. (2006): Scale Development for Consumer Confusion, in: Advances in Consumer Research, Vol. 33, pp. 184-190.

Schweizer, P. (2004): Consumer Confusion im Handel, Diss., Schesslitz.

Settersten, Jr. R. A./Mayer, K. U. (1997): The Measurement of Age, Age Structuring, and the Life Course, in: Annual Review of Sociology, Vol. 23, Issue 1, pp. 233-261.

Settle, R. B./Golden, L. L. (1974): Consumer Perceptions: Overchoice in the Market Place, in: Advances in Consumer Research, Vol. 1, Issue 1, pp. 29-37.

Sheth, J. N. (1981): Psychology of Innovation Resistance: The Less Developed Concept (LDC) in Diffusion Research, in: Research in Marketing, Vol. 4, pp. 273-282.

Shih, T.-H./Fan, X. (2008): Comparing Response Rates from Web and Mail Surveys: A Meta-Analysis, in: Field Methods, Vol. 20, No. 3, Aug, pp. 249-271.

Shimp, T. A./Bearden, W. O. (1982): Warranty and Other Extrinsic Cue Effects on Consumers' Risk Perceptions, in: Journal of Consumer Research, Vol. 9, Issue 1, June, pp. 38-46.

Siegel, A. F. (2012): Practical Business Statistics, 6. ed., Burlington MA; Oxford, UK.

Sigelman, C. K./Rider, E. A. (2012): Life-Span. Human Development, 7. ed., Belmont, CA.

Silvers, C. (1997): Smashing Old Stereotypes of 50-plus America, in: Journal of Consumer Marketing, Vol. 14, Issue 4/5, pp. 303-309.

Simcock, P./Sudbury, L./Wright, G. (2006): Age, Perceived Risk and Satisfaction in Consumer Decision Making: A Review and Extension, in: Journal of Marketing Management, Vol. 22, Issue 3/4, Apr, pp. 355-377.

Singer, T./Lindenberger, U. (2000): Plastizität, in: Wahl, H.-W./Tesch-Römer, C. (Eds.): Angewandte Gerontologie in Schlüsselbegriffen, Stuttgart; Berlin; Köln, pp. 39-43.

Sinha, I. (1994): A Conceptual Model of the Role of Situational Type on Consumer Choice Behavior and Consideration Sets, in: Advances in Consumer Research, Vol. 21, Issue 1, pp. 477-482.

Skiera, B./Albers, S. (2008): Regressionsanalyse, in: Herrmann, A./Homburg, C./Klarmann, M. (Eds.): Handbuch Marktforschung. Methoden - Anwendungen - Praxisbeispiele, 3. ed., Wiesbaden, pp. 467-497.

Smith, S. M./Albaum, G. S. (2005): Fundamentals of Marketing Research, Thousand Oaks (Calif.); London; New Delhi.

Solomon, M. R. (1999): Consumer Behavior. Buying, Having, and Being, 4. ed., Upper Saddle River, N.J.

Sood, A./Tellis, G. J. (2005): Technological Evolution and Radical Innovation, in: Journal of Marketing, Vol. 69, Issue 3, July, pp. 152-168.

Sorce, P. (1995): Cognitive Competence of Older Consumers, in: Psychology & Marketing, Vol. 12, Issue 6, Sep, pp. 467-480.

Sorokin, P. A./Zimmerman, C. C./Galpin, C. J. (1931): A Systematic Source Book in Rural Sociology, Vol. 2, Minneapolis.

Sproles, G. B./Kendall, E. L. (1986): A Methodology for Profiling Consumers' Decision-Marking Styles, in: Journal of Consumer Affairs, Vol. 20, Issue 2, Winter, pp. 267-279.

SPSS Inc. (2007): SPSS Base 16.0 User's Guide, Chicago.

Stadik, M. (2005): Medien rund um die Uhr. Mediennutzung im Wandel: "Online-Medien sind die neuen Massenmedien", in: TENDENZ (Bayerische Landeszentrale für neue Medien), Ausgabe 4, pp. 4-11.

Städtler, T. (2003): Selbstwirksamkeit, in: Lexikon der Psychologie, Stuttgart, pp. 971-972.

Statistisches Bundesamt (2006): Bevölkerung Deutschlands bis 2050. 11. koordinierte Bevölkerungsvorausberechnung, Wiesbaden.

Statistisches Bundesamt (2007a): Pressemitteilung Nr. 051 vom 07.02.2007, Wiesbaden.

Statistisches Bundesamt (2007b): Wirtschaftsrechnungen. Ausstattung privater Haushalte mit langlebigen Gebrauchsgütern; Fachserie 15 Reihe 2, Wiesbaden.

Statistisches Bundesamt (2009): Bevölkerung Deutschlands bis 2060. 12. koordinierte Bevölkerungsvorausberechnung, Wiesbaden.

Statistisches Bundesamt (2010a): Statistisches Jahrbuch 2010. Für die Bundesrepublik Deutschland mit "Internationalen Übersichten", Wiesbaden.

Statistisches Bundesamt (2010b): Wirtschaftsrechnungen. Einkommens- und Verbrauchsstichprobe. Aufwendungen privater Haushalte für den Privaten Konsum, Wiesbaden.

Statistisches Bundesamt (2010c): Wirtschaftsrechnungen. Einkommens- und Verbrauchsstichprobe. Einnahmen und Ausgaben privater Haushalte, Wiesbaden.

Statistisches Bundesamt (2010d): Wirtschaftsrechnungen. Einkommens- und Verbrauchsstichprobe. Geld- und Immobilienvermögen sowie Schulden privater Haushalte, Wiesbaden.

Statistisches Bundesamt (2011a): Ältere Menschen in Deutschland und der EU, Wiesbaden.

Statistisches Bundesamt (2011b): Bevölkerung und Erwerbstätigkeit. Bevölkerungsfortschreibung , Wiesbaden.

Statistisches Bundesamt (2011c): Bevölkerung und Erwerbstätigkeit. Entwicklung der Privathaushalte bis 2030. Ergebnisse der Haushaltsvorausberechnung, Wiesbaden.

Statistisches Bundesamt (2011d): Bevölkerung und Erwerbstätigkeit. Haushalte und Familien. Ergebnisse des Mikrozensus, Wiesbaden.

Statistisches Bundesamt (2011e): Modelle der Bevölkerungsvorausberechnungen, Wiesbaden.

Statistisches Bundesamt (2011f): Periodensterbetafeln für Deutschland. Allgemeine Sterbetafeln, abgekürzte Sterbetafeln und Sterbetafeln 1871/1881 bis 2007/2009, Wiesbaden.

Statt, D. A. (1997): Understanding the Consumer. A Psychological Approach, Houndmills [et al.].

Steenkamp, J.-B. E. M./Gielens, K. (2003): Consumer and Market Drivers of the Trial Probability of New Consumer Packaged Goods, in: Journal of Consumer Research, Vol. 30, Issue 3, Dec, pp. 368-384.

Steinhagen-Thiessen, E./Wrobel, N./Borchelt, M. (1999): Der Zahn der Zeit. Körperliche Veränderungen im Alter, in: Niederfranke, A./Naegele, G./Frahm, E. (Eds.): Funkkolleg Altern. Band 1: Die vielen Gesichter des Alterns, Opladen; Wiesbaden, pp. 239-276.

Stephens, N. (1991): Cognitive Age: A Useful Concept for Advertising?, in: Journal of Advertising, Vol. 20, Issue 4, Dec, pp. 37-48.

Stone, R. N./Gronhaug, K. (1993): Perceived Risk: Further Considerations for the Marketing Discipline, in: European Journal of Marketing, Vol. 27, Issue 3, pp. 39-50.

Strober, M. H./Weinberg, C. B. (1977): Working Wives and Major Family Expenditures, in: Journal of Consumer Research, Vol. 4, Issue 3, Dec, pp. 141-147.

Strodtbeck, F. L. (1973): Communication in Small Groups, in: Pool, I. d. S. et al. (Eds.): Handbook of Communication, Chicago, pp. 646-665.

Strutton, H. D./Lumpkin, J. R. (1993): The Influence of Determinant Product Attributes on the Elderly's Information Search for Health Care Innovations, in: Journal of Marketing Theory & Practice, Vol. 2, Issue 1; Fall, pp. 122-138.

Stuart-Hamilton, I. (2011): An Introduction to Gerontology, Cambridge [et al.].

Stumpf, H. (2008): Wenn das Wochenende 7 Tage hat. Berufsende - Rente - Älterwerden ...und alles, was Sie dazu wissen sollten, Norderstedt.

Swanson, D./Siegel, J. S. (2004): The Methods and Materials of Demography, 2. ed., San Diego (Calif.); London.

Szmigin, I./Carrigan, M. (2000): The Older Consumer as Innovator: Does Cognitive Age hold the Key?, in: Journal of Marketing Management, Vol. 16, Issue 5; June, pp. 505-527.

Szmigin, I./Carrigan, M. (2001a): Introduction to Special Issue on Cognitive Age and Consumption, in: Psychology & Marketing, Vol. 18 Issue 10, Oct, pp. 999-1002.

Szmigin, I./Carrigan, M. (2001b): Learning to Love the Older Consumer, in: Journal of Consumer Behaviour, Vol. 1, No. 1 pp. 22-34.

Szmigin, I./Carrigan, M. (2001c): Leisure and Tourism Services and the Older Innovator, in: Service Industries Journal, Vol. 21, Issue 3, Jul, pp. 113-129.

Tabachnick, B. G./ Fidell, L. S. (2010): Using Multivariate Statistics, 5. ed., Boston; München.

Tanny, S. M./Derzko, N. A. (1988): Innovators and Imitators in Innovation Diffusion Modelling, in: Journal of Forecasting, Vol. 7, Issue 4, Oct-Dec, pp. 225-234.

Tartler, R. (1961): Das Alter in der modernen Gesellschaft, Stuttgart.

Taylor, J. W. (1974): The Role of Risk in Consumer Behavior, in: Journal of Marketing, Vol. 38, Issue 2, Apr, pp. 54-60.

Taylor, J. W. (1977): A Striking Characteristic of Innovators, in: Journal of Marketing Research, Vol. 14, Issue 1, Feb, pp. 104-107.

Tellis, G. J./Yin, E./Bell, S. (2004): Global Consumer Innovativeness: Country Differences and Individual Commonalities, in: Marketing Science Institute and Journal of Marketing Research Joint Conference on Applied Research, Yale University, pp. 1-47.

Tesch-Römer, C./Andrick, R. (2011): Alter und Altern, Erfurt.

Tesch-Römer, C./Wiest, M./Wurm, S. (2010): Subjektives Wohlbefinden, in: Motel-Klingebiel, A./Wurm, S./Tesch-Römer, C. (Eds.): Altern im Wandel. Befunde des Deutschen Alterssurveys (DEAS), Stuttgart, pp. 263-283.

Tesch-Römer, C./Wurm, S. (2009): Lebenssituationen älter werdender und alter Menschen in Deutschland, in: Böhm, K./Tesch-Römer, C./Ziese, T. (Eds.): Gesundheit und Krankheit im Alter. Beiträge zur Gesundheitsberichterstattung des Bundes, Berlin, pp. 113-120.

Tews, H. P. (1990): Alternde Gesellschaft - Lebensbedingungen und -perspektiven älterer Menschen, in: WSI Mitteilungen, 43, No. 8, pp. 477-544.

Tews, H. P. (1993): Neue und alte Aspekte des Strukturwandels des Alters, in: Naegele, G./Tews, H. P. (Eds.): Lebenslagen im Strukturwandel des Alters, Opladen, pp. 15-42.

Thom, N. (1980): Grundlagen des betrieblichen Innovationsmanagements, 2. ed., Königstein (Ts.).

Thomae, H. (1970): Theory of Aging and Cognitive Theory of Personality, in: Human Development, Vol. 13, pp. 1-16.

Thompson, D. V./Hamilton, R. W./Rust, R. T. (2005): Feature Fatigue: When Product Capabilities Become Too Much of a Good Thing, in: Journal of Marketing Research, Vol. 42, Issue 4, Nov, pp. 431-442.

Tittel, K. (2003): Beschreibende und funktionelle Anatomie des Menschen, 14. ed., München; Jena.

Tongren, H. N. (1988): Determinant Behavior Characteristics of Older Consumers, in: Journal of Consumer Affairs, Vol. 22, No. 1, Summer, pp. 136-157.

Topritzhofer, E. (1969): Geldtheorie und Verbraucherverhalten, in: Der Markt, No. 29, pp. 64-66.

Tornatzky, L. G./Klein, K. J. (1982): Innovation Characteristics and Innovation Adoption-Implementation: A Meta-Analysis of Findings, in: IEEE Transactions on Engineering Management, Vol. 29, No. 1, Feb, pp. 28-45.

Trommsdorff, V. (2004): Konsumentenverhalten, 6. ed., Stuttgart.

Trommsdorff, V./Schneider, P. (1990): Grundzüge des betrieblichen Innovationsmanagement, in: Trommsdorff, V. (Ed.): Innovationsmanagement in kleinen und mittleren Unternehmen. Grundzüge und Fälle - ein Arbeitsergebnis des Modellversuchs Innovationsmanagement, München, pp. 1-26.

U.S. Census Bureau (2011a): International Data Base, Demographic Indicators, http://www.census.gov/population/international/data/idb/country.php, 7.9.2011.

U.S. Census Bureau (2011b): International Data Base, Population Pyramids, http://www.census.gov/population/international/data/idb/country.php, 7.9.2011.

U.S. National Institutes of Health (2011): National Institute on Aging. Longitudinal Studies (online), 3.10.2011.

Underhill, L./Cadwell, F. (1983): What Age Do You Feel: Age Perception Study, in: Journal of Consumer Marketing, Vol. 1, Issue 1, Summer, pp. 18-27.

Urban, D./Mayerl, J. (2006): Regressionsanalyse. Theorie, Technik und Anwendung, 2. ed., Wiesbaden.

VA (2007): VerbraucherAnalyse 2007/08 Klassik I Märkte.

Vacchiano, R. B./Strauss, P. S./Hochman, L. (1969): The Open and Closed Mind: A Review of Dogmatism, in: Psychological Bulletin, Vol. 71, No. 4, April, pp. 261-273.

van Eimeren, B./Frees, B. (2006): Zukünftige Medien: Praxistauglich für den Konsumenten? Eine Analyse auf Basis der Daten der ARD/ZDF-Online-

Studie und der ARD/ZDF Studie Massenkommunikation, in: Media Perspektiven (Ed.): Ausgabe 11, pp. 563-571.

van Eimeren, B./Frees, B. (2010): Fast 50 Millionen Deutsche online - Multimedia für alle?, in: Media Perspektiven (Ed.): Ausgabe 7-8, pp. 334-349.

Venkatesh, V. et al. (2003): User Acceptance of Information Technology: Toward a Unified View, in: MIS Quarterly, Vol. 27, Issue 3, Sep, pp. 425-478.

Venkatesh, V./Davis, F. D. (1996): A Model of the Antecedents of Perceived Ease of Use: Development and Test, in: Decision Sciences, Vol. 27, No. 3, Summer, pp. 451-481.

Venkatraman, M. P. (1991): The Impact of Innovativeness and Innovation Type on Adoption, in: Journal of Retailing, Vol. 67, Issue 1, Spring, pp. 51-67.

Venkatraman, M. P./Price, L. L. (1990): Differentiating Between Cognitive and Sensory Innovativeness. Concepts, Measurement, and Implications, in: Journal of Business Research, Vol. 20, Issue 4, June, pp. 293-315.

Veryzer, R. W. (1998): Key Factors Affecting Customer Evaluation of Discontinuous New Products, in: Journal of Product Innovation Management, Vol. 15, No. 2, Feb, pp. 136-150.

Victor, C. (2010): The Demography of Ageing, in: Dannefer, D./Phillipson, C. (ed.): The SAGE Handbook of Social Gerontology, London [et al.], pp. 61-74.

Vömel, T. (1991): Biologische Alternstheorien, in: Oswald, W. D. et al. (Eds.): Gerontologie. Medizinische, psychologische und sozialwissenschaftliche Grundbegriffe, 2. ed., Stuttgart; Berlin; Köln, pp. 47-53.

von Auer, L. (2005): Ökonometrie. Eine Einführung, 3. ed., Berlin; Heidelberg; New York (NY).

Voss, P. Jr. (1984): Status Shifts to Peer Influences, in: Advertising Age, May 17, pp. M-10.

VuMA (2007): VuMA Verbrauchs- und Medienanalyse – Basisauswertung.

Wahl, H.-W. (2004): Entwicklung gerontologischer Forschung, in: Kruse, A./Martin, M. (Eds.): Enzyklopädie der Gerontologie. Alternsprozesse in multidisziplinärer Sicht, Bern [et al.], pp. 29-48.

Wahl, H.-W./Heyl, V. (2004): Gerontologie - Einführung und Geschichte, Stuttgart.

Wahle, P. (1991): Erfolgsdeterminanten im Einzelhandel. Eine theoriegestützte, empirische Analyse strategischer Erfolgsdeterminanten, unter besonderer Berücksichtigung des Radio- und Fernsehfacheinzelhandels, Frankfurt (a. M.) [et al.].

Walker, A. (2010): Ageing and Quality of Life in Europe, in: Dannefer, D./Phillipson, C. (Eds.): The SAGE Handbook of Social Gerontology, London [et al.], pp. 573-586.

Wallach, M./Kogan, N. (1961): Aspects of Judgement and Decision-Making: Interrelationships and Changes with Age, in: Behavioral Science, Vol. 6, No. 1, Jan, pp. 23-36.

Walsh, G./Henning-Thurau, T. (2002): Wenn Konsumenten verwirrt sind. Empirische Analyse der Wirkungen eines vernachlässigten Konstruktes, in: Marketing ZFP (Zeitschrift für Forschung und Praxis), Heft 2, 24, pp. 95-109.

Wei, S.-C. (2005): Consumers' Demographic Characteristics, Cognitive Ages, and Innovativeness, in: Advances in Consumer Research, Vol. 32, Issue 1, pp. 633-640.

Weiber, R. (1992): Diffusion von Telekommunikation. Problem der kritischen Masse, Wiesbaden.

Weiber, R./Pohl, A. (1995): Nachfrageverhalten bei technologischen Innovationen: Herausforderung für das Marketing-Management, in: Zahn, E. (Ed.): Handbuch Technologiemanagement, Stuttgart, pp. 409-435.

Weick, S. (2004): Lebensbedingungen, Lebensqualität und Zeitverwendung, in: Statistisches Bundesamt (Ed.): Alltag in Deutschland. Analysen zur Zeitverwendung, Wiesbaden, pp. 412-430.

Weis, H. C. (2004): Marketing, 13. ed., Ludwigshafen (Rhein).

Wells, W. D./Gubar, G. (1966): Life Cycle Concept in Marketing Research, in: Journal of Marketing Research, Vol. III, Nov, pp. 355-363.

Wengler, T. (2005): Auswirkungen des Internet-Handels auf Shopping-Center. Eine empirische Analyse zu den Auswirkungen auf die Gestaltung und das Management von Shopping-Centern, Norderstedt 2005.

Wentura, D./Greve, W. (2000): Krise und Bewältigung, in: Wahl, H.-W./Tesch-Römer, C. (Eds.): Angewandte Gerontologie in Schlüsselbegriffen, Stuttgart; Berlin; Köln, pp. 49-53.

Weyerer, S. (2000): Epidemiologie, in: Wahl, H.-W./Tesch-Römer, C. (Eds.): Angewandte Gerontologie in Schlüsselbegriffen, Stuttgart; Berlin; Köln, pp. 27-32.

Wheaton, B. (1999): Social Stress, in: Aneshensel, C. S./Phelan, J. C. (Eds.): Handbook of the Sociology of Mental Health, New York (NY), pp. 277-300.

Wiedmann, K.-P./Walsh, G./Klee, A. (2001): Konsumentenverwirrtheit: Konstrukt und marketingpolitische Implikationen, in: Markteting ZFP (Zeitschrift für Forschung und Praxis), Heft 2, 2. Quartal pp. 83-99.

Wilkes, R. E. (1992): A Structural Modeling Approach to the Measurement and Meaning of Cognitive Age, in: Journal of Consumer Research, Vol. 19, Issue 2, Sep, pp. 292-301.

Wilkes, R. E. (1995): Household Life-Cycle Stages, Transitions, and Product Expenditures, in: Journal of Marketing Research, Vol. 22, No. 1, June, pp. 27-42.

Wilkie, W. L. (1986): Consumer Behavior, New York (NY) [et al.].

Wilmoth, J. M./Ferraro, K. F. (2007): Gerontology. Perspectives and Issues, 3. ed., New York (NY).

Witt, J. (1996): Grundlagen für die Entwicklung und Vermarktung neuer Produkte, in: Witt, J. (Ed.): Produktinnovation. Entwicklung und Vermarktung neuer Produkte, München, pp. 1-110.

Wöhe, G. (1996): Einführung in die allgemeine Betriebswirtschaftslehre, 19. ed., München.

Wood, S./Swait, J. (2002): Psychological Indicators of Innovation Adoption: Cross-Classification Based on Need for Cognition and Need for Change, in: Journal of Consumer Psychology, Vol. 12, Issue 1, pp. 1-13.

Wu, X. (2008): Population Policy and Human Capital Accumulation in China, Diss., College Park (University of Maryland).

Wurm, S./Schöllgen, I./Tesch-Römer, C. (2010): Gesundheit, in: Motel-Klingebiel, A./Wurm, S./Tesch-Römer, C. (ed.): Altern im Wandel. Befunde des Deutschen Alterssurveys (DEAS), Stuttgart, pp. 90-117.

Zacks, R. T./Hasher, L./Li, K. Z. H. (2000): Human Memory, in: Craik, F. I. M./Salthouse, T. A. (Eds.): The Handbook of Aging and Cognition, 2. ed., Mahwah, pp. 293-358.

Zahn, E. (1991): Innovation und Wettbewerb, in: Müller-Böling, D./Seibt, D./Winand, U. (Eds.): Innovations- und Technologiemanagement, Stuttgart, pp. 115-133.

Zahn, E./Weidler, A. (1995): Integriertes Innovationsmanagement, in: Zahn, E. (Ed.): Handbuch Technologiemanagement, Stuttgart, pp. 351-376.

Zaichkowsky, J. L. (1985): Measuring the Involvement Construct, in: Journal of Consumer Research, Vol. 12, No. 3, Dec, pp. 341–352.

Zank, S. (2000): Gesundheit und Krankheit, in: Wahl, H.-W./Tesch-Römer, C. (Eds.): Angewandte Gerontologie in Schlüsselbegriffen, Stuttgart; Berlin; Köln, pp. 44-48.

Zapf, W. (1989): Über soziale Innovationen, in: Soziale Welt, 40, Heft 1/2, pp. 170-183.

Zelan, J. (1985): Interviewing the Aged, in: Schewe, C. D. (Ed.): The Elderly Market: Selected Readings, Chicago, pp. 34-38 (Originally Published in: Public Opinion Quaterly, 33, (Fall 1969), pp. 420-424).

Ziamou, P. L. (1999): The Effect of the Degree of Newness of a 'Really New' Product on Consumers' Judgments, in: Advances in Consumer Research, Vol. 26, Issue 1, pp. 368-371.

Ziefle, M./Bay, S. (2005): How Older Adults Meet Complexity: Aging Effects on the Usability of Different Mobile Phones, in: Behaviour & Information Technology, Vol. 24, Issue 5, Sep/Oct 2005, pp. 375-389.

Zikmund, W. G. (2002): Essentials of Marketing Research, 2. ed., Mason (Ohio).

Zikmund, W. G. (2003): Business Research Methods, 7. ed., Mason (Ohio).

MARKTORIENTIERTE UNTERNEHMENSFÜHRUNG

Hrsg.: Prof. Dr. H. Freter

Band 1 Ulrich Döpke: Strategisches Marketing-Controllership. Eine empirische Untersuchung auf der Grundlage des situativen Ansatzes. 1986.

Band 2 Gerhard L. Laube: Betriebsgrößenspezifische Aspekte der Public Relations. Eine vergleichende theoretische und empirische Analyse der Public Relations in mittelständischen Unternehmungen und Großunternehmungen. 1986.

Band 3 Wolfgang Müller: Planung von Marketing-Strategien. Die Entwicklung von Marketing-Strategien im strategischen Marketing-Planungsprozeß. Eine theoretische Analyse unter besonderer Berücksichtigung eines markenbezogenen Produktpositionierungs-Modells. 1986.

Band 4 Friedhelm Klaes: Informationsverhalten kleiner und mittlerer Unternehmen der elektrotechnischen Investitionsgüterindustrie bei strategischen Entscheidungen. 1986.

Band 5 Walter Hinder / Sabine Bartosch: Strategisches Wettbewerbsverhalten in stagnierenden Märkten. Eine empirische Untersuchung der Damenoberbekleidungsindustrie der Bundesrepublik Deutschland. 1987.

Band 6 Wilfried R. Huber: Markenpolitische Strategien des Konsumgüterherstellers – dargestellt an Gütern des täglichen Bedarfs. 1988.

Band 7 Hans-Peter Walther: Erfolgreiches Strategisches Pharma-Marketing. Eine theoriegeleitete empirische Studie auf der Grundlage des situativen Ansatzes. 1988.

Band 8 Werner Gottschlich: Strategische Führung in mittleren Unternehmen. Konzepte, Operationalisierung und Messung. 1989.

Band 9 Dietmar Barzen: Marketing-Budgetierung. 1990.

Band 10 Volker Kuhl: Erfolgreiches Bier-Marketing nach dem Urteil des EUGH zum Reinheitsgebot. Marketing-Entscheidungen der deutschen Brauindustrie nach dem Verfahren des Europäischen Gerichtshofes um die Anwendung des Reinheitsgebotes auf Importbiere. 1990.

Band 11 Hans-Volker Eichler: Besonderheiten der Geschenksituation und ihre Auswirkungen auf das Konsumentenverhalten. 1991.

Band 12 Peter Wahle: Erfolgsdeterminanten im Einzelhandel. Eine theoriegestützte, empirische Analyse strategischer Erfolgsdeterminanten, unter besonderer Berücksichtigung des Radio- und Fernsehfacheinzelhandels. 1991.

Band 13 Stephan Georg Huth: Die marktorientierte Gestaltung der Logistik im Export der mittelständischen Unternehmung. Dargestellt am Beispiel zentraler Markenartikel. 1992.

Band 14 Jürgen Abel: Kooperationen als Wettbewerbsstrategien für Software-Unternehmen. 1992.

Band 15 Axel Eggert: Information und Innovation im industriellen Mittelstand. Eine theoriegeleitete empirische Untersuchung. 1992.

Band 16 Cornelia Meurer: Strategisches internationales Marketing für Dienstleistungen. Dargestellt am Beispiel des Management-Consulting. 1993.

Band 17 Frank Obbelode: Strategisches Marktauswahlverhalten mittelständischer Unternehmen auf internationalen Märkten. Eine theoretische Fundierung und empirische Untersuchung, dargestellt am Beispiel der Oberbekleidungsindustrie in der Bundesrepublik Deutschland. 1993.

Band 18 Harald Linné: Wahl geeigneter Kooperationspartner. Ein Beitrag zur strategischen Planung von F&E-Kooperationen. 1993.

Band 19 Michael Richter: Investitionsgütermarketing für CIM-Systeme. Unter besonderer Berücksichtigung von Integrationsdeterminanten. 1993.

Band 20 Frank Reintjes: Strategische Koordination von Beschaffung und Absatz. Entwicklung einer Konzeption zur Analyse und Gestaltung eines Koordinationssystems zur Abstimmung beschaffungs- und absatzwirtschaftlicher Ziel-, Strategie- und marktorientierter Instrumententenentscheidungen. 1995.

Band 21 Christian J. A. Glasmacher: Statische Preisentscheidungen im Rahmen eines wettbewerbsorientierten Krankenhausfinanzierungssystems. Eine Untersuchung vor dem Hintergrund des Gesundheitsstrukturgesetzes 1993 und der Bundespflegesatzverordnung 1995. 1996.

Band 22 Gabriele Barten: Marketingkommunikation im industriellen Anlagengeschäft. Zielorientierter Instrumenteinsatz in einem Beziehungsmanagement mit langfristiger Erfolgsperspektive. 1997.

Band 23 Carsten Baumgarth: Vertikale Marketingstrategien im Investitionsgüterbereich. Dargestellt am Beispiel von Einsatzstoffen. 1998.

Band 24 Kerstin Brockelmann: Multimedia im persönlichen Verkauf unter besonderer Berücksichtigung des Außendienstes. Eine empirische Untersuchung auf der Grundlage des situativen Ansatzes zur Erklärung der Akzeptanz und Effektivität von multimedialen Anwendungen. 1999.

Band 25 Henrike Sänger: Investor Relations im Internet. Eine Untersuchung auf der Grundlage funktionenorientierter und institutionenökonomischer Überlegungen. 2001.

Band 26 Florian Wecker: Kundenbindung im Markenportfolio. Ein Beitrag zur Erhöhung der Kundenbindung unter besonderer Berücksichtigung der Automobilindustrie. 2004.

Band 27 Marcus Stuhlert: Einsatz und Wirkung von Electronic Business-Anwendungen in Kunden-Lieferanten-Beziehungen. 2005.

Band 28 Sandra Feldmann: Bewertung von Sportmarken. Messung und Wirkungen der Markenstärke von Fußballbundesligavereinen. 2007.

Band 29 Hendrik Muth: Mehrmarkenhandel in der Automobilbranche. Eine Betrachtung aus Kundensicht. 2008.

Band 30 Marco Schmidt: Interne Markenstärke von B-to-B-Unternehmen. Konzeption – Determinanten – Erfolgsauswirkungen. 2009.

Band 31 Philipp Siebelt: Kooperation von Handelsunternehmen. Eine kritische Analyse am Beispiel des nicht filialisierten Einzelhandels mit Gebrauchsgütern auf Basis der Neuen Institutionenökonomik. 2010.

Band 32 Bastian Staub: Entscheidungsorientierte Marktsegmentbewertung mit dem Realoptionsansatz. 2010.

Band 33 Julia Naskrent: Verhaltenswissenschaftliche Determinanten der Spenderbindung. Eine empirische Untersuchung und Implikationen für das Spenderbindungsmanagement. 2010.

Band 34 Nikolaus A. D. Hohl: Nutzen als Basis von Kaufentscheidungen. Die Bedeutung von Bedürfnissen und Ressourcen für das Konsumentenverhalten. 2011.

Band 35 Antonia Barten: Rückgewinnungsmanagement öffentlicher Theaterbetriebe. Relevanz, Voraussetzungen, Handlungsempfehlungen. 2012.

Band 36 Armin Tank: Innovation Behavior of Older Consumers. An Empirical Analysis of Older and Younger Consumers' Determinants of Consumer Electronic Products Ownership. 2014.

www.peterlang.de